# THEORY AND APPLICATION OF LASER CHEMICAL VAPOR DEPOSITION

# LASERS, PHOTONICS, AND ELECTRO-OPTICS

Series Editor: H. Kogelnik

FUNDAMENTALS OF LASER OPTICS
Kenichi Iga

OPTICAL–THERMAL RESPONSE OF LASER-IRRADIATED TISSUE
Edited by Ashley J. Welch and Martin J. C. van Gemert

THEORY AND APPLICATION OF LASER CHEMICAL VAPOR DEPOSITION
Jyoti Mazumder and Aravinda Kar

A Continuation Order Plan is available for this series. A continuation order will bring delivery of each new volume immediately upon publication. Volumes are billed only upon actual shipment. For further information please contact the publisher.

# THEORY AND APPLICATION OF LASER CHEMICAL VAPOR DEPOSITION

**JYOTI MAZUMDER**
*University of Illinois at Urbana-Champaign*
*Urbana, Illinois*

**and**

**ARAVINDA KAR**
*University of Central Florida*
*Orlando, Florida*

PLENUM PRESS • NEW YORK AND LONDON

Library of Congress Cataloging-in-Publication Data

```
Mazumder, J.
    Theory and application of laser chemical vapor deposition / Jyoti
Mazumder and Aravinda Kar.
       p.   cm. -- (Lasers, photonics, and electro-optics)
    Includes bibliographical references and index.
    ISBN 0-306-44936-6
    1. Vapor-plating.  2. Lasers--Industrial applications.   I. Kar,
Aravinda.  II. Title.  III. Series.
TS695.M39   1995
671.7'35--dc20                                              95-25231
                                                                CIP
```

TS
·695
.M39
1995

ISBN 0-306-44936-6

© 1995 Plenum Press, New York
A Division of Plenum Publishing Corporation
233 Spring Street, New York, N. Y. 10013

10 9 8 7 6 5 4 3 2 1

All rights reserved

No part of this book may be reproduced, stored in a retrieval system, or transmitted in any form or by any means, electronic, mechanical, photocopying, microfilming, recording, or otherwise, without written permission from the Publisher

Printed in the United States of America

# PREFACE

With the advent of high-power lasers, laser-aided materials processing has grown into a new field in laser science and technology. The widespread use of laser has led to many innovations in applied optics and has emerged as optical engineering. Laser has become an important tool in novel manufacturing techniques that are efficient and cost-effective. Laser is also used to synthesize novel materials with improved microstructures and to deposit thin films of various shapes and sizes. Laser Chemical Vapor Deposition (LCVD) is a technique to deposit different types of thin films. For example, the LCVD technique can be used for surface modification by depositing thin films of high-temperature and high-strength materials to achieve improved oxidation, corrosion, and wear resistant properties that are useful for mechanical applications.

A proper understanding of the LCVD process requires an interdisciplinary approach, because the physics of this process involves a wide range of phenomena, including laser–matter interactions, heat transfer, fluid flow, chemical kinetics, and adsorption. The principles of these physicochemical phenomena are discussed in this book in a comprehensive way. The purpose of the volume is to show how these principles can be applied to develop theories about various types of LCVD processes in order to understand the basic mechanisms of the process, and to design and control an LCVD system in an optimum way.

The book is divided into five chapters. The first chapter presents an overview of the LCVD process, its applications, the interactions of lasers with various types of chemicals, and the safety issues related to laser handling. The second and third chapters explain the basic theories of the pyrolytic and photolytic LCVD processes, respectively. Finally, the fourth and fifth chapters respectively, show how these principles can be applied to model and understand various aspects of the pyrolytic and photolytic LCVD processes.

Aravinda Kar<br>Jyoti Mazumder

*Orlando and Urbana*

# CONTENTS

**Chapter 1. Introduction**

| | | |
|---|---|---|
| 1.1. | Laser Applications | 1 |
| 1.2. | Laser Beam Properties | 5 |
| | 1.2.1. Spot Size | 5 |
| | 1.2.2. Beam Divergence | 8 |
| | 1.2.3. Depth of Focus | 8 |
| | 1.2.4. Gaussian Beam Propagation | 10 |
| 1.3. | Fundamentals of Laser Chemical Vapor Deposition (LCVD) | 13 |
| 1.4. | Advantages of LCVD over Other Deposition Techniques | 14 |
| | 1.4.1. Deposition Rate Enhancement | 16 |
| | 1.4.2. Spatial Resolution and Control | 20 |
| | 1.4.3. Minimum Damage to the Substrate Film Materials | 21 |
| | 1.4.4. Ability to Process Temperature-Sensitive Materials | 21 |
| | 1.4.5. Isotope Separation | 23 |
| 1.5. | Literature Review | 24 |
| | 1.5.1. Laser Excitation of Polyatomic Molecules | 25 |
| | 1.5.2. Laser Excitation of Large Molecules | 27 |
| | 1.5.3. Laser-Induced Powder Generation | 32 |
| | 1.5.4. Thin Film Deposition | 36 |
| | 1.5.5. Film Characterization | 43 |
| 1.6. | Laser Safety | 44 |
| Nomenclature | | 47 |
| References | | 51 |

## Chapter 2. Pyrolytic LCVD

2.1. Introduction ................................................. 57
2.2. Process Variables in Pyrolytic LCVD ..................... 58
2.3. Laser Radiation Adsorption Phenomena ................. 59
2.4. Optical Properties .......................................... 60
2.5. Absorptivity of Metals at Normal Incidence ............. 66
2.6. Lorentz Model for Optical Properties of Dielectrics ...... 70
2.7. Drude–Zener Theory for Optical Properties of Metals .. 77
    2.7.1. Real Metals ......................................... 79
    2.7.2. Perfect Metals ...................................... 80
2.8. Effects of Temperature on the Absorptivity of Metals .... 81
2.9. Effects of Laser Intensity on Absorption .................. 84
    2.9.1. Two-Photon Absorption ........................... 85
    2.9.2. Free-Carrier Absorption ........................... 87
2.10. Thin Film Optics ........................................... 87
    2.10.1. Fresnel Coefficients ............................... 87
    2.10.2. Reflectivity and Transmissivity of a Multilayer Nonabsorbing ($k = 0$) Medium .................. 91
    2.10.3. Reflectivity and Transmissivity of a Multilayer Absorbing ($k \neq 0$) Medium ..................... 93
    2.10.4. Applicability of Thin Film Optics ............... 95
    2.10.5. Example–Calculation of Optical Properties at High Temperatures .............................. 96
2.11. Chemical Kinetics .......................................... 101
    2.11.1. Chemical Reaction and Reaction Rate .......... 101
    2.11.2. Types of Reactions ............................... 102
    2.11.3. Stoichiometric Coefficient, Molecularity, Order of a Reaction and Reaction Rate Constant ........................................... 103
    2.11.4. Thermodynamics and the Equilibrium Constant for Elementary Reactions ............. 105
    2.11.5. Reaction Quotient $Q$ ............................ 107
    2.11.6. Le Chatelier's Principle .......................... 107
    2.11.7. Effects of Temperature on the Reaction Rate Constant ........................................... 109
    2.11.8. Experimental Determination of the Pre–Exponential Factor $k_0$, Activation Energy $E$, and the Order of a Reaction ......... 113
    2.11.9. Example—Estimation of $k_0$ and Equations for Transport Coefficients in a Gaseous Medium ... 115

Nomenclature .................................................... 118
References ...................................................... 121

## Chapter 3. Photolytic LCVD

3.1. Introduction ............................................. 123
    3.1.1. Electromagnetic Radiation Spectrum .............. 125
    3.1.2. Interactions of Radiations with Chemical Reactants ......................................... 127
3.2. Definitions of Basic Photochemical Processes ............. 128
    3.2.1. Energy Level Diagrams ........................... 132
    3.2.2. Selection Rules ................................. 138
3.3. The Laws of Photochemical Reactions ..................... 142
3.4. Comparison of Photolytic and Pyrolytic LCVD ............. 143
3.5. Process Variables in Photolytic LCVD .................... 144
3.6. Photon Absorption Cross Section ......................... 145
    3.6.1. Absorption Cross Section ........................ 146
    3.6.2. The Beer–Lambert Law and the Attenuation Coefficient ...................................... 147
    3.6.3. Two-Photon Absorption Constant ................. 149
    3.6.4. Dissociation Cross Section ...................... 150
3.7. Chemical Bond Energy ................................... 152
    3.7.1. Calculation of Bond Dissociation Energies from Thermochemical Data ............................ 152
    3.7.2. Calculation of Bond Dissociation Energies from Spectroscopic Data .............................. 158
    3.7.3. Mean Metal–Carbon Dissociation Energy .......... 159
3.8. Example Problems ....................................... 161
3.9. Reaction Steps, Rates, and Correlation Rules in Photodissociation ....................................... 166
    3.9.1. Photophysical Processes in the Absence of Collisional Losses or Quenching of Excitation Energy .......................................... 167
    3.9.2. Photophysical Processes in the Presence of a Nonabsorbing Quenching Gas .................... 172
    3.9.3. Photodissociation and Predissociation of Diatomic Molecules .............................. 175
    3.9.4. Photodissociation Processes for Polyatomic Molecules ....................................... 176
    3.9.5. Correlation Rules for Photodissociation .......... 178

|       |        |                                                           |     |
|-------|--------|-----------------------------------------------------------|-----|
| 3.10. |        | Gas and Surface Phase Processes in Photolytic LCVD..      | 180 |
|       | 3.10.1.| Adsorption                                                | 181 |
|       | 3.10.2.| Langmuir (Ideal) Adsorption Isotherm                      | 184 |
|       | 3.10.3.| Adsorption with Dissociation                              | 185 |
|       | 3.10.4.| Adsorption of Two Types of Species                        | 186 |
|       | 3.10.5.| Nonideal Adsorption                                       | 187 |
| 3.11. |        | Concentration and Temperature Measurements                | 188 |
|       | 3.11.1.| Laser-Induced Fluorescence (LIF)                          | 188 |
|       | 3.11.2.| Advantages of LIF                                         | 189 |
|       | 3.11.3.| Basic Theory of LIF                                       | 189 |
|       | 3.11.4.| Linewidth Broadening Phenomena                            | 194 |
|       | 3.11.5.| Concentration Measurements Using LIF                      | 200 |
|       | 3.11.6.| Temperature Measurements Using LIF                        | 202 |
| 3.12. |        | Precursor Molecules                                       | 203 |
| Nomenclature |  |                                                           | 206 |
| References   |  |                                                           | 210 |

## Chapter 4. Pyrolytic LCVD Modeling

|      |        |                                                     |     |
|------|--------|-----------------------------------------------------|-----|
| 4.1. |        | Introduction                                        | 215 |
| 4.2. |        | Governing Equations and Boundary Conditions         | 218 |
|      | 4.2.1. | Mass Conservation Equation                          | 218 |
|      | 4.2.2. | Momentum Conservation Equation                      | 219 |
|      | 4.2.3. | Energy Conservation Equation                        | 220 |
| 4.3. |        | Modeling of the Temperature Field                   | 222 |
|      | 4.3.1. | Introduction                                        | 222 |
|      | 4.3.2. | Governing Equations and Boundary Conditions         | 223 |
|      | 4.3.3. | Method of Solution                                  | 226 |
|      | 4.3.4. | Results and Discussion                              | 235 |
|      | 4.3.5. | Summary                                             | 246 |
| 4.4. |        | Modeling of Metal (Ti) Film Deposition              | 246 |
|      | 4.4.1. | Introduction                                        | 247 |
|      | 4.4.2. | Governing Equations and Boundary Conditions         | 248 |
|      | 4.4.3. | Method of Solution                                  | 252 |
|      | 4.4.4. | Results and Discussion                              | 258 |
|      | 4.4.5. | Summary                                             | 267 |
| 4.5. |        | Modeling of Ceramic (TiN) Film Deposition           | 267 |
|      | 4.5.1. | Introduction                                        | 268 |
|      | 4.5.2. | Experimental Procedure                              | 269 |
|      | 4.5.3. | Governing Equations and Boundary Conditions         | 269 |
|      | 4.5.4. | Method of Solution                                  | 271 |

|       |        |                                                                 |     |
|-------|--------|-----------------------------------------------------------------|-----|
|       | 4.5.5. | Data Estimation                                                 | 274 |
|       | 4.5.6. | Results and Discusssion                                         | 275 |
|       | 4.5.7. | Summary                                                         | 283 |
| 4.6.  | Convection during LCVD                                                   | 284 |
| Nomenclature                                                                     | 286 |
| References                                                                       | 290 |

## Chapter 5. Photolytic LCVD Modeling

| | | | |
|---|---|---|---|
| 5.1. | Introduction | | 295 |
| 5.2. | Governing Equations and Boundary Conditions | | 296 |
| 5.3. | Uniform Source and Flux Model | | 302 |
| 5.4. | Identification of Gas and Adsorbed Phase Reactions | | 306 |
| 5.5. | Hemispheric Model | | 311 |
| | 5.5.1. | Single-Region ($w_0 \leqslant r < \infty$) Model with Hemispheric Surface Reaction | 311 |
| | 5.5.2. | Single-Region ($w_0 \leqslant r \leqslant R_i$) Model with Hemispheric Surface Reaction: Steady State Analysis | 316 |
| | 5.5.3. | Single-Region ($w_0 \leqslant r \leqslant R_i$) Model with Hemispheric Surface Reaction: Transient Analysis | 318 |
| | 5.5.4. | Two-Region ($0 \leqslant r \leqslant w_0$ and $w_0 \leqslant r < \infty$) Model with Hemispheric Volumetric Reaction | 323 |
| | 5.5.5. | Two-Region ($0 \leqslant r \leqslant w_0$ and $w_0 \leqslant r \leqslant R_i$) Model with Hemispheric Volumetric Reaction: Steady State Analysis | 328 |
| | 5.5.6. | Two-Region ($0 \leqslant r \leqslant w_0$ and $w_0 \leqslant r \leqslant R_i$) Model with Hemispheric Volumetric Reaction: Transient Analysis | 331 |
| 5.6. | Deposition Rate Expressions | | 337 |
| 5.7. | Kinetic Theory Model | | 338 |
| | 5.7.1. | Gas Phase Decomposition: Ballistic Deposition Model | 340 |
| | 5.7.2. | Gas Phase Decomposition: Diffusive Deposition Model | 344 |
| | 5.7.3. | Similarity between the Ballistic and Diffusive Deposition Models | 346 |
| | 5.7.4. | Determination of the Time-Independent Green Function | 349 |
| | 5.7.5. | Adsorbed Phase Decomposition: Effect of Chemisorbed Phase | 353 |

|       |       | 5.7.6. Adsorbed Phase Decomposition: Effect of Physisorbed Phase | 355 |
|-------|-------|-----|-----|

- 5.8. The Green Function Approach ............................. 357
  - 5.8.1. Effects of Surface Diffusion and Adsorption on Photolytic Surface Reactions ....................... 357
  - 5.8.2. Determination of the Time-Dependent Green Function ............................................. 359
- 5.9. Monte Carlo Method ....................................... 362
- 5.10. Multiphoton Processes .................................... 371
- Nomenclature ..................................................... 376
- References ....................................................... 381

## Appendix

- A.1. Definitions of Energy Density, Irradiance, and Intensity . 385
- A.2. Thermal Stress Analysis ................................... 386
- A.3. Volumetric Absorption Rate ................................ 389
- Nomenclature ..................................................... 391
- References ....................................................... 392

## Index ........................................................... 393

# ONE

# INTRODUCTION

## 1.1. LASER APPLICATIONS

In laser-aided materials processing, the heat, mass, and momentum transport processes depend on the intensity of the incident laser beam and the laser–materials interaction time, as shown in Fig. 1.1 [Mazumder (1991)]. The laser power density is about $10^3$ to $10^4$ W/cm$^2$ for laser surface hardening, where heat conduction and mass diffusion in the solid phase are important in order to determine the dwell time required for phase transformation. For surface melting and welding, the power density is about $10^5$ to $10^7$ W/cm$^2$, and convection in the melt pool becomes significant. During laser surface alloying and cladding, where the power density is about $10^5$ to $10^6$ W/cm$^2$, the convection heat and mass transfer processes affect the nonequilibrium microstructure and composition of the solidified material. Vaporization and plasma formation affect the surface contour, laser energy partitioning, and the depth of penetration during laser welding. When the power density exceeds $10^7$ W/cm$^2$, such as in laser drilling, vaporization and gas dynamical effects become important. For *l*aser *c*hemical *v*apor *d*eposition (LCVD) processes, the laser power density is about $10^3$ to $10^4$ W/cm$^2$, the substrate is not melted, only heat conduction occurs in the substrate, mass diffusion in the gas phase is important at low gas pressures, and convection in the gas phase becomes important at high gas pressures. Figure 1.1 [Mazumder (1991)] summarizes the transport phenomena associated with various laser-aided materials processing. Breinan *et al.* (1976), Ready (1978), and Roessler (1986) have also presented various laser parameter regimes similar to Fig. 1.1 for several laser-aided processes.

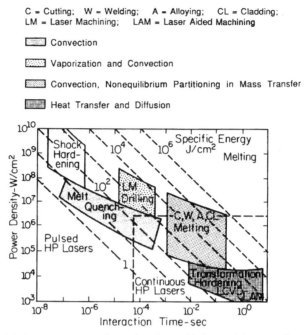

Figure 1.1. Laser parameters for various laser-aided materials processing.

Laser technology is so versatile that it can be used for various purposes. the LCVD technique provides a unique way of modifying the substrate surface by depositing thin films of desired optical, electrical, mechanical, or other properties. Boyd (1987) has summarized various applications of laser technology and its potential for future applications in *in-situ* thin film processing in the microelectronic and optoelectronic industries. Koebner (1984) has compiled several papers, which were contributed by various authors, to illustrate the applications of lasers. Some of the applications of laser technology are discussed below.

LASER HARDENING: lasers can be used to heat up a thin layer of the material near the substrate surface to a very high temperature within a very short time, and when the laser irradiation is stopped, the heated layer cools down so rapidly that the grain size is altered in the *heat-affected zone* (HAZ). Reduction in grain size usually improves the strength of the material. In transformation hardening of steel and cast iron, the material is heated above the austenitizing temperature for a sufficiently long time to accomplish the dissolution of C. When the irradiation is stopped, the dissipation of heat into the bulk of the material provides a rapid self-quenching mechanism, which leads to hardening of the material. By using the LCVD technique, thin films of hard materials, such as SiC, $Si_3N_4$, and TiN, can be deposited on the substrate to improve the surface strength of the material. Also, such films can improve the oxidation, corrosion, and wear-resistant properties of the material.

LASER CUTTING: Lasers can be used to cut materials at a rapid rate. Also, materials can be easily cut into various shapes by using lasers. Lasers are also useful to cut ceramics and other brittle materials.

LASER DRILLING: Lasers are used to drill holes with high precision and repeatability. Sometimes a laser is used to drill holes of very small sizes which cannot be achieved by other drilling techniques. Injection nozzles, coolant holes in automobile gears, cooling holes in turbine blades, watch bearings, screen plates, cells in gravure printing forms, and many other parts are drilled efficiently by lasers. Work on the development of automatic laser drilling machines for ruby watch-bearings began in 1966, and the first laser watch-bearing driller was used in production in 1968 [see Koebner (1984), p. 209]. The minimum diameter of the hole that can be drilled with a laser beam in percussion drilling is determined by the diffraction-limited laser spot size, and large-diameter holes are obtained by cutting out the material with a moving beam, which is known as trepanning. Since laser drilling produces holes with large aspect (depth-to-diameter) ratios, the electrical and optical interconnect technologies can be improved to achieve faster interchip or interboard communication by reducing the length of the interconnection pathways [see Boyd (1987), p. 285].

LASER WELDING: Laser welding is a widely accepted method for joining many different materials. Laser welds are found to be very strong, and laser welding produces a minimum HAZ compared to other welding techniques. Laser microsoldering and microbrazing are used for joining temperature-sensitive materials. For example, various components of an electronic board can be joined by local microsoldering, where the laser is used as a localized and well-controlled heat source to melt the solder without damaging any heat-sensitive component. Brazing is used to join thin metal parts because of the low heat-input requirement, and this technique usually requires heating of the entire part. For this reason, the conventional brazing technique becomes difficult to apply to a region that is surrounded by heat-sensitive materials, such as adhesively bonded connections, rubber seals, glass, or electronic components. Such problems arise in vacuum equipment, instrumentation, computers, communications, and electronics. In laser brazing, the filler metal is melted, and the molten metal distributes due to the capillary action along and between the parts that need to be joined. Also, rapid cooling and solidification improve the strength and microhardness of the joint. The LCVD technique can be used in such heat-sensitive environments to carry out microwelding.

LASERS IN MICROELECTRONICS INDUSTRIES: As noted earlier, LCVD is used extensively to deposit thin films of various semiconductor materials. Apart from this, laser is used to melt the surface of semiconductors for damage annealing. Due to rapid solidification, recrystallization occurs very fast, resulting in a high crystal perfection. Also, the implanted dopants are redistributed in the melt layer. Laser is used as well to trim resistors. Many different types of resistors are printed on electronic boards, and their resistivity is controlled by varying the following three parameters: (a) Area occupied by the resistor — the electrical resistance increases as the area decreases. (b) Type of paste — different pastes have different resistive properties. (c) Thickness of the paste deposit — for the same paste and resistor area, the resistance increases as the thickness of the deposit decreases. Usually, the dimensions of the printed resistors are larger than the required values. The correct size of the resistors is obtained by trimming them to attain the desired resistance. Laser is used to carry out this task with high precision.

OTHER APPLICATIONS: Laser surface alloying and laser cladding are used to improve the oxidation, corrosion, and wear-resistant properties of the substrate. Laser spectroscopy is used to determine temperature and concentration for understanding the physical and chemical phenomena of various processes. Laser is also used in medicine for surgery and other purposes. Laser holography produces three-dimensional images that can be used for turbine and generator construction. The hologram of a master-piece is developed, which serves as a standard for checking the shape and dimensional accuracy of other items that are similar to the master piece. The item whose deviations from the master-piece has to be checked is placed at the location where the master-piece was kept before, and the developed hologram plate is placed at the original place of shooting. The specimen is then exposed in the same way as the master-piece was exposed, and viewed through the developed hologram. The deviations in the shape of the specimen from that of the master-piece produce interference lines on the specimen. To carry out the safety analysis for fast breeder reactors in nuclear industries, various properties (such as the vapor pressure or other thermophysical properties) of liquid oxide fuel are required up to about 5000 K. Such high temperatures, preclude the use of conventional measurement techniques owing to the lack of suitable crucible materials. In such situations, laser can be used to design a containerless experiment for measuring

# INTRODUCTION

high-temperature properties of materials. To determine the vapor pressures of liquid urania and other oxide fuel materials, the specimens are heated locally with a laser beam, and the vapor pressure is calculated theoretically from the experimentally determined evaporation velocity or momentum of the vapor plume.

## 1.2. LASER BEAM PROPERTIES

Proper understanding of laser-induced processes requires experimental and theoretical studies of the process as well as knowledge concerning the properties of the laser beam. Various characteristics [Steen (1991), Boyd (1987)] of laser beams such as spot size, beam divergence, depth of focus, and beam propagation are reviewed briefly below.

A laser cavity can usually produce a laser beam with several *transverse electromagnetic modes* (TEM). However, only the fundamental (lowest-order) mode $TEM_{00}$ is obtained by using an aperture inside the cavity. The $TEM_{00}$ has a Gaussian profile and is most widely encountered in lasers. Several important properties of the fundamental mode are as follows.

1. The least amount of energy is required to support it.
2. The beam divergence is smaller than that of the higher-order modes.
3. The beam is symmetrical.
4. As the beam propagates through various optical systems, its "near-" and "far-" field Gaussian profiles do not alter, unlike the higher-order modes.

### 1.2.1. Spot Size

The distribution of electric field within the $TEM_{00}$ beam can be written as

$$E(r) = E_0 \exp\left(-\frac{r^2}{w_0^2}\right)$$

In mathematical modeling of laser-induced processes, the laser intensity is used instead of the electric field to carry out energy analyses. The intensity $I(r)$ is obtained from the square of the electric field, and can be written as

$$I(r) = I_0 \exp\left(-\frac{2r^2}{w_0^2}\right)$$

It should be noted that $w_0$, the radius of the beam at its waist, refers to that value of $r$ where $E(r) = E_0/e$, while at the same value of $r$, the intensity $I(r)$ decreases to $I_0/e^2$ as shown in Figs. 1.2a, b, and c.

The spot radius $w_0$ is defined as the distance from the beam center to the point where the electric field is $E_0/e$ or the intensity is $I_0/e^2$, and the spot

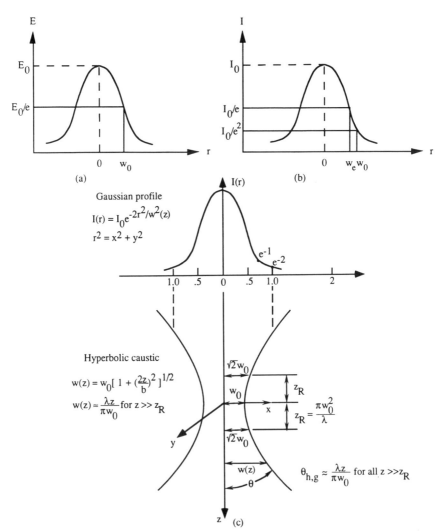

Figure 1.2. Various characteristics of a typical $TEM_{00}$ mode Gaussian laser beam: (a) Gaussian distribution of the electric field of the laser beam. (b) Gaussian distribution of the laser intensity. (c) A typical Gaussian beam with waist radius $w_0$, Rayleigh range $z_R$, confocal parameter $b$, and half-angle beam divergence $\theta_{h_G}$ in the far field. Note that $b = 2z_R$.

size is given by $2w_0$. The total power $P$ of the laser beam can be related to $I_0$ by noting that

$$P = \int_0^\infty I(r) 2\pi dr$$

which leads to the following expression:

$$P = \pi w_0^2 \frac{I_0}{2}$$

If the intensity distribution $I(r)$ is defined as

$$I(r) = I_0 \exp\left(\frac{-r^2}{w_e^2}\right)$$

where $w_e$ refers to the point at which $I(r) = I_0/e$, then the total laser power is given by

$$P = \pi w_e^2 I_0$$

By equating the above two expressions for $P$, it can be shown that $w_e = w_0/\sqrt{2}$.

For a short pulse of a laser with a Gaussian temporal distribution, the intensity distribution can be written as

$$I(r, t) = I_0 \exp\left(\frac{-r^2}{w_e^2}\right) \exp\left(\frac{-t^2}{t_p^2}\right)$$

and the energy of such a pulse is given by

$$E_p = I_0 \int_0^\infty \exp\left(\frac{-r^2}{w_e^2}\right) 2\pi r\, dr \int_{-\infty}^\infty \exp\left(\frac{-t^2}{t_p^2}\right) dt = \pi^{3/2} I_0 w_e^2 t_p$$

Also, the laser fluence is given by

$$F = \frac{E_p}{\pi w_e^2}$$

In materials processing, the above expressions for $I(r)$ or $I(r, t)$ refer to the intensity of the laser beam at normal incidence on the substrate surface. For oblique incidence, the intensity is given by Biyikli and Modest (1988):

$$I(x, y, z) = I_0'(\hat{k} + \hat{i}\tan\theta\cos\phi + \hat{j}\tan\theta\sin\phi)\frac{w_0^2}{w^2(z)}\exp\left[-\frac{(x^2+y^2)}{w^2(z)}\right]$$

where $I_0' = p/\pi w_0^2$.

### 1.2.2. Beam Divergence

The diffraction phenomenon plays an important role in determining the minimum divergence of a laser beam that can be obtained for a given optical system. Although the divergence of the laser beam can be minimized by designing the oscillator cavity properly, it cannot be reduced below the diffraction-limited divergence angle. In the far-field approximation, one-half of the divergence angle can be written as

$$\theta_{h_G} = \frac{\lambda}{\pi w_0}$$

for a Gaussian beam. For beams with uniform spatial distribution, the fraction $1/\pi$ in this expression should be replaced by 0.61. The radius of the focused spot is given by

$$w_0 = \frac{2\lambda f}{\pi D_l} = \frac{2\lambda}{\pi} F$$

According to this expression, lasers of shorter wavelengths can be focused to smaller spot sizes than those of longer wavelengths. Also, the smaller the $F$-number of the lens, the smaller the spot size.

### 1.2.3. Depth of Focus

Owing to diffraction, the focused laser beam does not remain focused as it propagates in free space (see Fig. 1.3), but diverges as rapidly as it was converged by the focusing lens. The radius of a Gaussian laser beam at any point after transmission by the lens is given by

$$w(z) = w_0\left[1 + \left(\frac{\lambda z}{\pi w_0^2}\right)^2\right]^{1/2}$$

# INTRODUCTION

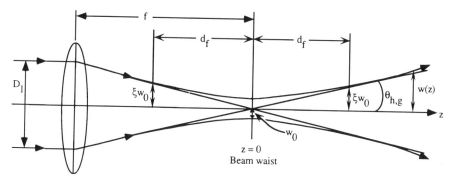

Figure 1.3. A schematic diagram for Gaussian laser beam propagation.

where the distance $z$ is measured from the center of the beam waist along the direction of the beam propagation. There is a small region on either side of the beam waist (focal plane) over which the beam radius $w(z)$ changes slowly and, therefore, the intensity of the beam is nearly constant. Letting $w(z) = \xi w_0$ at $z = \pm d_f$, we can write the above equation as

$$d_f = \pm \frac{\pi w_0^2}{\lambda}(\xi^2 - 1)^{1/2}$$

The physical significance of this expression is that at a distance $d$ on either side of the beam waist, the intensity of the laser beam decreases to $1/\xi^2$ of its maximum value because the beam radius increases to $\xi$ times its minimum value.

Ready (1971) has defined the depth of focus, $d_f$, as the distance on either side of the beam waist over which the laser intensity $I(0, z)$, drops to half its peak value $I(0, 0)$, that is, $I(0, z)/I(0, 0) = \frac{1}{2}$ at $z = \pm d_f$, where $I(r, z)$ represents the laser intensity at any radial and axial location $r$ and $z$, respectively. According to this definition, $\xi = \sqrt{2}$, and the depth of focus is given by

$$d_f = \pm \frac{\pi w_0^2}{\lambda}$$

at which the beam radius $w(z) = \sqrt{2} w_0$. At $d_f = \pm \pi w_0^2/\lambda$ the laser intensity $I(0, z)$ decreases to 50% of $I(0, 0)$, and the beam radius increases by 41.42% of the beam waist.

However, according to the Laser Institute of America (1977), the depth of focus is defined as the distance on either side of the beam waist at which

$\xi = 1.05$, which yields the following expression:

$$d_f = \pm 0.32 \frac{\pi w_0^2}{\lambda} = \pm \frac{w_0^2}{\lambda}$$

At $d_f = \pm w_0^2/\lambda$, which is the most widely accepted expression for the depth of focus, the laser intensity $I(0, z)$ decreases to 90.87% of $I(0,0)$, and the beam radius increases by 5% of the beam waist. The depth of focus given by the earlier expression for $d_f$, that is, $d_f = \pi w_0^2/\lambda$ is commonly known as the Rayleigh range, which is denoted by $z_R$; $b = 2z_R$ is referred to as the confocal parameter. The Rayleigh range is a measure of the length of the waist region over which the spot size is the smallest. The smaller the Rayleigh range, the smaller the spot size at the waist, which results in rapid growth of the beam radius from the waist. In practical applications, care must be taken to optimize the Rayleigh range, the effects of the fluctuations in $z_R$ on the process performance, and the required dimensions of the processed area on the workpiece.

### 1.2.4. Gaussian Beam Propagation

A Gaussian laser beam can be characterized completely by two geometrical properties [Kogelnik (1965), Li (1993)], the radius of curvature $R(z)$ of the phase fronts, and the spatial radius $w(z)$, which are given by

$$R(z) = z[1 + (z_R/z)^2] \qquad (1.1)$$

and

$$w^2(z) = w_0^2[1 + (z/z_R)^2] \qquad (1.2)$$

Note the change in notation in Eq. (1.2), where $w(z)$ refers to the same radial variable $r$ used in the above expression for the laser intensity $I(r)$. The above two parameters can be combined by defining a complex ray parameter $q(z)$ as

$$\frac{1}{q(z)} = \frac{1}{R(z)} - i\frac{\lambda}{\pi w^2(z)} \qquad (1.3)$$

Using Eqs. (1.1) and (1.2) in Eq. (1.3), one can obtain the following expression:

$$q(z) = q_0 + z \qquad (1.4)$$

where $q_0 = i(\pi w_0^2/\lambda) = iz_R$.

# INTRODUCTION

Figure 1.4 represents a typical Gaussian beam propagating in the $z$-direction, where $w_1$ and $w_2$ are the beam radii at distances $d_1$ and $d_2$ from the lens, respectively. The complex parameter of the beam at the incident surface of the lens is $q_1'$ with the radius of curvature $R_1'$ and $q_2'$ with the radius of curvature $R_2'$ at the transmitting surface of the lens; $q_1$ and $q_2$ represent the complex parameters at $z = -d_1$ and $z = d_2$, respectively. According to the laws of geometrical optics, a spherical wave with the radius of curvature $R_1'$ at the incident surface of the lens is transformed into a wave with a radius of curvature $R_2'$ at the transmitting surface of the lens. For a thin lens, $R_1'$ and $R_2'$ are related by

$$\frac{1}{R_2'} = \frac{1}{R_1'} - \frac{1}{f} \tag{1.5}$$

For a thin lens, $w(z)$ can be considered to be constant over the thickness of the lens, and under this assumption, $q_1'$ and $q_2'$ can be related by the following expression using Eqs. (1.3) and (1.5):

$$\frac{1}{q_2'} = \frac{1}{q_1'} - \frac{1}{f} \tag{1.6}$$

As the beam propagates, it is first transformed in the free space between plane $P_1$ and the lens, and according to Eq. (1.4), one can write

$$q_1' = q_1 + d_1 \tag{1.7}$$

where $q_1 = i(\pi w_1^2/\lambda)$.

As the beam passes through the lens, it is transformed according to the

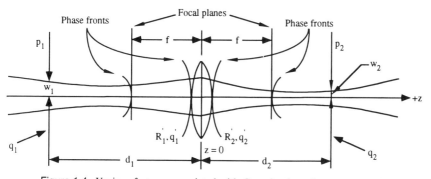

Figure 1.4. Various features associated with Gaussian laser beam propagation.

relation (1.6), which yields

$$q'_2 = \frac{q'_1}{1 - q'_1/f} \tag{1.8}$$

Before the beam reaches plane $P_2$, it is transformed in the free space between the lens and $P_2$, and one can write

$$q_2 = q'_2 + d_2 \tag{1.9}$$

Equations (1.7), (1.8), and (1.9) can be combined to obtain the *ABCD* law [Kogelnik (1965)] as follows:

$$q_2 = \frac{Aq_1 + B}{Cq_1 + D} \tag{1.10}$$

where $A = 1 - d_2/f$, $B = d_1 + d_2 - d_1 d_1/f$, $C = -1/f$, and $D = 1 - d_1/f$.

It should be noted that the spot size is usually assumed to be the same on either side of the lens, but a new beam waist is formed since the lens alters the radius of curvature of the beam. Expressions for the size of the waist and its location with respect to the lens are derived below. For this purpose, let $w_1$ and $w_2$ be the radii of two waists. Also, note that the radius of curvature of the phase front is infinite at the beam waist, which simplifies Eq. (1.3) to yield $q_1 = i\pi w_1^2/\lambda$ and $q_2 = i\pi w_2^2/\lambda$. Using these two values of $q_1$ and $q_2$ in Eq. (1.10), multiplying both sides of Eq. (1.10) by $Cq_1 + D$, and separating the real and imaginary parts, one can write

$$w_2^2(f - d_1) = w_1^2(f - d_2) \tag{1.11}$$

and

$$w_2^2 \left(\frac{\pi w_1}{\lambda}\right)^2 = f(d_1 + d_2) - d_1 d_2 \tag{1.12}$$

To determine the waist size and its location, that is, to calculate the values of $w_2$ and $d_2$, Eqs. (1.11) and (1.12) are rewritten in the following matrix form:

$$\begin{bmatrix} f - d_1 & w_1^2 \\ (\pi w_1/\lambda) & d_1 - f \end{bmatrix} \begin{bmatrix} w_2^2 \\ d_2 - f \end{bmatrix} = \begin{bmatrix} 0 \\ f^2 \end{bmatrix}$$

which can be solved to obtain

$$\frac{1}{w_2^2} = \frac{1}{w_1^2}\left(1 - \frac{d_1}{f}\right)^2 + \frac{1}{f^2}\left(\frac{\pi w_1}{\lambda}\right)^2$$

and

$$d_2 - f = (d_1 - f)\frac{f^2}{(d_1 - f)^2 + (\pi w_1^2/\lambda)^2}$$

## 1.3. FUNDAMENTALS OF LASER CHEMICAL VAPOR DEPOSITION (LCVD)

Laser chemical vapor deposition (LCVD) is a technique for depositing thin films of various materials on a substrate by inducing chemical reactions in a suitable reactant with the help of a laser beam. The LCVD technique can be classified into two categories, (1) pyrolytic and (2) photolytic, depending on the chemical reaction mechanisms. In pyrolytic LCVD, the laser beam interacts mainly with the substrate to produce a hot spot where thermally assisted chemical reactions take place resulting in the final product, which is the film material that eventually sticks to the substrate surface due to chemisorption. To model the pyrolytic LCVD processes, the chemical reactions are usually considered to occur only at the interface of the substrate and the chemical reactants, although a small volume just above the interface is also a possible region for chemical reaction. The reactants, that is, the precursors for pyrolytic LCVD must be selected in such a way that the chemical reactions can occur at a temperature below the melting temperature of the substrate because the substrate is not melted during LCVD. Usually infrared lasers, such as Nd:YAG and $CO_2$ lasers, are used for pyrolytic LCVD.

On the other hand, photolytic LCVD relies on the interactions of the laser beam with the chemical reactants. The precursor molecules absorb the photons of the laser beam, which causes the chemical bonds to break leading to the formation of the final product, which is the film material that deposits on the substrate. Usually, the visible and ultraviolet lasers are used for photolytic LCVD because the energy of each photon of these lasers is comparable to or exceeds the chemical bond energy of many chemical compounds. Table 1.1 shows the energy of a photon for various lasers to demonstrate that infrared lasers are unsuitable for photolytic LCVD because the energy of a photon of the infrared lasers is much smaller than the

Table 1.1. Energy of a Photon of Various Lasers

| Lasers | Regions of the electromagnetic spectrum | Wavelength (nm) | Energy of a photon (eV) |
|---|---|---|---|
| Nd:YAG | Infrared | 1060 | 1.17 |
| $CO_2$ | Infrared | 10600 | 0.117 |
| Cu Vapor | Yellow | 578 | 2.15 |
| Cu Vapor | Green | 511 | 2.43 |
| KrF | Ultraviolet | 248 | 5.00 |
| KrCl | Ultraviolet | 222 | 5.58 |
| XeCl | Ultraviolet | 308 | 4.03 |

usual chemical bond energy, which is approximately 5 eV. The laser and precursors for photolytic LCVD are so chosen that (1) the precursor molecules have a large absorption cross section for the laser beam, and (2) the chemical bond energy of the precursor molecules is less than or equal to the energy of a photon of the laser beam. In photolytic LCVD processes, the chemical reactions take place in the gas or vapor phase, which is made up of the chemical reactants and, therefore, the substrate does not have to be heated up to the chemical reaction temperature. This means that films can be deposited at a lower temperature by using the photolytic LCVD technique than the conventional deposition techniques, which is an attractive feature for the fabrication of semiconductor devices because thermally induced residual stress and impurity redistribution are minimized at lower temperatures.

## 1.4. ADVANTAGES OF LCVD OVER OTHER DEPOSITION TECHNIQUES

Besides the LCVD technique, there are several methods, such as the conventional *c*hemical *v*apor *d*eposition (CVD), *m*olecular *b*eam *e*pitaxy (MBE), *m*etal–*o*rganic *c*hemical *v*apor *d*eposition (MOCVD), and plasma-assisted deposition techniques, that are used to deposit thin films. In the conventional CVD technique, the whole substrate is heated, and the reactant gas or vapor is flown over the substrate surface, where chemical reactions occur and the film is deposited. This process is found to be very slow (the film growth rate is 100 to 1000 Å/min), and it is not energy-efficient since the whole substrate is heated, but only the hot surface is used to carry out the film deposition. In this technique, the substrate temperature can be as high as 1500 K depending on the chemical reaction temperature of the precursors and the properties of the film. Since the diffusion coefficients of the dopants become significant when the temperature exceeds about 1100 K, such high temperatures affect the quality of semiconductor films because of the change in the impurity concentration profiles due to the diffusion of dopants in the films. Also, the contaminants that may be present inside the CVD chamber or the wafer material can diffuse into the film grown at high temperatures. The volatilities of various species used for deposition compound semiconductor films are different, and therefore at high temperatures the species will evaporate during the film deposition process by different amounts, which can affect the properties of the film. Physical damage to the wafers at high temperatures can also cause defects in the devices produced by the conventional CVD method.

Prolonged exposure to high temperatures is harmful to almost any substrate material. Titanium alloys, which are used for aerospace applications, will deteriorate significantly due to the grain growth. Even ferrous alloys for automobile applications risk deterioration by grain growth and grain boundary diffusion.

In the MBE technique, molecular and/or atomic beams of the precursors impinge on the surface of a single-crystal sample which is heated suitably to facilitate the film growth process. The chemical reactions take place mainly at the substrate surface, and the choice of a single crystal as the substrate allows the film to grow epitaxially. The pressure inside the growth chamber is very low (usually about $10^{-11}$ Torr). Control of the MBE technique is so precise that the epitaxial film growth and composition of the film can be regulated to the monolayer level. The MBE technique is mostly used for producing semiconductors by using Group Periodic Table III and V elements such as Al, Ga, In, As, P, and Sb. A lot of work has also been done for the production of semiconductors with Si and Ge, II-VI, and IV-VI compounds, and for the fabrication of semiconductor devices by using this technique. However, the MBE process is extremely slow with a film growth rate of about 100–200 Å/min.

The MOCVD technique, which relies on the pyrolysis of various organometallic compounds and hydrides of the semiconductor materials, is similar to the conventional CVD process. Usually, low molecular weight metal alkyls, such as dimethyl cadmium ($[CH_3]_2 Cd$) or trimethyl gallium ($[CH_3]_3 Ga$), are used as the source for metals (Group II and III elements) to produce films of III-V and II-VI compounds and alloys. However, other types of organometallic compounds can also be used as the source for metals. For example, trimethyl indium-trimethyl phosphine ($[CH_3]_3 In-P[CH_3]_3$) is used as a source for In. Hydrides, such as $AsH_3$, $PH_3$, $H_2Se$, or $H_2S$; or alkyls, such as trimethyl antimony ($[CH_3]_3 Sb$) or diethyl tellurium ($[C_2H_5]_2 Te$) are usually used as a source for the nonmetal semiconductor materials which are Group V and VI elements. As in the conventional CVD case, the vapor of the organometallic compounds is flown through a reaction chamber, where pyrolysis occurs at the surface of the hot substrate and film deposition takes place. Since the substrate is heated, the high-temperature phenomena, such as the loss of film materials due to evaporation and redistribution of dopants because of diffusion, occur, which can affect the quality of the films grown by MOCVD. The MOCVD process is very slow, the reproducibility of some of the properties of the film grown by this technique is poor, and it is difficult to grow films with uniform layers over large areas.

The plasma-assisted or *p*lasma-*e*nhanced *c*hemical *v*apor *d*eposition (PECVD) technique was developed to grow films at a lower substrate temperature than that at which conventional CVD is carried out, in order to eliminate the diffusion and evaporation problems that are associated with high temperatures. In PECVD processes, a plasma is created by using an RF source (e.g., microwaves) or a high-frequency electric field across the reactants, which ionizes the reactants to induce chemical reactions in order to produce the film material that deposits on the substrate. The PECVD technique is mainly used to produce Si-based dielectrics and semiconducting materials, such as $SiO_2$, $Si_3N_4$, amorphous and polycrystalline Si. Amorphous hydrogenated Si (*a*-Si:H) is used for xerography, solar cells, and thin film field effect transistors. In telecommunication industries, optical fibers and intgrated optical structures are produced using the PECVD technique. The technique is also used to deposit thin films of many inorganic materials, such as GaAs, TiN, $TiO_2$, BN, and organic and organosilicon materials for various applications, such as reverse osmosis membranes, thin film capacitors, and laser guides in optoelectronic devices. In spite of the widespread applications of the PECVD technique, damage to the film from collisions between the highly energetic particles that are generated in the plasma and the substrate and depositing film surfaces is a serious problem with this technology.

The LCVD technique promises to overcome the above disadvantages of various film-depositing techniques. Photolytic LCVD is usually used for depositing electronic materials, and for isotope separation, while pyrolytic LCVD is used for applying oxidation, corrosion, and wear-resistant coatings of various materials. The advantages of pyrolytic and photolytic LCVD are listed below.

### 1.4.1. Deposition Rate Enhancement

Lydtin and Wilden (1973) reported the development of a new CVD technique at the Philips Research Laboratories in Aachen, Germany. They irradiated a substrate with a $CO_2$ laser beam for local heating, and the substrate was moved suitably to lay tracks of thin films. By using this new technique, which can be called pyrolytic LCVD, they deposited C, Si, and SiC films on Al substrates, and presented a figure to show a 0.35-mm-wide helical track of C resistor deposited on an Al rod 3.5 mm in diameter. Patterns of any shape can be deposited by using this technique and, therefore, the LCVD process can be used for manufacturing semiconductor devices. In many cases, the conventional CVD technique is used to deposit films on masked substrates, and then the pattern is obtained by machining or etching the unwanted material. However, such a method can become impracticable depending on the physicochemical properties of the film and substrate materials, and for this reason an electron beam is used to direction "write" patterns on substrates by heating the substrate locally. The vapor pressure inside the deposition chamber is very low in the electron-beam-assisted deposition technique, which leads to a very low deposition rate [Lydtin and Wilden (1973)] for all practical purposes. To overcome this problem, Lydtin and Wilden (1973) used the LCVD technique to deposit films for direct "writing" of patterns.

Steen studied the LCVD process both theoretically [Steen (1977)] and experimentally [Steen (1978)]. His theoretical studies involve the development of a mathematical model to determine the temperature distribution in a laser-heated spot on a substrate surface by considering that the laser beam operates with a coaxial nonreacting gas jet. Apart from the thermal history, he studied the variation in the temperature distribution due to various process parameters, such as the jet Reynolds number, the distance between the substrate and the nozzle mouth through which the gas jet exits, jet temperature, and reflectivity, which changes from the reflectivity value of the bare substrate to that of the film material as the deposition progresses. For experimental studies, he used a 30 W $CO_2$ laser to deposit CoO on a glass substrate by causing thermal decomposition of chelate (cobalt acetyl acetonate) vapor which was mixed with a $N_2$ carrier gas. The deposit was CoO because the jet impinging on the laser-heated spot was allowed to mix with air, and glass was used as the substrate because it allows measurement of the deposit thickness by using the color fringe and double-beam interference methods. He studied the effect of an increase in concentration on the deposition rate, and reported that A. Young [Steen (1978)] had deposited Si at the rate of 10 mm/min from silane vapor (200 mm Hg) and at the rate of 1 $\mu$m/min from 0.306 ppm silane in $N_2$ at atmospheric pressure, which Steen obtained from Young through a private communication. These results show that the film deposition rate increases as the concentration of the reactant is increased.

Allen (1980) carried out LCVD of Ni on $SiO_2$ (quartz), $TiO_2$ on $SiO_2$, TiC on $SiO_2$ and TiC on stainless steel according to the following overall reactions:

$$Ni(CO)_4 \xrightarrow{140°C} Ni + CO$$

$$TiCl_4 + H_2 + CO_2 \xrightarrow{900°C} TiO_2 + HCl + CO$$

$$TiCl_4 + CH_4 \xrightarrow{1200°C} TiC + HCl$$

where the quoted reaction temperatures refer to the minimum deposition temperatures for the equilibrium CVD process [Powell *et al.* (1955)]. A continuous wave (CW) $CO_2$ laser was used with a spot diameter of 0.6 mm at the $1/e^2$-point. For the LCVD of Ni, the deposition chamber was filled with $Ni(CO)_4$ at 40 Torr. It should be noted that the absorption spectrum of $Ni(CO)_4$ at the 9–11 μm wavelength range is dominated by a combination band at 920 cm$^{-1}$, according to Crawford and Cross (1938), and Allen (1980) reported that the absorption of a 1-cm pathlength at 10.6 μm (943 cm$^{-1}$) is less than 1% at 40 Torr of $Ni(CO)_4$, which means that the photolysis of $Ni(CO)_4$ at 10.6 μm is very small compared to its pyrolysis. Allen (1980) observed that the deposition rate of Ni film varies linearly with the deposition time during the initial stage of the deposition process, and reported that the Ni deposition rates are 1000 μm/min and 60 μm/min for incident laser powers 5 W and 0.5 W, respectively. For depositing $TiO_2$ film on $SiO_2$ substrates, Allen (1980) used $TiCl_4$, $H_2$, and $CO_2$ with partial pressures of 12.4, 205, and 205 Torr, respectively, noted that $TiCl_4$ is transparent to the $CO_2$ laser beam, and found that the deposition rates vary from 2 to 20 μm/min. The LCVD of TiC on quartz and stainless steel has been discussed in detail by Mazumder and Allen (1980). A pressure of 220 Torr of $CH_4$ was used in the deposition chamber which was attached to a container of vacuum-distilled liquid $TiCl_4$ (Allen, 1980). The lower absorptivity and higher thermal conductivity of stainless steel compared to those of quartz require the use of a higher (about 100 times larger) incident laser intensity for depositing TiC on stainless steel than on quartz, and for this reason, studies on the LCVD of TiC on stainless steel substrates have been carried out [Mazumder and Allen (1980)] by varying the incident laser power in the range from 240 to 750 W, and by focusing the laser beam on a spot of diameter 0.75 mm at the $1/e^2$-point. Mazumder and Allen (1980) observed that the deposited film thickness is directly proportional to the square root of the deposition time for a laser power of 400 and 600 W. Allen *et al.* (1983) have noted that very high deposition rates (> 100 μm/s) can be achieved by using the LCVD technique.

Bäuerle (1983a, 1983b, 1986) has discussed the LCVD of C from $C_2H_2$, $C_2H_4$, and $CH_4$, Si from $SiH_4$, Cd from $Cd(CH_3)_2$, and Ni from $Ni(CO)_4$, and noted that the LCVD rates are 100 to 1000 times greater than the maximal deposition rates obtained in the conventional CVD large-area flow reactors [Bäuerle (1983a)]. It should be noted that the LCVD process can be operated at higher partial pressures (several hundred mbar) of the reactants and at higher temperatures than the conventional CVD process, which could contribute to the deposition rate enhancement during LCVD. The conventional CVD reactors are usually operated at low partial pressures of the reactants to avoid film growth at the reactor wall. During the LCVD, the deposition occurs at the localized hot spot created by the laser beam on the substrate surface, and the rest of the LCVD reactor remains at a relatively low temperature. Therefore, the problem of deposition at the LCVD reactor wall does not arise even at high partial pressures (several hundred mbar) of the reactants. Bäuerle (1983b, 1986) has compared the deposition rates for Si film obtained from $SiH_4$ by the LCVD and the conventional CVD techniques, where the partial pressures of $SiH_4$ were 133 mbar and 1 mbar, respectively, and concluded that the LCVD rates are very high.

Moore *et al.* (1988) have reported that the conventional CVD rates are very low (100 to 1000 Å/min) because this technique relies on catalytic reactions at the hot substrate surface to dissociate the reactant molecules, whose binding energies are usually in the range of 1 to 3 eV, and also presented some data on the deposition rates, which are summarized in Table 1.2.

Table 1.2. Deposition Rates for Various Materials [after Moore *et al.* (1988)]

| Deposition method | Film material | Reactant | Deposition rate (Å/min) |
|---|---|---|---|
| LCVD using UV photons from excimer laser (0.248 μm) Photolytic process | Al<br>Mo<br>Cr<br>W | $Al(CH_3)_3$<br>$Mo(CO)_6$<br>$Cr(CO)_6$<br>$W(CO)_6$ | 500<br>2500<br>2000<br>1700 |
| Electron beam CVD. 4.7 kV, 16 mA/cm² | $SiO_2$ | Gas flow ratio: $N_2O/SiH_4/N_2 = 75/1/75$ Total pressure: 0.25 Torr | 500 |
| 2.3 kV, 13 mA/cm² | $Si_3N_4$ | Gas flow ratio: $NH_3/SiH_4/N_2 = 60/1/44$ Total pressure: 0.35 Torr | 200 |

## 1.4.2. Spatial Resolution and Control

The LCVD technique provides a very good means of controlling the film deposition process because the chemical reactions occur in a very small region at the localized hot spot on the substrate surface in the case of pyrolytic LCVD and along the path of the laser beam in the case of photolytic LCVD. Also, the dimension of the deposited film can be controlled in a very effective way. The diameter of a dot, which is deposited by keeping the laser beam and substrate stationary, and the width of a stripe, which is deposited by moving the substrate relative to the laser beam, depend on the size of the chemical reaction zone, which is defined as the region where the temperature is equal to or greater than the chemical reaction temperature at which the film material is produced. In the case of pyrolytic LCVD, the diameter of the dots can be controlled by adjusting the laser intensity and the irradiation time, and the width of the stripes can be controlled by regulating the laser intensity and the speed of the substrate relative to the laser beam.

For photolytic LCVD, the diameter of the dots and the width of the stripes depend on the diameter (spot size) of the laser beam on the substrate surface. The minimum laser beam diameter that can be attained theoretically for a given laser and a lens is the diffraction-limited spot size given by

$$D'_{min} = 2w_0 = \frac{4\lambda f}{\pi D_l}$$

for a Gaussian beam ($TEM_{00}$). In the far-field approximation ($z \to \infty$), the half-angle $\theta_{h_G} = \lambda/(\pi w_0)$ for a Gaussian beam ($TEM_{00}$), where the beam spread (divergence) is $2\theta_{h_G}$; $\theta_{h_G}$ can be related to the focal length of the focusing lens as

$$\theta_{h_G} = \frac{\lambda}{\pi w_0} = \frac{D_l}{2f}$$

to obtain the above expression for $D'_{min}$. However, in practice, multimode beams are usually encountered where the beam spread (divergence) $2\theta_{h_m}$ is always larger than the Gaussian beam spread $2\theta_{h_G}$. Accounting for this effect, we can write the expression for minimum spot size as

$$D_{min} = \frac{4M^2 f \lambda}{\pi D_l} \qquad (1.34)$$

where $M^2 = \theta_{h_m}/\theta_{h_G}$, and $\theta_{h_m}$ and $\theta_{h_G}$ refer to the half-angles of the multimode and Gaussian beams, respectively, which have the same radius at the waist. $M = 1$ and $M > 1$ refer to the Gaussian and non-Gaussian beams, respectively. It should be noted that in practice, the minimum spot size is about four times larger than the diffraction-limited spot size $D'_{min}$ because of the multimode characteristics of the beam and the spherical aberration caused by the focusing lens.

### 1.4.3. Minimum Damage to the Substrate and Film Materials

During LCVD, a very small area of the substrate surface is exposed to the laser beam to create a localized hot spot and the rest of the substrate remains at a relatively low temperature. Consequently, a small portion of the substrate experiences thermal stress resulting in limited substrate distortion.

Due to localized heating in LCVD, an extremely high heating rate (about $10^{10}$ K/s) [Allen et al. (1983)] is achieved, and the heat is transferred from the hot spot to the rest of the substrate very rapidly. As a result, the HAZ in the substrate is very small, and the grain size of the depositing film can be controlled since it depends on the cooling rate. Also, due to the rapid cooling rate, the film and substrate cool down so rapidly that there is not enough time for the substrate atoms to diffuse into the film, and so the film is not contaminated by the substrate atoms. Therefore, clean films can be deposited by the LCVD technique.

### 1.4.4. Ability to Process Temperature-Sensitive Materials

Materials processing at low temperatures is very important [Ibbs and Osgood (1989)], especially in microelectronics industries for fabricating very large-scale integrated (VLSI) circuits and electronic devices, because the performance of these devices and the high-density circuits depends on the dimensions and purity of the material. At high temperatures, the atoms diffuse from one layer of the material to another, which alters the concentration profile in the device, and thus the electronic properties of the device changes. Also, many other materials such as the compound semiconductor HgCdTe–GaAs–InP, which as the high vapor pressure elements Cd, As, P, decompose at high temperatures, and their composition changes due to the evaporative loss of the volatile elements. Even rapid changes in temperature (high heating and cooling rates) [Appleton and Allen (1982)] can generate defects in these types of compound semiconductors. These problems can be overcome by processing the temperature-sensitive materials at low temperatures.

Both pyrolytic and photolytic LCVD techniques provide means of achieving low processing temperatures. Photolysis can be carried out in two different ways, (1) photolysis by single-photon absorption, and (2) photolysis by multiphoton absorption. In the first method, the reactant is chosen in such a way that the reactant molecules absorb photons from an ultraviolet (or sometimes a visible) laser beam that leads directly to the breakage of the molecular bonds to yield the film material. In the second method, an intense infrared laser beam is used to induce multiple excitation of the vibrational modes of the reactant molecules by a multiphoton absorption process [Steinfeld *et al.* (1980)], which eventually leads to the dissociation of the reactant molecules to produce the film material. The pyrolytic low-temperature processing technique uses a short-pulse or a rapidly scanning laser beam to heat a small region on the substrate surface, where chemical reactions and film deposition take place. The optical absorption length of many semiconductors is less than 10 nm for ultraviolet excimer lasers, and due to the short irradiation time achieved with a short-pulse or a rapidly scanning laser beam, the thermal diffusion length is so small that the heated zone is about 1000 nm for a Si wafer [Ibbs and Osgood (1989)].

Although low-temperature processing has many advantages, it must be used with care, because in many cases, especially in photolytic deposition of metal films from vapor phases or metal–alkyl precursors, the deposited film is found to be nonuniform with poor electrical and mechanical properties. Such problems arise because at low temperatures, the desorption of the species adsorbed at the substrate surface is not complete, and the surface diffusion of the deposited atoms is very low, which causes columnar growth and the incorporation of impurities into the film. However, film quality can be improved by thermal treatment, which causes coalescence of the particulates of the deposited film [Ibbs and Osgood (1989)].

The LCVD technique has been used successfully for materials processing at low temperatures. Metal has been deposited by carrying out the photolysis of organic metallic precursors with a frequency-doubled Ar-ion laser [Ehrlich *et al.* (1982)], and metal lines have been drawn on very-low-melting-point plastics without damaging the plastic layer [Gilgen *et al.* (1984)]. Ehrlich *et al.* (1980) used a dual-wavelength approach to dope $p$-type InP compound semiconductors, which are difficult to dope with other techniques because of the dissociation of such materials near their melting points. They used an Ar-ion laser as a two-wavelength source and generated an UV wavelength (257 nm) laser of a few milliwatts by frequency-doubling the fundamental wavelength of the Ar-ion laser. The original visible (green) light of the Ar-ion laser source was used to heat a small region of the substrate surface; and near that hot spot, the dopant atoms were produced by photodissociating an organometallic precursor using the UV laser beam

# INTRODUCTION

to carry out doping by the solid state diffusion of the dopant atoms into the substrate.

## 1.4.5. Isotope Separation

A photolytic decomposition technique can be used to separate isotopes. It can lead to an efficient and economical process, especially for separating U isotopes, compared to other separation processes, such as gaseous diffusion, gas centrifuge, the Becker jet nozzle process, and electromagnetic separation [Gross (1974), Lamarsh (1983)]. The laser isotope separation technique relies on the dependence of the energies of the excited states of atoms and molecules on their atomic masses. Owing to the differences in the atomic masses of the isotopes, the atomic and molecular absorption spectra of the elements exhibit peaks, that is, resonant excitation energies, for various isotopes; which is called the isotope shift given by the difference between the resonant excitation energies of any two isotopes. The isotope that needs to be separated must have a considerable isotope shift so that it can be excited preferentially, for which a highly monochromatic light source, usually a laser, is required. The separation proceeds in two steps. In the first step, the isotope absorbs the photon of the laser beam and becomes excited due to the increase in its internal energy, and, in the second step, the desired isotope is separated from the mixture by a chemical or physical process.

The absorption spectrum of U shows that the energies of the excitation resonances for $^{235}$U and $^{238}$U are approximately at 5027.3 and 5027.4 Å, respectively, with an isotope shift of 0.1 Å [Lamarsh (1983)]. It should be noted that 5027 Å corresponds to the wavelength of green light with a photon energy of about 2.5 eV and, for this reason, the green light of the Cu vapor laser can be used to separate the $^{235}$U isotope. Zuber (1935) obtained a mixture of $^{200}$Hg and $^{202}$Hg isotopes by separating them from the $^{198}$Hg, $^{199}$Hg, $^{201}$Hg and $^{204}$Hg isotopes using the light of a mercury resonance lamp. The absorption peaks of $^{200}$Hg and $^{202}$Hg are much higher than those of the other mercury isotopes near the Hg 2537 Å line. The absorption spectrum of mercury near the 2537 Å wavelength has been given by Gross (1974), who has discussed as well the separation of $^{50}$Ti isotopes from TiCl$_4$ by using a CO$_2$ laser and a N$_2$ laser. An infrared CO$_2$ laser can be used to separate $^{50}$Ti because the isotope shift for $^{48}$Ti and $^{50}$Ti isotopes is well-defined near 10 μm. Also, $^{50}$Ti$^{35}$Cl$_4$ attains an excited state after absorbing an infrared photon, which exhibits a strong absorption peak at 3374 Å that corresponds to the wavelength of the N$_2$ laser. The thermal neutron absorption cross sections of pure $^{50}$Ti, $^{50}$TiO$_2$, $^{50}$Ti-alloy (92 wt.% $^{50}$Ti + 4 wt.% Al + 1 wt.% V + 3 wt.% Mo), pure Zr, and zircaloy are 140,

140, 265, 182, and 190 mbarns (1 barn = $10^{-24}\,cm^2$) [Gross (1974)]. In the same work, Gross also reported a few thermophysical and mechanical properties of these materials, noted that Ti is a good corrosion resistant material with excellent mechanical properties at high temperatures, and concluded that $^{50}$Ti can be used as a cladding material for nuclear reactor fuel elements instead of Z, and as a structural shell material for fast breeder and fusion reactors to minimize the radioactivity and after-heats produced in the shell as a result of high neutron fluxes.

Apart from the above isotopes, those of Cl and Ca [Kuhn and Martin (1932) and Kuhn et al. (1941)]; O [Liuti et al. (1966), Dunn et al. (1973)], and $H_2$ [Bazhin et al. (1974)] atoms have also been separated using a nonlaser photochemical technique in which the light of a resonance lamp containing mostly the isotope of interest is used. Gunning (1963), Gerard (1966), and Farrar and Smith (1972) have reviewed the nonlaser photochemical separation of isotopes in detail. However, laser is an efficient source of light because of its narrow spectral bandwidth which enables the separation of isotopes having a very small isotope shift. Due to the coherence of laser, the laser beam can be manipulated efficiently by using large $F$-number (ratio of the focal length of a lens to its diameter) lenses to maximize the photolytic reaction volume, as well as by focusing the beam on a tiny spot to increase its intensity, in order to induce nonlinear processes such as multiphoton absorption [Aldridge et al. (1976)]. Aldridge *et al.*, (1976), apart from their own results, included information about laser isotope separation that had been presented by several other authors.

Although the laser isotope separation technique was originally developed to separate U isotopes, and as noted above $^{235}$U and $^{50}$Ti isotopes ahve nuclear applications, the technique can also be used to separate various isotopes for other applications by suitable selection of the laser beam. For example, the isotopes of $N_2$, S, and C can be used as nonradioactive tracers for production control, agriculture, and environmental protection. Apart from these applications, laser isotope separation has scientific merit because it provides a tool to understand the chemical kinetics of the nonthermal processes encountered during laser-induced photochemical reactions. Such understanding is essential in order to synthesize new materials, develop catalysts, and improve management of radioactive waste [Aldridge *et al.* (1976)].

## 1.5. LITERATURE REVIEW

There is a vast amount of information about LCVD in the literature, including books, conference proceedings, journal articles, and reports. Some of these works are reviewed in this section, and the others will be discussed in the subsequent chapters.

### 1.5.1. Laser Excitation of Polyatomic Molecules

Steinfeld (1981) has four chapters, contributed by several authors, concerning various aspects of laser-induced chemical processes. In the first chapter of this reference, H. Galbraith and J. Ackerhalt discussed the multiphoton excitation and dissociation of polyatomic molecules such as $SF_6$. The most important dependent variable in the experimental studies of multiphoton excitation is the photon absorption rate, that is, the number of photons absorbed by the gas or vapor molecules per unit volume per unit time, and it is given by [Lyman et al. (1986), Allmen (1987)]

$$\dot{R} = N\sigma\phi^n$$

In the case of single-photon absorption, the order of the photon absorption rate is unity, that is $n = 1$. The important independent variables are laser fluence ($F = \int_0^\tau I\,dt$), laser pulselength, laser wavelength, and the gas temperature and pressure, whose effects on the average number of photons absorbed per unit volume, $\phi_a = \int_0^\tau \dot{R}\,dt$, can be summarized as follows:

1. Photon absorption $\phi_a$ increases as the laser fluence (J/cm$^2$) increases, and there is no threshold fluence for multiphoton excitation [Akhmanov et al. (1978)]. Lyman et al. (1986) have found that deviation from linear absorption ($n = 1$) occurs in $SF_6$ at very low fluences of magnitudes in the range of $\mu J/cm^2$.
2. The dependence of photon absorption $\phi_a$ on the laser pulselength is very weak [Lyman et al. (1986)].
3. Stafast et al. (1977) and Nowak and Lyman (1975) studied the effects of $CO_2$ laser frequencies on photon absorption in $SF_6$ gas. It is found that the temperature at which the absorption coefficient $[(mol/cm^3)^{-1} cm^{-1}]$ is a maximum decreases as the laser frequency increases, and the maximum absorption coefficient increases as the laser frequency increases [Nowak and Lyman (1975)]. The variation in the average number of absorbed photons with laser frequency shows that the absorption peak is slightly red-shifted; that is, the absorption peak shifts toward the longer wavelengths as the laser energy is increased.
4. Tsay et al. (1979) found that an increase in temperature leads to a substantial increase in the absorption of photons, especially for

photons with frequencies lower than the resonant frequency. Quigley (1978) found that an increase in pressure increases the absorption. Note that experiments are usually carried out at low pressures because the mean free path is very large at low pressures, which reduces the probability of intermolecular interactions, so that the behavior of an individual molecule in the photon field of the laser beam can be studied.

The effects of the above independent variables on multiphoton dissociation can be summarized as follows:

1. The dissociation of molecules proceeds only above a threshold value of the laser fluence because dissociation cannot occur below the activation energy. During the dissociation process, the molecules are excited to a level where the normal mode density of states is so large that a statistical rate model such as that of Rice, Ramsperger, Kassel and Marcus (RRKM) can be used to study the dissociation kinetics [Robinson and Holbrook (1972), Thiele et al. (1980)].
2. The dependence of the dissociation rate on the laser intensity is weak [Woodin et al. (1978)].
3. The effect of laser frequency on multiphoton dissociation is similar to the case of multiphoton excitation [Gower and Billman (1977)].
4. Dissociation usually decreases as the gas temperature increases [Duperrex and Van den Bergh, (1979a)]. As the gas pressure is increased, dissociation increases [Duperrex and Van den Bergh, (1979b)].

Besides the above experimental studies, mathematical models have also been developed to understand multiphoton absorption phenomena through classical and quantum mechanics [Walker and Preston (1977)]. Some of the difficulties of the classical models are as follows:

1. The classical models require knowledge of the potential energy surface, which is sometimes difficult to obtain [Truhlar (1981)].
2. In the classical models, Newton's equations of motion are integrated with respect to time over the laser pulse, which usually corresponds to a large number of oscillations (about $10^6$) of the normal modes. This requires that integration time steps be very small, which can introduce errors into the calculations, and lead to a long computational time.
3. For accurate statistics, the motion of each particle has to be studied for various initial phases of the particle. This also increases the computational time.

# INTRODUCTION

4. To reduce the computational time and error, the simulations are usually carried out for short times by using a high-power laser, bearing in mind that such calculations artificially introduce a large degree of coherence in the excitation. Actually, the results of such simulations must be compared with the experimental data obtained by using a high-power laser with short pulses to avoid any confusion that might arise due to the degree of coherence noted above [Black et al. (1979)].

Similar to the classical models, the quantum models also have a few difficulties, such as:

1. The distribution of energy over the normal modes is difficult to determine.
2. The intramolecular V-V relaxation phenomena are difficult to account for correctly in the calculations.

The quantum models can be classified into the following four categories:

1. Fully coherent Schrödinger equation dynamics [Ackerhalt and Galbraith (1978)].
2. Incoherent rate equation dynamics [Emanuel (1979)]. In this approach, no attempt is made to be consistent with spectroscopy.
3. Molecular quasi-continuum approach [Isenor et al. (1973)].
4. "Complete" model — a combination of the first-order spectroscopic and the quasi-continuum models — [Ackerhalt and Galbraith (1979)] that describes the multiphoton absorption phenomena from the ground state through dissociation.

## 1.5.2. Laser Excitation of Large Molecules

Danen and Jang [Steinfeld (1981), Chapter 2] have discussed multiphoton excitation and the reaction of "large" organic molecules, which they define as any organic compound that has five to six or more atoms other than hydrogen or a halogen, and a density of vibrational and rotational states higher than $10^3$ to $10^4 \, cm^{-1}$ at room temperature. According to this definition, $SF_6$ can be considered as a small inorganic molecule. The laser-induced excitation and dissociation phenomena can be described as follows. The small polyatomic molecules have discrete vibrational and rotational states at low energies, and an intense laser beam is required to excite the molecules through these discrete states via resonant absorption. Owing to the absorption of photons, the vibrational levels become anharmonic; that is, the spacings between the vibrational energy levels become unequal, leading to closer spacings between higher vibrational levels. This hinders the multiphoton excitation process, because the laser, which is tuned for the lowest vibrational energy transition, falls out of resonance for higher transitions. Excitation through these discrete states depends on the laser intensity rather than on the laser fluence. As the laser intensity increases, the density of states increases, multiphoton (about 3 to 4 photons) absorption takes place, and the molecules reach the vibrational quasi-continuum state. Molecules do not exhibit anharmonicity in this state, because the density of states is so large in the quasi-continuum regime that there is always a rotational–vibrational level in resonance with the laser light.

The laser fluence is found to be more important than the laser intensity for exciting the molecules up to and beyond the reaction threshold level from the quasi-continuum state. After the molecules attain the reaction threshold, either reactions can occur or the molecules become further excited, depending on the reaction rate constant and the laser pulselength. At this stage of photon absorption, the laser intensity again becomes more important than the laser fluence in determining the ultimate excited state that the molecules can achieve before the reaction becomes so fast that further excitation is impossible. Typically, for a laser of pulselength 100 ns ($10^{-7}$ s) and a molecule with a reaction rate constant of $10^7$ to $10^8$ s$^{-1}$, the reaction competes with the excitation of the molecule from the reaction threshold to higher levels. The photon absorption process is shown schematically in Fig. 1.5.

In infrared multiphoton absorption studies, difficulty sometimes arises in classifying the reaction as a thermal or a nonthermal process. Studies on infrared laser isotope separation and mode-selective chemical reactions [Brenner (1978), Kaldor et al. (1979)] suggest that a pulsed infrared laser can produce results different from conventional heating. In conventional heating, the molecules are excited in the following three steps:

1. Energy is first transferred from the hot walls of the reaction vessel into the translational motion of the molecules when a chemical is heated, which means that the molecules begin to travel faster.
2. Part of the translational energy is converted into rotational energy as a result of collisions among the translationally excited molecules, which means that the molecules begin to rotate (spin) in space.
3. Owing to further collisions among the molecules, part of the transitional and rotational energies is converted into vibrational energy, which means that the atoms of the molecule begin to vibrate faster,

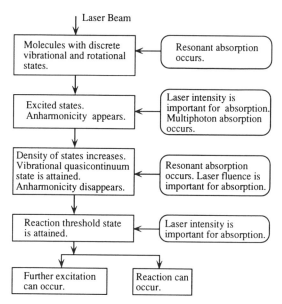

Figure 1.5. A schematic representation of multiphoton excitation and reaction processes.

and that the molecule is now translationally, rotationally, and vibrationally excited. It should be noted that although the total energy of the molecule, which is the sum of its various energies, is larger than the vibrational energy, only the vibrational energy induces unimolecular reactions because the interatomic bonds break only when the vibrational energy exceeds the threshold energy.

In the case of pulsed infrared laser-induced reactions, the photons of the laser beam can import energy directly to the atoms of a molecule to excite it vibrationally, while keeping it translationally and rotationally unexcited. Since vibrational excitation is necessary and sufficient to induce unimolecular reactions, such reactions are classified as nonthermal or thermal depending on the intramolecular and intermolecular distributions of the vibrational energy. A unimolecular reaction is considered a nonthermal or nonequilibrium process if the vibrational energy is statistically distributed intramolecularly throughout the molecule, which usually occurs when the molecule is excited by raising only its vibrational energy. On the other hand, a unimolecular reaction is considered a thermal process if the vibrational energy is distributed intermolecularly among various molecules. It should be noted that this definition of a thermal process differs from the usual concept of thermal equilibrium, which refers to the situation where there is an equilibrium distribution of the translational, rotational, and vibrational energies among various molecules.

To determine whether a unimolecular reaction is thermal or nonthermal, thermal monitor molecules are used in the reaction chamber [Danen et al. (1977), Tsang et al. (1978)]. Thermal monitor molecules can undergo a unimolecular thermal reaction and not absorb the laser radiation, and for this reason are referred to as chemical thermometers. Upon irradiation with a pulsed laser, if the laser light absorbing molecules undergo a reaction but the monitor molecules remain unreacted, then the reaction can be identified as a laser-specific nonequilibrium process. Many molecules are difficult to excite to the reaction threshold by direct absorption of light and, according to the first law of photochemistry [Calvert and Pitts (1966)], only light that is absorbed by a molecule can cause photochemical change in the molecule. For such cases, photosensitization [Shaub and Bauer (1975), Garcia and Keehn (1978)] is used to carry out photochemical reactions. In this technique, the mixture of the reactant and sensitizer is irradiated with a laser beam, where the sensitizer refers to a chemical agent that absorbs the laser light, transfers a large amount of the absorbed energy to the reactant molecules, remains unreacted, and does not participate chemically in the reaction process. If $S$ and $R$ represent the sensitizer and reactant molecules, respectively, then the main steps of the sensitizer-assisted reactions are as follows:

$$S \xrightarrow{\text{infrared laser}} S^{**}$$

$$S^{**} + R \xleftarrow{\text{many collisions}} S^{*} + R^{*}$$

$$R^{*} \longrightarrow \text{Products}$$

where $R^*$, $S^*$ and $S^{**}$ refer to the excited states of $R$ and $S$, respectively.

In view of the above discussion on multiphoton excitation and reaction, the following features of large molecules should be noted:

1. Due to the high density of vibrational and rotational states, large molecules are very close to their quasi-continuum state at room temperature. For this reason, they can undergo chemical reactions at a low reaction threshold temperature.
2. In the presence of a high-intensity ($W/cm^2$) laser, the rotational–vibrational state that is being pumped, can become depleted of small molecules. Such "hole burning" phenomena are not usually exhibited by large molecules.

3. Due to the high density of rotational and vibrational states and the number of degrees of freedom of large molecules, the probability of reactions occurring at a temperature higher than the reaction threshold temperature is very low, because the population of excited molecules above the reaction threshold energy is very low. For this reason, the vibrationally excited large molecules usually become deactivated due to collisions with unexcited reactant molecules or carrier gases, which decreases the reaction rate, resulting in long decomposition lifetimes. This phenomenon is often observed at low or moderate laser fluences ($J/cm^2$). Complete reactions of large molecules can be achieved at low pressures with a high-intensity laser.
4. Since the heat capacity of large molecules is due mainly to vibrational energy, the vibrational temperature of the molecules is very little affected by intramolecular relaxation of vibrational energy into translational and rotational energy.
5. Large molecules can be excited to various vibrational energy levels with a laser beam. The population of these vibrationally excited molecules is so broad that it can be approximated by a Boltzmann distribution.

On the other hand, small polyatomic molecules, such as $SF_6$, have: (a) a low density of low-energy vibrational states, which makes coherent excitation between the discrete levels possible, (b) high reaction threshold energies, and (c) high reaction rates, that is, short decomposition lifetimes.

Danen et al. (1977) and Danen (1980) studied multiphoton excitation and reaction of various esters experimentally. Ethyl acetate can be considered as a representative large molecule with a broad absorption band centered around $1055\,cm^{-1}$. For their studies, ethyl acetate was kept in small cells at pressures of 0.1 Torr or less and irradiated with a Lumonics Model 103 TEA (*t*ransversely *e*xcited *a*tmospheric) grating-tuned, multimode $CO_2$ laser by varying the laser fluences from 0.2 to $10\,J/cm^2$. The reaction probability for 0.05 Torr ethyl acetate was found to increase as the laser fluence increases; when the fluence is $10\,J/cm^2$ or more, the reaction probability becomes almost unity, that is, all of the irradiated ethyl acetate molecules undergo reaction. Such trends were observed for various $CO_2$ laser lines such as $P(12) = 1053.9\,cm^{-1}$, $P(16) = 1050.4\,cm^{-1}$, $P(18) = 1048.7\,cm^{-1}$, $P(20) = 1046.9\,cm^{-1}$, and $P(38) = 1029.4\,cm^{-1}$. However, the absorption cross section of ethyl acetate was found to decrease for the P(20) line and increase for the P(38) line as the laser fluence increases.

It should be noted that the P(20) and P(38) lines are near the single-photon absorption maximum and on the low-frequency side of the

O–CH$_2$ stretching mode of ethyl acetate molecules, respectively. Due to red-shifting of the absorption band (that is, the shifting of the absorption band toward longer wavelengths) at higher fluences, the absorption cross section for the P(38) line increases as the laser fluence increases. For the P(20) line, the decrease in absorption cross section as the laser fluence increases could be due to several factors, such as: (a) an anharmonic red-shifting of the absorption maximum; (b) more molecules undergo reactions at the early stages of the laser pulse and, consequently, fewer molecules are available to absorb the laser beam during the latter part of the pulse; and (c) saturation effects, that is, the molecules become saturated after absorbing a quantity of photons, and cannot absorb any more. However, the last two effects were found to be unimportant for ethyl acetate in the fluence range 0 to 3 J/cm$^2$. For the P(20) line, the average number of photons absorbed, the average number of photons absorbed per reacted molecule, and the reaction probability were found to increase, decrease, and increase, respectively, as the laser fluence increases for 0.05 Torr ethyl acetate. In addition to experimental studies, mathematical models [Black et al. (1977), Baldwin et al. (1979), Stephensen et al. (1979), Quack (1978), Robinson and Holbrook (1972)] that were originally developed for small polyatomic molecules have been extended to large molecules to study their multiphoton excitation and reaction phenomena.

### 1.5.3. Laser-Induced Power Generation

J. S. Haggerty and W. R. Cannon [Steinfeld (1981), Chapter 3] used laser-induced chemical reactions to produce ceramic powders of Si$_3$N$_4$ and SiC. Improved properties of ceramic materials (such as the hardness; erosion, oxidation and corrosion resistances; low density; high-temperature strength; and electrical and optical properties) make them excellent materials for many applications.

Ceramic materials can be used to manufacture high-performance engine parts, such as the all-ceramic turbine, caps and rings for pistons, rotors for turbochargers, injector nozzles, and inner walls of combustion chambers. However, ceramic materials are brittle and fail catastrophically. The strength of such materials can be improved if they are sintered properly to consolidate them into dense parts, which can be achieved if the powder has the following properties:

1. The particle size distribution of the powder must be uniform with a size of less than 0.5 μm.
2. The particles must not have any agglomerates and must be free of any contaminants and multiple polymorphic phases.

3. The microstructure of the particles must be equiaxed; that is, the particles must have spherical morphology.

When these requirements are fulfilled, the powder can be sintered to the theoretical density without using pressure or additives, and the final grain size and microstructure can be controlled to achieve the desired properties. Conventionally, $Si_3N_4$ and SiC are synthesized by using: (a) DC arc plasma techniques, (b) vapor phase reactions in heated tube furnaces, and (c) nitriding or carbiding of Si metal. However, these processes do not produce particles with the above properties. Although the arc plasma and furnace-heated vapor phase reactions produce very small particles with a narrow particle size distribution, the thermal profile in the reaction zone is such that the nucleation and growth times vary from point to point within the reaction zone, leading to the formation of agglomerates. In the nitriding or carbiding process, $Si_3N_4$ or SiC particles are formed on the surface of Si. The material has to be ground to separate the Si core, and the size distribution of nonagglomerated particles is found to be nonuniform. However, a dilute gas-phase reaction is considered to be a very good technique to synthesize ideal powders.

By using laser-induced gas phase reactions, Haggerty and Cannon [Steinfeld (1981), Chapter 3] produced $Si_3N_4$ and SiC particles with sizes mostly in the range of 15–20 nm. They used a horizontal $CO_2$ laser beam, and the reactant gases, such as $SiH_4$ and $NH_3$ for $Si_3N_4$ and $SiH_4$ and $CH_4$ or $SiH_4$ and $C_2H_4$ for SiC, were passed vertically through the laser beam. The chemical reactions proceed as follows:

$$3SiH_4(g) + 4NH_3(g) \xrightarrow{h\nu} Si_3N_4(s) + 12H_2(g)$$

$$SiH_4(g) + CH_4(g) \xrightarrow{h\nu} SiC(s) + 4H_2(g)$$

$$2SiH_4(g) + C_2H_4(g) \xrightarrow{h\nu} 2SiC(s) + 6H_2(g)$$

For the P(18) and P(20) $CO_2$ laser lines, the absorption coefficient ($cm^{-1}$ $atm^{-1}$) of $SiH_4$ reaches a maximum at about 0.025 atm, and the absorption coefficient of $NH_3$ reaches a minimum at about 0.25 atm. In a mixture having the ratio of $NH_3$ to $SiH_4$ as 10:1 with a total pressure of 1 atm, about 70% of the laser energy is absorbed in the first centimeter of the mixture, which makes the powder synthesis process very efficient. It should be noted that although $CH_4$ can be used to synthesize SiC from $SiH_4$, the pyrolysis of $CH_4$ proceeds through the formation of a $C_2H_4$ intermediate [Powell (1966)]. Also, the absorptivity of $C_2H_4$ for the P(20)

CO$_2$ laser line is high [Patty et al. (1974)] and, therefore, C$_2$H$_4$ would be a better choice to produce SiC from SiH$_4$.

Haggerty and Cannon [Steinfeld (1981), Chapter 3] carried out experimental studies using a reaction chamber as shown in Fig. 1.6. The laser beam, which can be focused or unfocused, enters into the reaction chamber horizontally through a KCl window that is cooled with Ar gas. The reactants are flown into the chamber vertically through a nozzle at a controlled pressure. Owing to the interactions between the laser beam and the reactant gases, the powder is produced and is collected at a microfiber filter between the reaction chamber and the vacuum pump. With a laser intensity of 760 W/cm$^2$, a reaction chamber pressure of 0.2 atm, SiH$_4$ and NH$_3$ flow rates of 11 and 110 cm$^3$/min, respectively, Ar flow rates to the window and to the annulus of 600 and 400 cm$^3$/min, respectively, Si$_3$N$_4$ powder was produced at the rate of 1 g/h. In addition to this type of result, Haggerty and Cannon [Steinfeld (1981), Chapter 3] presented a mathematical analysis of the laser-induced powder synthesis problem. The energy balance in any volume element of the reaction zone, where the reactants interact with the laser beam, can be written as

$$I\Delta A \exp\left(-\sum \alpha_i p_i x\right)\left[1 - \exp\left(-\sum \alpha_i p_i \Delta x\right)\right]$$
$$= C_p N_m \Delta V \frac{dT}{dt} + \Delta H \Delta V \frac{dN_m}{dt} + \text{heat transfer losses}$$

Assuming that all of the absorbed energy is converted to sensible heat, using $\exp(-\alpha p \Delta x) \approx 1 - \alpha p \Delta x$, $N_m = P/RT$, and taking the effect of temperature on absorption into account, that is, $\alpha p(T) = \alpha_0 p_0 (T_0/T)$,

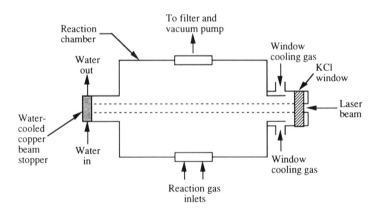

Figure 1.6. A typical laser-induced gas-phase reaction chamber for powder synthesis.

one can write

$$\Delta t = \frac{C_p(T_R - T_0)P}{I\alpha_0 RT_0 p_0}$$

for $x = 0$. Here $\Delta t$ is the threshold laser pulselength needed to heat the reactants to the reaction temperature $T_R$.

Assuming that the absorbed energy is lost to the surrounding walls by conduction through the gas, one can write the rate of heat loss per unit length from the hot cylindrical column of gas to the wall as

$$\frac{Q}{L} = 2\pi k \frac{T_i - T_0}{\ln(r_0/r_i)}$$

Also, the amount of heat produced per unit length of the absorbing gas column is given by

$$\frac{Q}{\Delta x} = I\pi r_i^2 \alpha p \exp(-\alpha p x)$$

Equating these two expressions, and setting $x = 0$ and $T_i = T_R$, one can write

$$p_{crit} = \frac{2k}{\alpha I r_i^2} \frac{T_R - T_0}{\ln(r_0/r_i)}$$

which represents the critical minimum pressure needed to induce the chemical reaction. An X-ray line broadening technique was used to estimate $Si_3N_4$ and SiC particle sizes using the following expression [Cullity (1956)]:

$$d = \frac{0.9\lambda_x}{B(p)\cos\theta_B}$$

where the particle size broadening $B(p)$ can be related to the machine broadening by

$$B^2(p) = B^2(h) - B^2(g)$$

It should be noted that the strength of a material depends on its grain size. For example, according to the Hall–Petch relation [Dieter (1986)], the yield stress of a material is given by

$$\sigma_0 = \sigma_i + \frac{k_d}{\sqrt{d_g}}$$

and the Herring–Coble [Coble (1963)] model leads to the following deformation rate $\dot{\varepsilon}$ of diffusional creep as a function of the grain size:

$$\dot{\varepsilon} = \frac{\sigma_a \Omega}{d_g^2 k_B T_s} \left( 10 D_v + \frac{148 \delta D_b}{d_g} \right)$$

Therefore, particle size is an important variable affecting the mechanical properties of any component fabricated from these powders.

To understand the agglomeration process, Mason's (1977) theory can be used under two different assumptions:

1. *Perikinetic solution:* Assume Brownian motion for the particles. For noninteracting spheres of equal size, the frequency of collision of a single sphere with any other sphere is

$$f_1 = \frac{8}{3} \frac{N_1 k_B T}{\mu}$$

2. *Orthokinetic solution:* Assume that the shearing action of the flowing medium brings the particles together. For equisized noninteracting spheres, the frequency of collision is

$$f_1 = \frac{32}{3} b^3 NG$$

However, only a fraction of the collisions lead to the agglomeration of particles where the particles are held together by surface forces [Easterling and Thölen (1972)]. Steinfeld (1981, Chapter 4) has presented a survey of laser-induced chemical reactions.

### 1.5.4. Thin Film Deposition

Eden (1992) has summarized various reactants used for depositing thin films. Al is the most commonly used conductor metal in integrated circuit technology. Cacouris *et al.* (1988) carried out laser direct writing of Al interconnects by photodissociation of dimethylaluminumhydride (DMAlH) with an Ar-ion laser of wavelengths 514, 350, and 275 nm, and a frequency-doubled $Ar^+$ laser (257 nm). They varied the laser fluence from 114 to 227 kW/cm$^2$, the spot size was 2.5 $\mu$m, the pressure inside the deposition chamber was 1 Torr, and the scan speed was varied from 0.4 to 100 $\mu$m/s. The film growth rate was found to be 0.09 $\mu$m/s for photolytic deposition at 114 kW/cm$^2$, and 7 $\mu$m/s for pyrolytic deposition at 227 kW/cm$^2$. If trimethylaluminum is used as the reactant instead of DMAlH, then C is produced, which leads to the formation of carbide.

Uesugi *et al.* (1988) studied the pyrolytic deposition of Mo lines from $Mo(CO)_6$ using a CW $Ar^+$ laser (515 nm wavelength, 6 $\mu$m beam diameter, 0.3 MW/cm$^2$ intensity) The $Mo(CO)_6$ vapor (1 Torr) was introduced into the deposition chamber at a pressure of 1 atm in pure Ar gas. In large-scale integrated circuit technology, the contact resistance between the direct written Mo lines and Al interconnections should be as low as possible (several tens of ohms or less), and the width of the Mo lines should be very small (2–3 $\mu$m or less). If a CW $Ar^+$ laser is used, heat is dissipated through Al interconnections due to conduction, which lowers the deposition temperature, degrades the deposited film quality, and makes the Mo lines wide. To overcome these problems, Uesugi *et al.* (1988) used short and high-peak-power pulsed laser (Q-switched Nd:YAG laser, 20-ns, 8-kHz pulses, 532 nm wavelength, 4 $\mu$m beam diameter, 1.1 MW/cm$^2$ intensity) to deposit Mo lines, and were able to reduce the contact resistance to several tens of ohms, which was about one-third of that obtained by the CW $Ar^+$ laser. Also, they were able to produce Mo lines of widths very close to the beam diameter. Lyons *et al.* (1988) carried out direct writing of conductive Ca lines in polymeric substrates, such as phenolformaldehyde paper laminate, using an ArF-pulsed excimer laser (193 nm), a Nd:YAG laser (1.06 $\mu$m), and a $CO_2$ laser (10.6 $\mu$m).

The temperature of the polymer substrate increases as it absorbs the laser, which decomposes the polymer to form C, and eventually a conductive C pattern is formed as the substrate is moved relative to the laser beam. The advantages of C patterns are that the same material can be used for both interconnections and contacts, and that the pattern can be formed easily by the laser direct writing technique. In the usual telephone keypads, Cu patterns, produced by plating and lithographic techniques, are used for interconnections; to meet the noncorrosive requirement of switches, gold or other noble metals, applied by screen printing of metal filled inks, are used for contacts. All these manufacturing steps and the use of costly metals, such as Cu and Au, can be eliminated if the laser technology for direct writing of C is used, because C is less expensive and noncorrosive.

Petzold *et al.* (1988) studied laser-induced localized deposition of metal on Si membranes with a view toward repairing defects on X-ray marks with films having high aspect ratios and high X-ray opacity. They compared the results of laser-pyrolytic deposition of W film from $W(CO)_6$ with those of photolytic deposition of Sn from $Sn(CH_3)_4$, and found that the latter process is better in terms of process cleanliness, deposition rates, X-ray opacity, and aspect ratio. The pyrolytic process led to mask distortions, back side deposition, and the diameter of the deposited W spots was larger than that of the laser spot size due to heat conduction. They used an $Ar^+$ laser (514 nm) and a frequency-doubled $Ar^+$ laser (257 nm) for the pyrolytic and photolytic experiments, respectively.

Natzle (1988) deposited Cr film along with some contaminants on the inner surface of a quartz window of a reaction chamber containing 110–150 mTorr $Cr(CO)_6$ by using a 325-nm CW He–Cd laser. He observed a dual morphology of the film, that is, the quartz surface was found to have a wide deposit area with a film thickness of less than $0.16\,\mu m$, and on the top of this film, the material deposited up to several microns in thickness over an area smaller than that of the $0.16\,\mu m$ thick deposit. Such dual morphology was also observed by other investigators [Arnone et al. (1986), Yokoyama et al. (1984), Froidevaux et al. (1982)]. Natzle (1988) used the dual morphology of the deposited film to distinguish the photolytic and pyrolytic deposition mechanisms. For this purpose, a transparent substrate is used to deposit a film of an opaque material by illuminating the back side of the substrate. As a laser does not heat up transparent substrates, the initial film growth must be due to the photolytic reaction. However, care must be taken to arrive at this conclusion, because although the photons of the laser beam can be absorbed by the reactant molecules to undergo photolytic dissociation, the laser beam can also heat up the reactants to induce thermal decomposition of the molecules. In any case, when the opaque film reaches a certain thickness, it prevents the transmission of the laser beam into the gas phase. If the laser intensity is strong enough to raise the temperature of the film to the thermal decomposition temperature of the reactant molecules, a pyrolytic reaction will occur at the interface of the film and the gas phase.

Nishizawa and Kurabayashi (1988) investigated GaAs vapor phase epitaxial growth, and found that the growth rate increases when the substrate zone is irradiated with an excimer (KrF) laser of wavelength 249 nm, and that the growth rate decreases when the Ga source zone is irradiated with a 249 nm (KrF) or 222 nm (KrCl) excimer laser at temperatures below 650°C. The effects of temperature and irradiation are shown in Fig. 1.7. It should be noted that III-V compound semiconductors are important for fabricating high-speed integrated circuits and microwave and optoelectronic devices for which the epitaxy of the film has to be controlled

INTRODUCTION 39

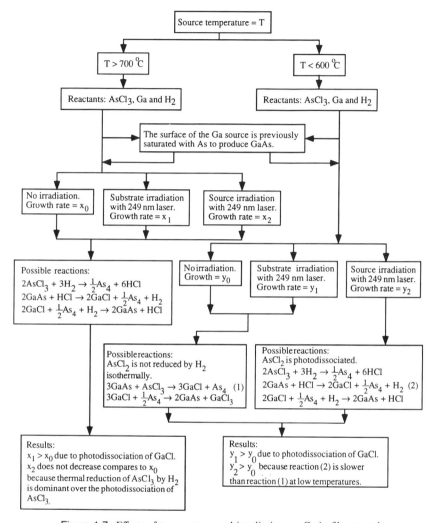

Figure 1.7. Effects of temperature and irradiation on GaAs film growth.

as exactly as the atomic dimension, and with the stoichiometry being accurately preserved. Also, low crystal growth temperature produces high-quality material and devices, because high-temperature processing introduces interstitials, vacancies, and other defects, and causes deviations in the stoichiometry. The photolytic reactions of trimethylgallium and arsine in the vapor phase have been found to occur at low temperatures in the presence of a 249-nm excimer laser:

$$Ga(CH_3)_3 \xrightarrow{h\nu} Ga + \tfrac{3}{2}C_2H_6 \text{ at room temperature}$$

$$Ga(CH_3)_3 + \tfrac{3}{2}H_2 \xrightarrow{h\nu} Ga + CH_4 \text{ above } 400°C$$

$$AsH_3 \xrightarrow{h\nu} As + \tfrac{3}{2}H_2 \text{ at room temperature}$$

$$Ga(CH_3)_3 + AsH_3 \xrightarrow{h\nu} Ga + As + 3CH_4 \text{ at room temperature}$$

Karam et al. (1988) studied selective epitaxial growth of III-V compounds based on GaAs using an $Ar^+$ LCVD on GaAs substrates. They used two techniques, conventional LCVD and *laser-assisted atomic layer* epitaxy (LALE) to grow GaAs films. In the conventional LCVD technique, the substrate is exposed to trimethylgallium (TMG), arsine ($AsH_3$), and the laser beam simultaneously. However, in the case of LALE, the substrate is exposed to TMG and the laser beam simultaneously, then the fluxes of TMG and laser irradiation are stopped, the $H_2$ is purged from the deposition chamber, and $AsH_3$ is introduced into the chamber. The film growth rate can be controlled to a few angstroms per second at a substrate temperature as low as 250°C by the conventional LCVD technique, but this usually produces Gaussian film thickness profiles. Flat top thickness profiles can be produced by direct writing of a GaAs monolayer using the LALE technique. The advantages of LCVD to produce III-V compound semiconductors are as follows:

1. It allows maskless selective epitaxial growth.
2. It allows *in-situ* multicomponent device integration on the same wafer.
3. It requires low substrate temperature.
4. It provides high spatial resolution for selective area deposition.
5. It can be used for a wide variety of materials.

In the MBE and MOCVD techniques for selective epitaxy, the entire wafer is heated up to the film growth temperature (600–800°C), which may damage other devices that are already present on the chip. Also, the wafers have to be cleaned before putting them into the reactor chamber to take care of the contamination problem arising from the organic materials used in photolithographic processes.

Zuhoski et al. (1988) deposited InSb film on GaAs substrates at room temperature by inducing photolysis of a mixture of trimethylindium (TMIn) and trimethylantimony (TMSb) in a $H_2$ carrier gas with an excimer laser of wavelength 193 nm (ArF) or 248 nm (KrF) Strained-layer superlattices of $InSb/InAs_xSb_{1-x}$ are promising materials for long-wavelength detectors in the 8–12 μm range. The MOCVD technique is not suitable for growing InSb film, because the melting temperature of InSb is 525°C and it is also used as the substrate material. Therefore, the film deposition has to be carried out below 525°C. At growth temperatures of 475 to 500°C, the rate of reaction of the reactant source chemicals is very low and, consequently, the film growth rate is extremely low.

Zinck et al. (1988) used a 248-nm (KrF) excimer laser-assisted *m*etal–*o*rganic *v*apor *p*hase *e*pitaxy (MOVPE) to deposit CdTe(111) films on a GaAs(100) substrate from a mixture of dimethylcadmium (DMC) and diethyltellurium (DET). CdTe and CdTe/HgTe superlattices are used for infrared detectors. Since the substrate temperature and reactant dissociation can be controlled independently in the photoassisted MOVPE technique, it is better than the thermal MOVPE. Also, photoassisted MOVPE allows the growing of films at low temperatures, which is important for growing CdTe/HgTe superlattices, where the formation of Hg vacancies and interdiffusion between layers have to be prevented. Although thermal MOVPE can be used to grow films at 250°C, this temperature is still so high that well-defined interfaces between the layers cannot be obtained. Besides the growth of II–VI superlattices at low temperatures, the growth of CdTe on other substrates is important for the integration of high-speed electronics and optical materials. Eres et al. (1988) deposited amorphous Si thin films at low temperatures by photodissociating disilane ($Si_2H_6$) with a pulsed ArF (193 nm) excimer laser. Amorphous Si is a good material for producing photovoltaic cells. Tachibana et al. (1988) produced amorphous C films at low temperatures by photodissociating $C_2H_3Cl$ or $CCl_4$ gas using a pulsed ArF (193 nm) excimer laser. Amorphous C has several important properties, such as extreme hardness, electrical resistivity, chemical inertness, optical transparency, and a large energy gap. Other techniques, such as low-energy C ion-beam deposition, electron beam deposition, sputtering, $H_2$ gas reactive RF sputtering, and RF plasma deposition can be used to generate amorphous C. However, the main advantages of the LCVD technique are as follows:

1. It is a low-temperature process.
2. It does not require ion bombardment.
3. It allows selective area deposition.

Padmanabhan et al. (1988) passivated GaAs MESFET (*m*etal *s*emiconductor *f*ield *e*ffect *t*ransistor) by depositing $Si_3N_4$ film using Hg as the photosensitizer and a set of UV lamps for photochemical activation of Hg and the reactants ($NH_3$ and $SiH_4$). GaAs MESFETs are used for signal processing, communications, displays, remote controls, and other high-frequency systems. involving the microwave region of the electromagnetic spectrum. However, the degradation in device characteristics due to the changes in channel thickness and channel surface is the major problem with such transistors. The presence of deep levels or traps at the interfaces, buffer layer, and substrate affects the channel thickness, and the trapping of surface charges forms an unstable depletion region on the channel surface. Photo-assisted CVD is considered to minimize the degradation arising due to the trapping of surface charges. Itoh and Yanai (1980) have investigated the effects of deep level traps on the channel thickness.

Nakamatsu et al. (1988) synthesized epitaxial SiC films on sapphire or $\alpha$-$Al_2O_3$ substrate by using the LCVD technique. Si carbides have many polytypes with different electronic properties, and can be used as semiconductors. However, it is difficult to grow SiC crystals because SiC decomposes at 2830°C and does not melt. Nakamatsu et al. (1988) used an ArF (193-nm) excimer laser and a mixture of $Si_2H_6$ and $C_2H_2$ gases, and were able to grow epitaxial films at 980°C. The ArF laser beam can be absorbed by $Si_2H_6$ and $C_2H_2$ in the gas phase and in the adsorbed layer, intermediate products, by-products, and the SiC film. The substrate was also irradiated with UV light from a deuterium ($D_2$) lamp. The light from the $D_2$ lamp was selected by sharp cut filters. Epitaxial and adherent film was found to grow when the wavelength of the light from the $D_2$ lamp was shorter than about 310 nm.

Irvine (1989) has summarized various aspects of the epitaxial growth and the importance of LCVD, which is termed photoepitaxy in this context, for growing films of temperature-sensitive materials. In the conventional thermally assisted epitaxial processes, hydrides are decomposed for depositing Si films, or metal–organics are used for III-V and II-VI semiconductors to achieve MOVPE, OMVPE, or MOCVD. However, photoepitaxy has several advantages over the conventional processes, because it allows selective area deposition, low processing temperature, in-situ processing, and independent control of the substrate temperature and precursor decomposition. Photoepitaxy is classified into three categories:

1. Photomodified epitaxy — refers to laser-enhanced doping or laser-induced change in alloy composition [Kisker and Feldman (1985), Roth et al. (1983)]. In this technique, the film growth rate is not enhanced compared to that of conventional methods.
2. Photoenhanced epitaxy — the film deposition rate is enhanced compared to that of conventional methods. Kisker and Feldman (1985) obtained enhanced epitaxial growth of CdTe by photolytic decomposition, and Roth et al. (1983) demonstrated enhanced low-temperature growth of GaAs by gas phase pyrolytic decomposition using a Nd:YAG laser.

3. Photoinduced epitaxy — refers to the process where deposition does not occur outside the laser-irradiated region.

Boyd (1989) has discussed the role of LCVD in doping semiconductors and producing $SiO_2$ for fabricating microelectronic devices. Usually, ion implantation is used for doping purposes, but this technique requires a high-temperature annealing cycle to rebuild the sample surface to its original crystalline perfection. Such high temperatures can affect the properties of the underlayers and other devices that might be present in the wafer. As submicron features are used to produce devices in VLSI technology, the high-temperature processing time should be minimized in the production line. Also, photolithography is used to define the areas to be doped. Laser processing can be used to overcome these problems. Apart from laser doping, laser oxidation is also of considerable importance, especially for the production of $SiO_2$, which is an important material in *m*etal *o*xide *s*emiconductor (MOS) technology. It has been found [Boyd (1987)] that optical radiation can not only oxidize various materials, but can also enhance the oxidation reaction. Usually, RF-heated quartz-tube furnaces are used to carry out oxidation at temperatures between 700 and 1200°C. However, optical radiation can be used to produce $SiO_2$. Heating rates can be as high as $10^{15}$ K/s when a picosecond laser is used, and it can be about 500 K/s when an incoherent light source is used. Laser allows localized processing, and the production of fine-scale patterns by using direct writing or projection methods.

## 1.5.5. Film Characterization

As discussed above, the dimensions of the deposited films are very small and, therefore, proper experimental techniques have to be used to characterize the properties and structures of these films. Murphy and Brueck (1983) used Raman spectroscopy for microanalysis of semiconductor structures. Raman scattering from optical phonons in semiconductor materials can be utilized to study the following characteristics:

STRESS: An applied stress causes lattice strain, which alters the phonon energy resulting in a shift of the optical phonon line.

DEFECT DENSITY: Lattice defects increase phonon scattering, which shortens phonon lifetime resulting in a broadened Raman linewidth.

SURFACE MORPHOLOGY: The radiation scattered from a roughened surface is depolarized and, in many cases, the intensity of the scattered radiation is more than that observed from the bulk material.

CRYSTALLINITY: The phonon lineshape and linewidth, and the polarization of the scattered radiation depend on the crystallite size and orientation in polycrystalline materials.

TEMPERATURE: The phonon population, which is related to the ratio of the intensity of the Stokes to anti-Stokes lines, can be used for lattice temperature measurement.

Kirillov and Merz (1983) used Raman scattering to measure lattice temperature inside the laser spot during CW laser heating of Si. Polycrystalline Si is important for photovoltaic devices and integrated circuit technology, but the grain boundaries affect many properties of the material. Poon et al. (1983) irradiated the grain boundary with a laser beam and measured the photoconductivity of the sample as a function of temperature to determine various grain boundary parameters, such as the barrier height, interface charge density, trap energy, and thermal capture cross section. Salathe and Gilgen (1983) measured laser-induced luminescence of direct gap semiconductors with high spatial ($1\,\mu m$) and temporal ($<1\,ns$) resolution. This technique relies on the radiative transitions of direct gap semiconductors between the states which are located near the conduction and valence band edges, and it allows the determination of temperature, carrier density, and band gap energy within the laser-irradiated zone.

## 1.6. LASER SAFETY

During LCVD or any other laser-aided materials processing, the safety aspects of handling the laser beam must be considered carefully [Steen (1991)]. Exposure to any laser beam can cause damage to the human body. The eyes are the most vulnerable, because they can be exposed to the beam directly during the beam-alignment, or the scattered beam might go into the eyes. To understand the safety features, lasers are classified into three regimes in the electromagnetic spectrum: ultraviolet, ocular focus, and infrared; and two major components of the eyes, that is, the cornea, which is an exterior component, and the retina, which is an interior component, are considered. The lens of the eye is located between the cornea and the retina. The ocular focus region spans the wavelength in the range from 0.4 to $1.4\,\mu m$, which contains the visible portion of the spectrum with wavelength 0.4 to $0.7\,\mu m$, and a portion of the infrared region with wavelength 0.7 to 1.4 μm. It should be noted that the eye's components can focus radiations of wavelengths 0.4 to 1.4 μm with a power of about $10^5$ times, which means that in this wavelength range, the intensity of the radiation that falls on the exterior surface of the eye will be amplified by the eye's lens by a factor of about $10^5$ on the retina. Due to this increase in intensity at the retina, even low-energy lasers of wavelengths from 0.4 to 1.4 μm can damage retinal tissues. Wavelengths lower than 0.4 μm or higher than 1.4 μm are not hazardous to the eye, because the cornea and other exterior parts of the eye absorb essentially the entire radiation at such wavelengths, and consequently the radiation is neither transmitted to the retina nor focused on the retina. For example, 90% of the energy of a 10.6-μm $CO_2$ laser is absorbed in a 25 μm thick layer of water [Meyer-Arendt (1984)]. Protective eyewear should be used whenever working with any laser beam.

The absorptive or filtering property of protective eyewear is expressed in terms of the optical density, which represents the ability of a material to absorb the wavelength of interest [Winburn (1990)]. According to the ANSI Z136.1 standard, the optical density $D_\lambda$ is defined as

$$D_\lambda = -\log_{10} \tau_\lambda$$

However, the following expression is also useful to define the optical density $O_d$:

$$O_d = \log_{10}(I_i/I_t)$$

For protective eyewear, the value of $O_d$ does not usually exceed 6. Based on the results of Meyer-Arendt (1984), Table 1.3 shows the threshold laser energy levels at wavelengths from 0.4 to 1.4 μm that can damage eyes that look directly into a laser beam. It should be noted that proper safety care

Table 1.3. Maximum Permissible Exposure at Wavelengths 0.4 to 1.4 μm [after Meyer-Arendt (1984)]

| Exposure condition | Exposure time for very short exposures | Maximum permissible exposures |
|---|---|---|
| Looking directly into a laser beam | 1 ns to $2 \times 10^{-5}$ s | 0.0005 mJ/cm² |
|  | $2 \times 10^{-5}$ to 10 s | $1.8t^{3/4}$ mJ/cm² (t is in seconds) |
|  | Exposures longer than 10 s | 10 mJ/cm² |
| Looking at diffusely reflected laser light | 1 ns to 10 s | $10t^{1/3}$ J/cm² sr, (t is in seconds) |

must be always taken even the laser energy is lower than the values given in Table 1.3. Maximum permissible exposures for other cases have been calculated by Winburn (1990).

BRIGHTNESS AND TEMPERATURE: To avoid any damage to the eyes, one should not look directly into a laser beam even if the laser power is very low. When the brightness of a laser source is compared with that of a thermal radiation source, the laser source exhibits the characteristics of a very high-temperature thermal radiation source. The brightness of a source is defined as the power emitted by the source at a frequency $v$ in a particular direction of interest, per unit solid angle about this direction, and per unit frequency interval $\Delta v$ about $v$ [Das (1991)], that is,

$$B_s = \frac{P_s}{(\Delta\Omega)(\Delta v)} \tag{1.13}$$

Brightness is essentially equal to the radiance per unit frequency interval $\Delta v$ about $v$. Noting that $\Delta\Omega \approx \lambda^2/A$ for a laser beam, one can express the brightness of a laser source with output power $P$ by [Das (1991)]

$$B_{\text{laser}} = \frac{P}{(\lambda^2/A)(\Delta v)} \tag{1.14}$$

Here $A$, which is the cross-sectional area (projected area) of the radiation source, is perpendicular to the direction of interest.

To compare $B_{\text{laser}}$ with the brightness $B_{\text{thermal}}$ of a thermal radiation source, let us consider a thermal radiation source at temperature $T$ K, and filter it temporally and spatially to obtain a brightness for the thermal source equal to that of the laser source. Using Planck's theory of blackbody radiation, the intensity $I_v$, that is, the energy of the radiation emitted by a blackbody per unit time, per unit area, per unit solid angle (steradian, sr), per unit frequency interval, is given by

$$I_v = \frac{2}{\lambda^2} \frac{hv}{\exp(hv/k_B T) - 1} \ [\text{W}/(\text{m}^2 \text{ sr s}^{-1})] \tag{1.15}$$

To determine the emissive power of the surface of area $A$, we consider a hemisphere whose base and center are in the plane and center of $A$, respectively. The emissive power $E_v$, that is, the energy emitted by the area $A$ in all the directions of the hemisphere per unit time, per unit area of the

# INTRODUCTION

emitting surface, per unit frequency interval, is given by

$$E_v = \pi I_v = \frac{2\pi}{\lambda^2} \frac{hv}{\exp(hv/k_B T) - 1} \, [\text{W}/(\text{m}^2 \text{s}^{-1})] \quad (1.16)$$

$E_v$ is sometimes referred to as the hemispherical spectral emissive power of a black surface. The surface of area $A$ emits radiation in infinitely many directions, with each ray penetrating through a hemisphere of unit radius. As the hemispherical solid angle is $2\pi$ sr, the total solid angle above the surface of area $A$ is $2\pi$ sr, and the surface radiates into this solid angle. Therefore, the brightness of the thermal radiation source can be written as

$$B_{\text{thermal}} = A \frac{1}{\lambda^2} \frac{hv}{\exp(hv/k_B T) - 1} \quad (1.17)$$

Equating equations (1.14) and (1.17), we obtain

$$T = \frac{hv}{k_B} \frac{1}{\ln(1 + hv \, \Delta v / P)} \approx \frac{P}{k_B \Delta v} \quad (1.18)$$

For a He–Ne laser with output power $P = 1 \, \text{mW}$ and $\Delta v \approx 10^9 \, \text{sr}^{-1}$, Equation (1.18) yields

$$T = \frac{10^{-3} \, \text{W}}{(1.3805 \times 10^{-23} \, \text{J/K})(10^9 \, \text{s}^{-1})} = 7.24 \times 10^{10} \, \text{K}$$

which illustrates that even a low-power laser source will appear to the eyes as a very high-temperature thermal radiation source and, for this reason, one should never look directly at a laser even if the laser power is very low.

## NOMENCLATURE

| | |
|---|---|
| $A$ | Cross-sectional area (projected area) of the radiation source. This area is perpendicular to the direction of interest (see Section 1.6.1) |
| $\Delta A$ | Cross section of the elemental volume |
| $b$ | Radius of the particles |
| $B(g)$ | Machine broadening |
| $B(h)$ | Peak breadth at half-height |
| $B(p)$ | Peak breadth of half-intensity |

$B_{laser}$ — Brightness of the laser source
$B_s$ — Brightness of any source $s$
$B_{thermal}$ — Brightness of the thermal radiation source
$C_p$ — Specific heat of the gas
$d$ — Particle size
$d_f$ — Depth of focus
$d_g$ — Grain diameter
$d_1$ — Distance between the lens and the waist of radius $w_1$
$d_2$ — Distance between the lens and the waist of radius $w_2$
$D_b$ — Boundary diffusivity
$D_v$ — Volumetric diffusivity
$D_l$ — Laser beam diameter on the lens
$D_\lambda$ — Optical density according to the ANSI Z136.1 standard
$D_{min}$ — Diffraction-limited multimode laser beam diameter (spot size), see p. 20
$D'_{min}$ — Diffraction-limited Gaussian laser beam diameter (spot size), see p. 20
$E(r)$ — Electric field at any location $r$
$E_0$ — Peak of the amplitude of the electric field
$E_p$ — Energy of a pulse
$f$ — Focal length of the lens
$f_1$ — Frequency of collision of a single sphere with any other sphere
$F$ — F number of the lens, $F = f/D_l$ (in Section 1.4.2).
$F$ — Laser fluence (J/m$^2$)
$G$ — Gradient of the fluid flow velocity
$h$ — Planck constant
$\Delta H$ — Heat of reaction (J/mole)
$i, j,$ and $k$ — Unit vectors in the $x$, $y$, and $z$ directions, respectively
$I$ — Laser intensity (W/m$^2$)
$I(x, y, z)$ — Laser intensity at a point $(x, y, z)$ in Cartesian coordinates
$I_i$ — Intensity of the incident light
$I_0$ — Laser intensity at $r = 0$
$I_t$ — Intensity of the transmitted light
$k$ — Thermal conductivity of the gas
$k_B$ — Boltzmann constant
$k_D$ — "Locking parameter" which measures the contribution of the grain boundaries to the relative hardening
$M$ — Laser beam quality
$n$ — Order of the photon absorption rate
$N$ — Number density of the gas or vapor molecules absorbing photons

# INTRODUCTION 49

| | |
|---|---|
| $N_1$ | Number density of the particles |
| $N_m$ | Molar density of the gas |
| $dN_m/dt$ | Moles of gas reacting per unit volume per unit time |
| $O_d$ | Optical density |
| $p_{crit}$ | Minimum critical pressure needed to induce chemical reactions |
| $p_i$ | Partial pressure of the $i$th species |
| $p_0$ | Pressure at temperature $T_0$ |
| $P$ | Laser power (in Sections 1.2.1 and 1.6.1) |
| $P$ | Pressure at temperature $T$ |
| $P_s$ | Output power of any source $s$ |
| $q_1$ | Complex ray parameter at the waist before the beam passes through the lens |
| $q_2$ | Complex ray parameter at the waist after the beam passes through the lens |
| $q'_1$ | Complex ray parameter at the incident surface of the lens |
| $q'_2$ | Complex ray parameter at the transmitting surface of the lens |
| $q(z)$ | Complex ray parameter |
| $Q/\Delta x$ | Heat produced per unit length per unit time |
| $Q/L$ | Heat loss per unit length per unit time |
| $r_i$ | Radius of the inner cylindrical column of heated gas |
| $r_0$ | Radius of the outer wall of the reaction chamber |
| $R$ | Universal gas constant |
| $\dot{R}$ | Photon absorption rate per unit volume per unit time |
| $R(z)$ | Radius of the curvature of the laser phase front (in Section 1.2.4) |
| $R'_1$ | Radius of curvature of the laser phase front at the incident surface of the lens |
| $R'_2$ | Radius of curvature of the laser phase front at the transmitting surface of the lens |
| $t$ | Time variable |
| $\Delta t$ | Threshold laser pulse length to induce chemical reactions |
| $t_p$ | Half-width of the pulse duration between the $1/e$-point of the intensity varying with time |
| $T$ | Temperature of the particles in Kelvin |
| $T$ | Temperature of the thermal radiation source (in Section 1.6.1) |
| $T_i$ | Temperature of the inner cylindrical column of the heated gas |
| $T_R$ | Chemical reaction threshold temperature |
| $T_s$ | Temperature of the material |
| $T_0$ | Initial temperature of the outer wall of the reaction chamber |
| $dT/dt$ | Rate of change of temperature |
| $\Delta V$ | Volume of the elemental volume |
| $w(z)$ | Spatial variation of the beam radius along the $z$-axis |

| | |
|---|---|
| $w_1$ | Radius of the waist of the beam before it passes through the lens |
| $w_2$ | Radius of the waist of the beam after it passes through the lens |
| $w_e$ | Radius of the beam where $I(r) = I_0/e$ |
| $w_0$ | Radius of the waist of the laser beam |
| $x$ | Distance of the elemental volume from the KCl window (in Section 1.5.3) |
| $x$ | $x$-coordinate of a point |
| $\Delta x$ | Thickness of the elemental volume |
| $y$ | $y$-coordinate of a point |
| $z$ | Distance from the focusing lens along the direction of the laser beam propagation |
| $z_R$ | Rayleigh range, or confocal parameter |

GREEK SYMBOLS

| | |
|---|---|
| $\alpha$ | Absorption coefficient at temperature $T$ |
| $\alpha_i$ | Absorption coefficient of the $i$th species |
| $\alpha_0$ | Absorption coefficient at temperature $T_0$ |
| $\delta$ | Thickness of the grain boundary |
| $\dot{\varepsilon}$ | Deformation rate |
| $\theta$ | Angle between a laser ray and the $z$-axis |
| $\theta_B$ | Bragg angle in radians |
| $\theta_{h_G}$ | Half-angle for a Gaussian (TEM$_{00}$) beam with spot size $2\omega_0$ |
| $\theta_{h_m}$ | Half-angle for a multimode beam with spot size $2\omega_0$ |
| $\lambda$ | Wavelength of the laser |
| $\lambda_x$ | Wavelength of the X-ray beam |
| $\mu$ | Viscosity of the fluid medium containing particles of concentration $N_i$ |
| $\nu$ | Frequency of the emitted radiation |
| $\Delta \nu$ | Frequency interval about the frequency $\nu$ |
| $\sigma$ | Photon absorption cross section with dimension $m^{2n}s^{n-1}$ |
| $\sigma_a$ | Applied tensile stress |
| $\sigma_i$ | Friction stress, which represents the resistance of the crystal lattice to dislocation movement |
| $\sigma_0$ | Yield stress |
| $\tau$ | Time for irradiation with a laser beam |
| $\tau_\lambda$ | Spectral transmittance |
| $\phi$ | Azimuthal angle for the laser ray measured from the $x$-axis in the $x$-$y$ plane |
| $\phi$ | Photon flux |
| $\phi_a$ | Absorption of photon per unit volume |
| $\Omega$ | Vacancy volume |
| $\Delta \Omega$ | Solid angle into which the source emits radiation |

# REFERENCES

Ackerhalt, J., and Galbraith, H. (1978), Collisionless Multiple Photon Excitation of $SF_6$: A Comparison of Anharmonic Oscillators with and without Octahedral Splitting in the Presence of Rotational Effects, *J. Chem. Phys.* **69**: 1200.

Ackerhalt, J., and Galbraith, H. (1979), Collisionless Multiple Photon Excitation in $SF_6$: Thermal or Not?, in *Laser Spectroscopy IV*, Proc. 4th International Conference, Rottach-Egern, Springer Series in Optical Sciences, Walther, H., and Rothe, K., eds., Springer-Verlag, New York.

Akhmanov, S., Gordienkv, V., Mikhunko, A., and Panchenko, V. (1978), Dependence of the Rate of Vibrational-Translational Relaxation in $SF_6$ on the Intensity of Selective Laser Excitation, *JETP Lett.* **26**, 453.

Aldridge III, J. P., Birely, J. H., Cantrell III, C. D., and Cartwright, D. C. (1976), Experimental and Theoretical Studies of Laser Isotope Separation, in: *Laser Photochemistry, Tunable Lasers, and Other Topics*, Jacobs, S. F., Sargent III, M., Scully, M. O., and Walker, C. T., eds., Addison-Wesley, Reading, pp. 57–144.

Allen, S. D. (1980), Laser Chemical Vapor Deposition—Applications in Materials Processing, in: *Laser Applications in Materials Processing*, Society of Photo-Optical Instrumentation Engineers (SPIE), Vol. 198, August 27–28, 1979, San Diego, CA, Ready, J. F., ed., Soc. Photo-Optical Instrumentation Engineers, Washington, pp. 49–56.

Allen, S. D., Trigubo, A. B., and Jan, R. Y. (1983), Direct Writing Using Laser Chemical Vapor Deposition, In: *Laser Diagnostics and Photochemical Processing for Semiconductor Devices*, Materials Research Soc. Symp. Proc., Vol. 17, Osgood, R. M., Brueck, S. R. J., and Schlossberg, H. R., eds., Elsevier, New York, pp. 207–214.

Allmen, M. V. (1987), *Laser-Beam Interactions with Materials: Physical Principles and Applications*, Springer-Verlag, New York, p. 34.

Appleton, B. R., and Allen, G. K., eds. (1982), *Laser and Electron Beam Processing*, Elsevier, New York.

Arnone, C., Rothschild, M., Black, J. G., and Ehrlich, D. J. (1986), *Appl. Phys. Lett.* **48**, 1018.

Baldwin, A. C., Barker, J. R., Golden, D. M., Duperrex, R., and Van den Bergh, H. (1979), Infrared Multiphoton Chemistry: Comparison of Theory and Experiment, Solution of the Master Equation, *Chem. Phys. Lett.* **62**, 178.

Bäuerle, D. (1983a), Production of Microstructures by Laser Pyrolysis, in: *Laser Diagnostics and Photo-Chemical Processing for Semiconductor Devices*, Materials Research Soc. Symp. Proc., Vol. 17, Osgood, R. M., Brueck, S. R. J. and Schlossberg, H. R., eds., Elsevier, New York, pp. 19–28.

Bäuerle, D. (1983b), Laser Induced Chemical Vapor Deposition, in: *Surface Studies with Lasers*, Proc. Int. Conf., Mauterndorf, Austria, March 9–11, Aussenegg, F. R., Leitner, A., and Lippitsch, M. E., eds., Springer-Verlag, New York, pp. 178–188.

Bäuerle, D. (1986), *Chemical Processing with Lasers*, Springer-Verlag, New York, pp. 42 and 93.

Bazhin, H. M., Skubnevskaya, G. I., Sorokin, N. I., and Molin, Y. N. (1974), *JETP Lett.*, **20**, 18.

Biyikli, S., and Modest, M. F. (1988), Beam Expansion and Focusing Effects on Evaporative Laser Cutting, *ASME J. Heat Transfer* **110**, 529.

Black, J., Kolodner, P., Shultz, M., Yablonovitch, E., and Bloembergen, N. (1979), Collisionless Multiphoton Energy Deposition and Dissociation of $SF_6$, *Phys. Rev. A* **19**, 704.

Black, J., Yablonovitch, E., Bloembergen, N., and Mukamel, M. (1977), Collisionless Multiphoton Dissociation of $SF_6$: A Statistical Thermodynamics Process, *Phys. Rev. Lett.* **38**, 1131.

Boyd, I. W. (1989), Doping and Oxidation, Chapter 10, in: *Laser Microfabrication: Thin Film Processes and Lithography*, Ehrlich, D. J., and Tsao, J. Y., eds., Academic, New York, pp. 559–580.

Boyd, I. W. (1987), *Laser Processing of Thin Films and Microstructures*, Springer-Verlag, New York.

Breinan, E. M., Kear, B. H., and Banas, C. M. (1976), Processing Materials with Lasers, *Physics Today* **29**, 44–50.

Brenner, D. M. (1978), Infrared Multiphoton-Induced Chemistry of Ethyl Vinyl Ether: Dependence of Branching Ratio on Laser Pulse Duration, *Chem. Phys. Lett.* **57**, 357.

Cacouris, T., Scelsi, G., Scarmozzino, R., Osgood, Jr., R. M., and Krchnavek, R. R. (1988), Laser Direct Writing of Aluminum, in: *Laser and Particle-Beam Chemical Processing for Microelectronics*, Materials Research Soc. Proc., Vol. 101, December 1–3, 1987, Boston, MA, Ehrlich, D. J., Higashi, G. S., and Oprysko, M. M. eds., Materials Research Society, Pittsburgh, pp. 43–48.

Calveert, J. G., and Pitts, J. N., Jr. (1966), *Photochemistry*, Wiley, New York, p. 19.

Coble, R. L. (1963), A Model for Boundary Diffusion-Controlled Creep in Polycrystalline Materials, *J. Appl. Phys.* **34**, 1679.

Crawford, B. L., and Cross, P. C. (1938), *J. Chem. Phys.* **6**, 535.

Cullity, B. D. (1956), *Elements of X-Ray Diffraction*, Addison-Wesley, Reading, 261 pp.

Danen, W. C. (1980), Pulsed Infrared Laser Induced Organic Chemical Reactions, *Opt. Eng.* **19**, 21.

Danen, W. C., Munslow, W. D., and Setser, D. W. (1977), Infrared Laser Induced Organic Reactions: (1) Irradiation of Ethyl Acetate With a Pulsed $CO_2$ Laser. Selective Inducement vs. Thermal Reaction, *J. Am. Chem. Soc.* **99**, 6961.

Das, P. (1991), Lasers and Optical Engineering, Springer-Verlag, New York, pp. 237–240.

Dieter, G. E. (1986), *Mechanical Metallurgy*, 3rd Ed., McGraw-Hill, New York, 189 pp.

Dunn, O., Harteck, P., and Dondes, S. (1973), *J. Phys. Chem.* **77**, 878.

Duperrex, R., and Van den Bergh, H. (1979a), Temperature Dependence in the Multiphoton Dissociation of $^{32}SF_6$, *J. Chem. Phys.* **70**, 5672.

Duperrex, R., and Van den Bergh, H. (1979b), Competition Between Collisions and Optical Pumping in Unimolecular Reactions Induced by Monochromatic Infrared Radiation, *J. Chem. Phys.* **40**, 275.

Easterling, K. E., nd Thölen, A. R. (1972), *Acta Met.* **20**, 1001.

Eden, J. G. (1992), *Photochemical Vapor Deposition*, Wiley, New York.

Ehrlich, D. J., Osgood, R. M., and Deutsch, T. F. (1980), *Appl. Phys. Lett.* **36**, 916.

Ehrlich, D. J., Osgood, R. M., and Deutsch, T. F. (1982), *J. Vac. Sci. Technol.* **21**, 23.

Emanuel, G. (1979), A Simple Model for Multiphoton Molecular Absorption, *J. Quant. Spectros. and Radiat. Transfer* **21**, 147.

Eres, D., Lowndes, D. H., Geohegan, D. B., and Mashburn, D. N. (1988), Laser Photochemical Growth of Amorphous Silicon at Low Temperatures and Comparison with Thermal Chemical Vapor Deposition, in: *Laser and Particle-Beam Chemical Processing for Microelectronics*, Materials Research Soc. Proc., Vol. 101, Symposium held December 1–3, 1987, Boston, MA, Ehrlich, D. J., Higashi, G. S., and Oprysko, M. M., eds., Materials Research Society, Pittsburgh, pp. 355–360.

Farrar, Jr., R. L., and Smith, D. F. (1972), ORGDP Report KL 3054, Rev. 1, March.

Froidevaux, Y. R., Salathe, R. P., Gilgen, H. H., and Weber, H. P. (1982), Cadmium Deposition on Transparent Substrates by Laser-Induced Dissociation of $Cd(CH_3)_2$ at Visible Wavelengths, *Appl. Phys.* **A.27**, pp. 133.

Garcia, D., and Keehn, P. M. (1978), Organic Chemistry by Infrared Lasers: (2) Retro-Diels-Alder Reactions, *J. Am. Chem. Soc.* **100**, 6111.

Gerard, M. (ed), (1966), ONRL-TR-1045, *Isotopes and Radiation Technology*, Vol. 3, p. 200.

Gilgen, H. H., Chen, C. J., Krchnavek, R., and Osgood, R. M. (1984), *Laser Processing and Diagnostics*, Bäerle, D., ed., Springer-Verlag, New York.

Gower, M., and Billman, K. (1977), Collisionless Dissociation and Isotopic Enrichment of $SF_6$ Using High Powered $CO_2$ Laser Radiation, *Opt. Comm.* **20**, 123.

Gross, R. W. F. (1974), Laser Isotope Separation, *Opt. Eng.* **13**, 506.

Gunning, H. E. (1963), *J. Chem. Phys.* **60**, 197.

Ibbs, K. G., and Osgood, R. M. (1989), Applications of Laser Chemical Techniques to Microelectronics Fabrication, in: *Laser Chemical Processing for Microelectronics*, Ibbs, K. G., and Osgood, R. M., eds., Cambridge University Press, New York, pp. 5–7.

Irvine, S. J. C. (1989), Epitaxy, Chapter 9, in: *Laser Microfabrication: Thin Film Processes and Lithography*, Ehrlich, D. J., and Tsao J. Y., eds., Academic, New York, pp. 503–538.

Isenor, N., Merchant, V., Hallsworth, R., and Richardson, M. (1973), $CO_2$ Laser-Induced Dissociation of $SiF_4$ Molecules into Electronically Excited Fragments, *Canad. J. Phys.* **51**, 1281.

Itoh, T., and Yanai, H. (1980), *IEEE Trans. Electron. Devices*, **ED-27** (6), 1037.

Kaldor, A., Hall, R. B., Cox, D. M., Horsley, J. A., Rabinowitz, P., and Kramer, G. M. (1979), Infrared Laser Chemistry of Large Molecules, *J. Am. Chem. Soc.* **101**, 4465.

Karam, N. H., Liu, H., Yoshida, I., Katsuyama, T., and Bedair, S. M. (1988), Laser-Enhanced Selective Epitaxy of III-V Compounds, in: *Laser and Particle-Beam Chemical Processing for Microelectronics*, Materials Research Soc. Proc., Vol. 101, Symposium held December 1–3, 1987, Boston, MA, Ehrlich, D. J., Higashi, G. S., and Oprysko, M. M., eds., Materials Reseach Society, Pittsburgh, pp. 285–290.

Kirillov, D., and Merz, J. L. (1983), Raman Scattering as a Temperature Probe for Laser Heating of Si, in: *Laser Diagnostics and Photochemical Processing for Semiconductor Devices*, Materials Research Soc. Symp. Proc., Vol. 17, Osgood, R. M., Brueck, S. R. J., and Schlossberg, H. R., eds., Elsevier, New York, pp. 95–102.

Kisker, D. W., and Feldman, R. D. (1985), Photon-Assisted OMVPE Growth of CdTe, *J. Cryst. Growth* **72**, 102.

Koebner, H., (ed.), (1984), *Industrial Applications of Lasers*, Wiley, New York.

Kogelnik, H. (1965), Imaging of Optical Modes — Resonators with Internal Lenses, *Bell. Sys. Tech. J.* **44**, 455.

Kuhn, W., and Martin, H. (1932), *Naturwissen.* **20**, 772.

Kuhn, W., Martin, M., and Eldau, K. H. (1941), *Z. Phys. Chem. Abt.* **50B**, 213.

Lamarsh, J. R. (1983), *Introduction to Nuclear Engineering*, Addison-Wesley, Reading, MA, pp. 168–175.

Laser Institute of America (1977), *Guide for Material Processing by Lasers*, Paul M. Harrod, Baltimore, p. 4–6.

Li, Y. (1993), Accurate Approximation of the Focal Shift in the Transformation of Truncated Gaussian Beams, *Opt. Eng.* **32**, 774.

Liuti, G., Dondes, S., and Harteck, P. (1966), *J. Chem. Phys.* **44**, 4052.

Lydtin, H., and Wilden, R. (1973), Deposition of Metal: Laser-Aided Technique, *Metals and Materials* **7**, 159.

Lyman, J. L., Quigley, G. P., and Judd, O. P. (1986), Single-Infrared-Frequency Studies of Multiple-Photon Excitation and Dissociation of Polyatomic Molecules, in: *Multiple-Photon Excitation and Dissociation of Polyatomic Molecules*, Cantrell, C. D., ed., Vol. 35 of Topics in Current Physics, Springer-Verlag, New York, pp. 9–94.

Lyons, A. M., Wilkins, Jr., C. W., and Mendenhall, F. T. (1988), Direct Writing of Carbon Interconnections, in: *Laser and Particle-Beam Chemical Processing for Microelectronics*, Materials Research Soc. Proc., Vol. 101, Symposium held December 1–3, 1987, Boston, MA, Ehrlich, D. J., Higashi, G. S., and Oprysko, M. M., eds., Materials Research Society, Pittsburgh, pp. 67–73.

Mason, S. C. (1977), *J. Colloid. and Interf. Sci.* **58**, 275.
Mazumder, J. (1991), Overview of Melt Dynamics in Laser Processing, *Opt. Eng.* **30**, 1208.
Mazumder, J., and Allen, S. D. (1980), Laser Chemical Vapor Deposition of Titanium Carbide, in: *Laser Applications in Materials Processing*, San Diego, CA, Proc. Society of Photo-Optical Instrumentation Engineers (SPIE), Vol. 198, August 27–28, 1979, San Diego, CA, Ready, J F., ed., Soc. Photo-Optical Instrumentation Engineers, Washington, pp. 73–80.
Meyer-Arendt, J. R. (1984), *Introduction to Classical and Modern Optics*, Second Edition, Prentice-Hall, Englewood Cliffs, pp. 516–517.
Moore, C. A., Yu, Z. Q., Thompson, L. R., and Collins, G. J. (1988), Laser and Electron Beam Assisted Processing, *Handbook of Thin-Film Deposition Processes and Techniques: Principles, Methods, and Equipment and Applications*, Schuegraf, K. K., ed., Noyes Publications, Park Ridge, N.J., pp. 318–343.
Murphy, D. V., and Brueck, S. R. J. (1983), Optical Microanalysis of Small Semiconductor Structures, in: *Laser Diagnostics and Photochemical Processing for Semiconductor Devices*, Materials Research Soc. Symp. Proc., Vol. 17, Osgood, R. M., Brueck, S. R. J., and Schlossberg, H R., eds., Elsevier, pp. 81–94.
Nakamatsu, H., Hirata, K., and Kawai, S. (1988), Synthesis of Epitaxial Silicon Carbide Films by Laser CVD, in: *Laser and Particle-Beam Chemical Processing for Microelectronics*, Materials Research Soc. Proc., Vol. 101, December 1–3, 1987, Boston, MA, Ehrlich, D. J., Higashi, G. S., and Oprysko, M. M., eds., Materials Research Society, Pittsburgh, pp. 397–402.
Natzle, W. C. (1988), Distinguishing Laser-Induced Thermal and Photochemical Surface Reactions by Photodeposit Morphology, in: *Laser and Particle-Beam Chemical Processing for Microelectronics*, Materials Research Soc. Proc., Vol. 101, December 1–3, 1987, Boston, MA, Ehrlich, D. J., Higashi, G. S., and Oprysko, M. M., eds., Materials Research Society, Pittsburgh, pp. 213–216.
Nishizawa, J. I. and Kurabayashi, T. (1988), Photo-Assisted Molecular Layer Epitaxy, in: *Laser and Particle-Beam Chemical Processing for Microelectronics*, Materials Research Soc. Proc., Vol. 101, December 1–3, 1987, Boston, MA, Ehrlich, D. J., Higashi, G. S., and Oprysko, M. M., eds., Materials Research Society, Pittsburgh, pp. 275–284.
Nowak, A. V., and Lyman, J. L. (1975), The Temperature-Dependent Absorption Spectrum of the $v_3$ Band of $SF_6$ at 10.6 $\mu$m, *J. Quant. Spectros. Radiat. Transfer* **15**, 945.
Padmanabhan, R., Miller, B. J., and Tam, G. (1988), Photo-CVD Silicon Nitride for GaAs MESFET Passivation, in: *Laser and Particle-Beam Chemical Processing for Microelectronics*, Materials Research Society Proc., Vol. 101, December 1–3, 1987, Boston, MA, Ehrlich, D. J. Hagashi, G. S., and Oprysko, M. M., eds., Materials Research Society, Pittsburgh, pp. 379–384.
Patty, R. R., Russwarm, G. M., and Morgan, D. R. (1974), *Appl. Opt.* **13**, 2850.
Petzold, H. C., Putzar, R., Weigmann, U., and Wilke, I. (1988), Laser-Induced Metal Deposition on Silicon Membranes for X-Ray Lithography, in: *Laser and Particle-Beam Chemical Processing for Microelectronics*, Materials Reseach Soc. Proc., Vol. 101, December 1–3, 1987, Boston, MA, Ehrlich, D. J., Higashi, G. S., and Oprysko, M. M., eds., Materials Research Society, Pittsburgh, pp. 75–80.
Poon, E., Evans, H. L. Wang, W. H., and Osgood, Jr., R. M., and Yang, E. S. (1983), Measurement of Grain Boundary Parameters by Laser-Spot Photoconductivity, in: *Laser Diagnostics and Photochemical Processing for Semiconductor Devices*, Materials Research Soc. Symp. Proc., Vol. 17, Osgood, R. M. Brueck, S. R. J., and Schlossberg, H. R., eds., Elsevier, New York, pp. 103–108.
Powell, C. F. (1966), *Chemically Deposited Nonmetals in Vapor Deposition*, Powell, C. F., Oxley, J. H., and Blocher, J. M., eds., Wiley, New York, 343 pp.

Powell, C. F., Campbell, I. E., and Gonser, B. W. (1955), *Vapor Plating*, Wiley, New York.
Quack, M. (1978), Theory of Unimolecular Reactions Induced by the Monochromatic Infrared Radiation, *J. Chem. Phys.* **69**, 1282.
Quigley, G. (1978), Collisional Effects in Multiple Photon IR Absorption in: *Advances in Laser Chemistry*, Springer Series in Chemical Physics, Zewail, A. Z., ed., Springer-Verlag, New York.
Ready, J. F. (1971), *Effects of High-Power Laser Radiation*, Academic, New York, pp. 19–20.
Ready, J. F. (1978), *Industrial Applications of Lasers*, Academic, New York, p. 356.
Robinson, P. J., and Holbrook, K. A. (1972), *Unimolecular Reactions*, Wiley, New York.
Roessler, D. M. (1986), An Introduction to the Laser Processing of Materials, *The Industrial Laser Annual Handbook*, Belforte, D., and Levitt, M., eds., SPIE Vol. 629, PennWell Books, Tulsa, pp. 16–30.
Roth, W., Kräutle, H., Krings, H., and Beneking, H., 1983, Laser-Stimulated Growth of Epitaxial GaAs, *Mat. Res. Soc. Symp. Proc.* **17**, 193–198.
Salathe, R. P., and Gilgen, H. H. (1983), High Resolution Laser Diagnostics for Direct Gap Semiconductor Materials, in: *Laser Diagnostics and Photochemical Processing for Semiconductor Devices*, Materials Research Soc. Symp. Proc., Vol. 17, Osgood, R. M., Brueck, S. R. J., and Schlossberg, H. R., eds., Elsevier, New York, pp. 109–116.
Shaub, W. M., and Bauer, S. H. (1975), Laser-Powered Homogeneous Pyrolysis, *Int. J. Chem. Kinet.* **7**, 509.
Stafast, H., Schmid, W. E., and Kompa, K. L. (1977), Absorption of $CO_2$ Laser Pulses at Different Wavelengths by Ground-State and Vibrationally Heated $SF_6$, *Opt. Comm.* **21**, 121.
Steen, W. M. (1977), The Thermal History of a Spot Heated by a Laser, *Lett. Heat and Mass Transfer* **4**, 167.
Steen, W. M. (1978), Surface Coating with a Laser, *Int. Conf. Advances in Surface Coating Technology*, Vol. 1, The Welding Institute, February 13–15, Cambridge, London, pp. 175–187.
Steen, W. M. (1991), *Laser Material Processing*, Springer-Verlag, London, pp. 52–61.
Steinfeld, J. I. (ed.), (1981), *Laser-Induced Chemical Processes*, Plenum, New York.
Steinfeld, J. I., Anderson, T. G., Reisner, C., Denison, D. R., Hartsough, L. D., and Hollahan, J. R. (1980), *J. Electrochem. Soc.* **127**, 514.
Stephensen, J. C., King, D. S., Goodman, M. F., and Stone, J. (1979), Experiment and Theory for $CO_2$ Laser-Induced $CF_2$ HCl Decomposition Rate Dependence on Pressure and Intensity, *J. Chem. Phys.* **70**, 4496.
Tachibana, H., Nakaue, A., and Kawate, Y. (1988), Deposition of Amorphous Carbon Films by Laser-Induced CVD, in: *Laser and Particle-Beam Chemical Processing for Microelectronics*, Materials Research Soc. Proc. Vol. 101, December 1–3, 1987, Boston, MA, Ehrlich, D. J., Higashi, G. S., and Oprysko, M. M., eds., Materials Research Society, Pittsburgh, Pennsylvania, pp. 367–370.
Thiele, E., Goodman, M., and Stone, J. (1980), Can Lasers Be Used to Break Chemical Bonds Selectively? (Laser Application to Chemistry Issue) *Opt. Eng.*, **19**, 10.
Truhlar, D. G. (ed.), (1981), *Potential Energy Surfaces and Dynamics Calculations*, Plenum, New York.
Tsang, W., Walker, J. A., Braun, W., and Herron, J. T. (1978), Mechanisms of Decomposition of Mixtures of Ethyl Acetate and Isopropyl Bromide Subjected to Pulsed Infrared Laser Irradiation, *Chem. Phys. Lett.* **59**, 487.
Tsay, W., Riley, C., and Ham, D., 1979, Thermal Enhancement of Multiple Photon Absorption by $SF_6$, *J. Chem. Phys.* **70**, 3558.
Uesugi, F., Morishige, Y., Shinzawa, T., Kishida, S., Hirata, M., Yamada, H., and Matsumoto, K. (1988), "Low Resistivity Contact Formation for LSI Interconnection with Short-Pulse-Laser Induced MOCVD, in: *Laser and Particle-Beam Chemical Processing for Microelectronics*,

Materials Research Soc. Proc., Vol. 101, December 1-3, 1987, Boston, MA, Ehrlich, D. J., Higashi, G. S., and Oprysko, M. M., eds., Materials Research Society, Pittsburgh, pp. 61-65.

Walker, R. B., and Preston, R. K. (1977), Quantum versus Classical Dynamics in the Treatment of Multiple Photon Excitation of the Anharmonic Oscillator, *J. Chem. Phys.* **67**, 2017.

Winburn, D. C. (1990), *Practical Laser Safety*, 2nd Ed., Marcel Dekker, New York.

Woodin, R. L., Bomse, D. S., and Beauchamp, J. L. (1978), Multiphoton Dissociation of Molecules with Low Power Continuous Wave Infrared Laser Radiation, *J. Am. Chem. Soc.* **100**, 3248.

Yokoyama, H., Kishida, S., and Washio, K. (1984), *Appl. Phys. Lett.* **44**, 755.

Zinck, J. J., Brewer, P. D., Jensen, J. E., Olson, G. L., and Tutt, L. W. (1988), Excimer Laser-assisted MOVPE of CdTe on GaAs (100): Crystal Growth and Mechanisms, in: *Laser and Particle-Beam Chemical Processing for Microelectronics*, Materials Research Soc. Proc., Vol. 101, December 1-3, 1987, Boston, MA, Ehrlich, D. J., Higashi, G. S., and Oprysko, M. M., eds., Materials Research Society, Pittsburgh, pp. 319-326.

Zuber, K., 1935, *Nature* **136**, 796.

Zuhoski, S. P., Killeen, K. P., and Biefeld, R. M. (1988), Photolytic Deposition of InSb Film, in: *Laser and Particle-Beam Chemical Processing for Microelectronics*, Materials Research Soc. Proc., Vol. 101, December 1-3, 1987, Boston, MA, Ehrlich, D. J., Higashi, G. S., and Oprysko, M. M., eds., Materials Research Society, Pittsburgh, pp. 313-318.

# TWO

# PYROLYTIC LCVD

## 2.1. INTRODUCTION

Laser can be used to heat up chemical reactants directly or indirectly to induce chemical reactions for depositing thin films of materials, and such processes are called pyrolytic LCVD. Various aspects of LCVD have been discussed by Bäuerle (1986). In the case of indirect heating of the chemical reactants, the substrate absorbs the laser energy to produce a localized hot spot at its surface, and collisions between the reactant molecules and the hot spot transfer thermal energy to the reactants. Sometimes, the energy radiated by the hot spot can heat up the reactants to induce chemical reactions above the substrate surface. It should be noted that the width of a depositing film can be controlled better during pyrolytic LCVD, if the chemical reactions occur only at the hot spot on the substrate surface. In the case of direct heating, the reactants, which can be in the gas, liquid, or solid phase, are chosen in such a way that the reactant molecules absorb the laser energy to reach an excited state. The intramolecular and intermolecular collisions cause this excitation energy to be redistributed within the translational, rotational, and vibrational modes of the same molecule and among other molecules, respectively, within $10^{-12}$ to $10^{-7}$. When the temperature of the molecules reaches the reaction threshold temperature, the chemical reactions occur. As the reactants can be raised to a very high temperature in a small volume over a short time by using a laser beam, novel reaction products due to different reaction paths are expected in pyrolytic LCVD processes compared to the reactions induced by conventional heating.

## 2.2. PROCESS VARIABLES IN PYROLYTIC LCVD

There are several parameters that can affect the LCVD process in many different ways:

INCIDENT LASER POWER DISTRIBUTION: The power of a laser beam can vary spatially and temporally. These spatial and temporal variations in laser power determine the total amount of energy supplied for the LCVD process. For a Gaussian beam, the laser power varies radially; for a top hat beam, the power remains unchanged along the radius of the laser beam. Depending on the variation of laser power with time, the lasers are classified as continuous wave or pulsed beam.

LASER BEAM DIAMETER: Laser beam diameter is twice the radius of that part of the beam where the power is equal to $1/e^2$ of the power at the beam center. The beam diameter affects the laser intensity, which determines the heat flux to the substrate or reactants.

LASER WAVELENGTH: Since absorptivity depends on the wavelength of the incident radiation, the choice of laser of a particular wavelength determines whether the reactants will undergo photolytic or pyrolytic reactions. Also, the laser wavelength affects the fraction of the input laser energy utilized in the LCVD process.

RELATIVE SPEED BETWEEN THE LASER BEAM AND THE SUBSTRATE: This parameter affects the laser–materials interaction time, and the size (thickness and width) and shape of the deposited film. If the beam does not move with respect to the substrate, the film is expected to deposit in the form of a dot (circular spot). On the other hand, if the beam moves relative to the substrate, the film will be deposited in the form of a stripe. The width of the stripe can be controlled by controlling the relative speed of the beam and the substrate.

REACTANT CONCENTRATION: The concentrations of the reactants determine the chemical reaction rate, which, in turn, affects the film deposition rate. As the concentration increases, the collisions between the molecules will increase, and more molecules will reach the activated state to eventually undergo chemical reaction. However, at high concentrations, the probability of the activated molecules becoming deactivated will also increase, and for this reason, there is a critical concentration at which the reaction rate is optimum.

PHYSICOCHEMICAL PROPERTIES OF THE REACTANTS: These properties affect the laser-induced heating of the reactants and determine whether the LCVD process will involve photolytic or pyrolytic reactions for a given laser wavelength.

THERMOPHYSICAL PROPERTIES OF THE SUBSTRATE: These properties control the flow of laser energy in the substrate, and the temperature and size of the localized hot spot on the substrate surface. It should be noted that the dimensions of the depositing film depend on the temperature and size of the hot spot.

In mathematical modeling of pyrolytic LCVD, the absorptivity of the substrate, and the reaction rate constants, such as the activation energy and the Arrhenius constant, play important roles. The absorptivity is used to determine the amount of laser energy being utilized to heat the substrate, in order to calculate the temperature distribution within it. The reaction rate constants allow the modeling of the reaction yields and the film deposition rate.

## 2.3. LASER RADIATION ABSORPTION PHENOMENA

In laser-aided materials processing, the laser–materials interactions [Bass (1991)] should be analyzed carefully because they determine how and how much of the laser energy is utilized in a given process. The electric and magnetic fields of the lasers interact with the electrons of other materials according to the classical theory of electrodynamics. In quantum mechanics, the photons of light are considered to interact with the electrons of other materials in the presence of the lattice of nuclei. It should be noted that the lattice must be presumed to be present in order to conserve momentum [Bass (1991)]. Both classical and quantum-mechanical treatments of such interactions can be found in Sokolov (1967). As the laser wavelengths are much larger than the interatomic dimensions, the classical theory of electrodynamics, that is, Maxwell's equations of electromagnetism, can be used to study the laser–materials interactions, and such interactions can be described in terms of macroscopic optical properties defined by the refractive index, extinction coefficient, and absorption index, and the absorption coefficient of the material for a given wavelength [see Boyd (1987), p. 37]. The interactions of lasers with various types of materials are discussed briefly below.

METALS: The electrons of the conduction band of metals acquire energy by interacting with the electromagnetic field of the laser beam. The energy of these energetic electrons is transferred to the vibrating lattice by collisions, causing the material to become heated.

SEMICONDUCTORS: In semiconductors, the electrons of the conduction band and the holes of the valence band acquire energy by interacting with the laser beam.

DIELECTRICS OR ELECTRICAL INSULATORS: In dielectrics, the electrons are strongly bound in the atoms or molecules of the material and, consequently, the electromagnetic field of the laser beam polarizes the material. During relaxation of the polarized states, some of the energy of polarization is transferred to the lattice, causing the material to become heated.

GASEOUS MEDIUM: The atoms and molecules of a gaseous medium have fewer optically excitable states than those of the solid or liquid materials [see Boyd (1987), p. 17]. Laser irradiation usually causes gaseous atoms and molecules to undergo rotational, vibrational, and electronic transitions. The ratio of energies required for electronic, vibrational, and rotational transitions is approximately $10^6:10^3:1$. Only those molecules which have a permanent dipole moment exhibit the rotational spectra, and a change in dipole during the motion results in the vibrational spectra. However, all molecules exhibit the electronic spectra. The optical excitations of gaseous atoms and molecules can produce reactive species or lead to chemical reactions.

## 2.4. OPTICAL PROPERTIES

Optical properties characterize the interactions of lasers with materials, and the expressions for such properties can be obtained by analyzing Maxwell's equations. Various aspects of the optical properties have been discussed in Born and Wolf (1980). For a homogeneous and isotropic medium, Maxwell's equations can be written in SI units as follows

$$\nabla \cdot \mathbf{D} = \rho \tag{2.1}$$

$$\nabla \cdot \mathbf{B} = 0 \tag{2.2}$$

$$\nabla \times \mathbf{E} = -\frac{\partial \mathbf{B}}{\partial t} \tag{2.3}$$

$$\nabla \times \mathbf{H} = -\frac{\partial \mathbf{D}}{\partial t} + \mathbf{J} \tag{2.4}$$

Table 2.1. Maxwell's Equations in SI and Gaussian Units

| SI units | Gaussian units |
|---|---|
| $\nabla \cdot \mathbf{D} = \rho$ | $\nabla \cdot \mathbf{D} = 4\pi\rho$ |
| $\nabla \cdot \mathbf{B} = 0$ | $\nabla \cdot \mathbf{B} = 0$ |
| $\nabla \times \mathbf{E} = -\dfrac{\partial \mathbf{B}}{\partial t}$ | $\nabla \times \mathbf{E} = -\dfrac{1}{c}\dfrac{\partial \mathbf{B}}{\partial t}$ |
| $\nabla \times \mathbf{H} = -\dfrac{\partial \mathbf{D}}{\partial t} + \mathbf{J}$ | $\nabla \times \mathbf{H} = \dfrac{1}{c}\dfrac{\partial \mathbf{D}}{\partial t} + \dfrac{4\pi}{c}\mathbf{J}$ |
| $\mathbf{D} = \varepsilon_r\varepsilon_0\mathbf{E} = \varepsilon\mathbf{E}$ | $\mathbf{D} = \varepsilon_r\mathbf{E}$ |
| $\mathbf{B} = \mu_r\mu_0\mathbf{H} = \mu\mathbf{H}$ | $\mathbf{B} = \mu_r\mathbf{H}$ |
| $\varepsilon = \varepsilon_r\varepsilon_0,\ \mu = \mu_r\mu_0$ | |

These equations are supplemented by the following constitutive relations:

$$\mathbf{J} = \sigma\mathbf{E} \tag{2.5}$$

$$\mathbf{D} = \varepsilon_r\varepsilon_0\mathbf{E} \tag{2.6}$$

$$\mathbf{B} = \mu_r\mu_0\mathbf{H} \tag{2.7}$$

Equations (2.1) and (2.2) describe the sources that generate the electric and magnetic fields, respectively. Equation (2.1) represents Coulomb's law, which states that the force on a charge due to another charge is proportional to the product of the charges and inversely proportional to the square of the distance between them. Equation (2.2) implies that there are no magnetic monopoles. Since only magnetic dipoles exist, the magnetic field lines are always closed and they do not spread out radially with spherical symmetry from a point pole, like the electric field lines emanating from a point charge. A magnetic line of force that starts from a north pole ends at a south pole and, therefore, there is no divergence of magnetic field lines. Equation (2.3) represents Faraday's law of induction, which states that an electric field can be produced by changing a magnetic field. Equation (2.4) stands for Ampére's law, according to which an electric current passing through a circular loop of wire produces a magnetic field at the center of the loop and along the normal to the plane of the loop.

Sometimes Maxwell's equations are used in Gaussian units and, for this reason, they are presented in both units in Table 2.1 for comparison purposes. Using the constitutive relations, Eqs. (2.1)–(2.4) can be written as

$$\nabla \cdot \mathbf{E} = \dfrac{1}{\varepsilon_r\varepsilon_0}\rho \tag{2.8}$$

$$\nabla \cdot \mathbf{H} = 0 \tag{2.9}$$

$$\nabla \times \mathbf{E} = -\mu_r \mu_0 \frac{\partial \mathbf{H}}{\partial t} \qquad (2.10)$$

$$\nabla \times \mathbf{H} = \varepsilon_r \varepsilon_0 \frac{\partial \mathbf{E}}{\partial t} + \sigma \mathbf{E} \qquad (2.11)$$

To determine the effect of $\rho$ on $\mathbf{E}$, let us take divergence of Eq. (2.11) and use Eq. (2.8) in the resulting expression to obtain

$$\frac{\partial \rho}{\partial t} + \frac{\sigma}{\varepsilon_r \varepsilon_0} \rho = 0 \qquad (2.12)$$

The solution of Eq. (2.12) is given by

$$\rho = \rho_0 \exp(-t/\tau) \qquad (2.13)$$

where the relaxation time $\tau = \varepsilon_r \varepsilon_0 / \sigma$. However, $\tau$ is very small for many materials that have some conductivity. For metals, $\tau \approx 10^{-18}$ s [see Born and Wolf (1980), p. 612]. Thus, $\rho$ can be assumed to be zero for metals and, therefore, Eq. (2.8) can be simplified as

$$\nabla \cdot \mathbf{E} = 0 \qquad (2.14)$$

Taking the curl of Eq. (2.10), and using Eqs. (2.11), (2.14), and the identity $\nabla \times (\nabla \times \mathbf{E}) \equiv \nabla(\nabla \cdot \mathbf{E}) - \nabla^2 \mathbf{E}$, we obtain

$$\nabla^2 \mathbf{E} = \varepsilon_r \varepsilon_0 \mu_r \mu_0 \frac{\partial^2 \mathbf{E}}{\partial t^2} + \mu_r \mu_0 \sigma \frac{\partial \mathbf{E}}{\partial t} \qquad (2.15)$$

which holds good for a conducting medium. The presence of $\partial \mathbf{E}/\partial t$ in Eq. (2.15) suggests that the wave will be attenuated as it propagates through the medium.

For a dielectric medium, $\tau$ need not be very small and, consequently, Eq. (2.8) will hold good. In this situation, taking the curl of Eq. (2.10), and using Eqs. (2.11), (2.8), and the identity $\nabla \times (\nabla \times \mathbf{E}) \equiv \nabla(\nabla \cdot \mathbf{E}) - \nabla^2 \mathbf{E}$, we obtain

$$\nabla^2 \mathbf{E} = \varepsilon_r \mu_r \varepsilon_0 \mu_0 \frac{\partial^2 \mathbf{E}}{\partial t^2} + \mu_r \mu_0 \sigma \frac{\partial \mathbf{E}}{\partial t} + \frac{\nabla \rho}{\varepsilon_r \varepsilon_0} \qquad (2.16)$$

Since laser is a monochromatic beam of radiation, $\mathbf{E}$ can be written in the form

$$\mathbf{E} = \mathbf{E}_0 \exp(-i\omega t) \tag{2.17}$$

to solve Eqs. (2.15) and (2.16).

SOLUTION FOR DIELECTRIC MEDIUMS

Although $\rho \neq 0$ for dielectric mediums (electrical insulators), we know from Eq. (2.13) that the space charge $\rho$ will decay in a finite time. Thus, the term containing $\nabla \rho$ in Eq. (2.16) can be neglected for steady state conditions attained after the decay of the space charge. Also, noting that $\sigma = 0$ for dielectrics, and $c = 1/\sqrt{\mu_0 \varepsilon_0}$, Eq. (2.16) can be written as

$$\nabla^2 \mathbf{E} = \frac{\varepsilon_r \mu_r}{c^2} \frac{\partial^2 \mathbf{E}}{\partial t^2} \tag{2.18}$$

which represents a wave motion without any damping. The wave speed $v$ is given by

$$\frac{1}{v^2} = \frac{\varepsilon_r \mu_r}{c^2} \tag{2.19}$$

and the refractive index $n$ is given by

$$n = c/v = \sqrt{\varepsilon_r \mu_r} \tag{2.20}$$

SOLUTION FOR CONDUCTING MEDIUMS

Noting that $c = 1/\sqrt{\mu_0 \varepsilon_0}$, Eq. (2.15) can be written as

$$\nabla^2 \mathbf{E} = \frac{\varepsilon_r \mu_r}{c^2} \frac{\partial^2 \mathbf{E}}{\partial t^2} + \frac{\mu_r \sigma}{\varepsilon_0 c^2} \frac{\partial \mathbf{E}}{\partial t} \tag{2.21}$$

Substituting Eq. (2.17) in Eq. (2.21), we obtain

$$\nabla^2 \mathbf{E}_0 + \hat{k}^2 \mathbf{E}_0 = 0 \tag{2.22}$$

where the wave number $\hat{k}$ is given by

$$\hat{k}^2 = \frac{\mu_r \omega^2}{c^2} \left( \varepsilon_r + i \frac{\sigma}{\varepsilon_0 \omega} \right) \tag{2.23}$$

Defining a complex dielectric constant $\hat{\varepsilon}_r$ as

$$\hat{\varepsilon}_r = \varepsilon_r + i\frac{\sigma}{\varepsilon_0 \omega} \qquad (2.24)$$

and noting that the wave number, $\hat{k} = 2\pi/\lambda$ and $\omega = 2\pi v/\lambda$, Eq. (2.23) can be rewritten as

$$c/v = \sqrt{\mu_r \hat{\varepsilon}_r} = \hat{n} \qquad (2.25)$$

which defines the complex refractive index $\hat{n}$. Letting

$$\hat{n} = n + ik \qquad (2.26)$$

Combining Eqs. (2.25) and (2.26), and separating the real and imaginary parts, we obtain

$$n^2 - k^2 = \mu_r \varepsilon_r \qquad (2.27)$$

and

$$nk = \frac{\sigma \mu_r}{2\varepsilon_0 \omega} \qquad (2.28)$$

Equations (2.27) and (2.28) can be solved to write

$$n^2 = \frac{\mu_r}{2}\left[\left(\varepsilon_r^2 + \frac{\sigma^2}{\varepsilon_0^2 \omega^2}\right)^{1/2} + \varepsilon_r\right] \qquad (2.29)$$

and

$$k^2 = \frac{\mu_r}{2}\left[\left(\varepsilon_r^2 + \frac{\sigma^2}{\varepsilon_0^2 \omega^2}\right)^{1/2} - \varepsilon_r\right] \qquad (2.30)$$

To understand the attenuation of the wave propagating through a conducting medium as given by Eq. (2.21), we need to solve Eq. (2.21). The simplest solution for Eq. (2.21) is the plane, time-harmonic wave solution, which is given by

$$\mathbf{E} = \mathbf{e}_0 \exp\{i[\hat{k}(\mathbf{r} \cdot \mathbf{s}) - \omega t]\} \qquad (2.31)$$

PYROLYTIC LCVD

Using Eqs. 2.23–2.26), we can show that

$$\hat{k} = \frac{\omega}{c}\hat{n} = \frac{\omega}{c}(n + ik) \tag{2.32}$$

which modifies Eq. (2.31) to the following form:

$$\mathbf{E} = \mathbf{e}_0 \exp\left[-\frac{\omega}{c}k(\mathbf{r}\cdot\mathbf{s})\right]\exp\left\{i\omega\left[\frac{n}{c}(\mathbf{r}\cdot\mathbf{s}) - t\right]\right\} \tag{2.33}$$

Taking the real part of Eq. (2.33), we obtain

$$\mathbf{E} = \mathbf{e}_0 \exp\left[-\frac{\omega}{c}k(\mathbf{r}\cdot\mathbf{s})\right]\cos\left\{\omega\left[\frac{n}{c}(\mathbf{r}\cdot\mathbf{s}) - t\right]\right\} \tag{2.34}$$

where the exponential term indicates that the wave will attenuate as it propagates through the medium. Using Eq. (2.34), we can write the energy density of the radiation, which is proportional to the time average of $\mathbf{E}^2$, as

$$I = I_0 \exp\left[-\frac{2\omega}{c}k(\mathbf{r}\cdot\mathbf{s})\right] \tag{2.35}$$

which determines the attenuation of the beam as it propagates through the medium.

The absorption coefficient $\eta$ is defined as

$$\eta = \frac{2\omega}{c}k = \frac{4\pi k}{\lambda_0} \tag{2.36}$$

Equation (2.35) shows that $I$ will decrease to $1/e$ of $I_0$ at a depth

$$d = \frac{1}{\eta} = \frac{\lambda_0}{4\pi k} \tag{2.37}$$

When the depth of penetration $d$ is very small compared to the wavelength, it is referred to as the "skin depth," and the penetration phenomenon is known as the "skin effect." For long wavelengths, that is, when the wavelength is 100 μm or more [see Sokolov (1967), p. 13], $\omega$ will be small and $1/\omega$ will be large, and Eqs. (2.29), (2.30), and (2.37) can be written as

$$n \approx k \approx \sqrt{\frac{\mu_r \sigma}{2\varepsilon_0 \omega}} \qquad (2.38)$$

and

$$d \approx \frac{c\sqrt{\varepsilon_0}}{\sqrt{2\mu_r \sigma \omega}} \qquad (2.39)$$

## 2.5. ABSORPTIVITY OF METALS AT NORMAL INCIDENCE

Metals have a complex refractive index whose imaginary part is associated with the attenuation of the beam propagating through the medium. The attenuation is mainly due to the reflection of the light wave by the electrons of metals. Only a small fraction of the incident radiation is absorbed by the metal and is utilized in heating the metal. To determine the amount of energy absorbed by the metal, let us first calculate the amount of energy reflected from a metal surface. For this purpose, consider a plane light wave falling normally on a metal surface that lies in the $yz$-plane. It should be noted that a plane wave refers to a wave train for which a series of parallel planes can be constructed in such a way that the magnitude and direction of **E** and **H** are constant on these planes. These parallel planes are called wave fronts or planes of equal phase, and the wave travels in a direction perpendicular to these planes. Since we are considering plane waves, $E_y = E_y(x, t)$, $H_z = H_z(x, t)$, and all other components of **E** and **H** are zero. Thus, Eqs. (2.10) and (2.11) can be written as

$$\frac{\partial E_y}{\partial x} = -\mu_r \mu_0 \frac{\partial H_z}{\partial t} \qquad (2.40)$$

and

$$-\frac{\partial H_z}{\partial x} = \varepsilon_r \varepsilon_0 \frac{\partial E_y}{\partial t} + \sigma E_y \qquad (2.41)$$

We can consider the solution of these equations in the form of the following plane waves:

$$E_y = A_1 \exp\left[i\omega\left(\frac{\hat{n}x}{c} - t\right)\right] \qquad (2.42)$$

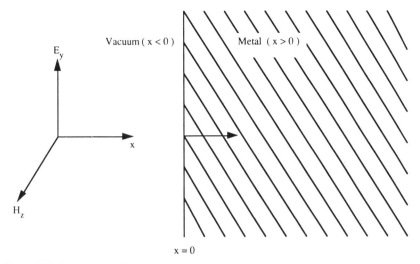

Figure 2.1. A schematic diagram for electric and magnetic fields propagating in the positive direction of the x-axis

$$H_z = B_1 \exp\left[i\omega\left(\frac{\hat{n}x}{c} - t\right)\right] \tag{2.43}$$

Substituting these solutions into Eqs. (2.40) and (2.41), we obtain

$$B_1 = \frac{\hat{n}}{\mu_r \mu_0 c} A_1$$

and

$$\hat{n}^2 = \mu_r \hat{\varepsilon}_r$$

and $H_2$ becomes

$$H_z = \frac{\hat{n}}{\mu_r \mu_0 c} E_y \tag{2.44}$$

Now, in vacuum, that is, for $x < 0$, the equations for the incident waves are

$$E_{yi} = A_i \exp\left[i\omega\left(\frac{x}{c} - t\right)\right] \tag{2.45}$$

and

$$H_{zi} = \frac{1}{\mu_0 c} A_i \exp\left[i\omega\left(\frac{x}{c} - t\right)\right] \tag{2.46}$$

and the equations for reflected waves are

$$E_{yr} = A_r \exp\left[i\omega\left(\frac{x}{c} + t\right)\right] \quad (2.47)$$

and

$$H_{zr} = -\frac{1}{\mu_0 c} A_r \exp\left[i\omega\left(\frac{x}{c} + t\right)\right] \quad (2.48)$$

In a metal, that is, for $x > 0$, the equations for the transmitted waves are

$$E_{yt} = A_t \exp\left[i\omega\left(\frac{\hat{n}x}{c} - t\right)\right] \quad (2.49)$$

and

$$H_{zt} = \frac{\hat{n}}{\mu_r \mu_0 c} A_t \exp\left[i\omega\left(\frac{\hat{n}x}{c} - t\right)\right] \quad (2.50)$$

Equations (2.45)–(2.50) must satisfy the boundary conditions at the vacuum–metal boundary, which is located at $x = 0$. According to the boundary conditions, the tangential components of **E** and **H**, which are $E_y$ and $H_z$ in this case, must be continuous across the boundary at $x = 0$.

The continuity of $E_y$ leads to

$$A_i - A_r = A_t \quad (2.51)$$

and the continuity of $H_z$ leads to

$$A_i + A_r = \frac{\hat{n}}{\mu_r} A_t \quad (2.52)$$

From Eqs. (2.51) and (2.52), the amplitudes of the reflected and transmitted waves can be written as

$$A_r = \frac{\hat{n} - \mu_r}{\hat{n} + \mu_r} A_i \quad (2.53)$$

$$A_t = \frac{2\mu_r}{\hat{n} + \mu_r} A_i \quad (2.54)$$

Since the intensity of a wave is proportional to the square of the modulus of its amplitude, and the reflectivity is given by the ratio of the intensities of the reflected and incident waves, we can write

$$R = \frac{|A_r|^2}{|A_i|^2} = \frac{A_r A_r^*}{A_i A_i^*} = \frac{(\hat{n} - \mu_r)(\hat{n}^* - \mu_r)}{(\hat{n} + \mu_r)(\hat{n}^* + \mu_r)}$$

which can be rewritten as the following expression by substituting $\hat{n} = n + ik$.

$$R = \frac{(n - \mu_r)^2 + k^2}{(n + \mu_r)^2 + k^2} \tag{2.55}$$

It should be noted that $\mu_r$ is almost equal to unity for wavelengths ranging from the long-wave infrared region to the visible and ultraviolet regions [see Sokolov (1967), p. 15]. Thus, we can write

$$R = \frac{(n - 1)^2 + k^2}{(n + 1)^2 + k^2} \tag{2.56}$$

which is the Fresnel relation for interface reflectivity at normal incidence.

For a transparent dielectric, that is, for a nonabsorbing medium, $k = 0$, and Eq. (2.56) can be written as

$$R = \frac{(n - 1)^2}{(n + 1)^2} \tag{2.57}$$

For long wavelengths, $n \approx \eta$ as shown in Eq. (2.38) and, therefore, Eq. (2.56) can be approximated as

$$\begin{aligned} R &= \left(1 - \frac{2n - 1}{2n^2}\right)\left(1 + \frac{2n + 1}{2n^2}\right)^{-1} \\ &= \left(1 - \frac{2n - 1}{2n^2}\right)\left(1 - \frac{2n + 1}{2n^2} - \cdots\right) \\ &= 1 - \frac{2}{n} + \frac{1}{n^2} + \cdots \end{aligned} \tag{2.58}$$

It should be noted that $R + A + T = 1$ due to the conservation of energy, because the energy of the incident light wave is divided into reflected, absorbed, and transmitted components. Also, the transmittivity $T$ can be neglected for metals, and thus $A = 1 - R$. Taking the first two terms on the right-hand side of Eq. (2.58), using Eq. (2.38), and noting that $\mu_r \approx 1$ for long wavelengths as discussed above, we obtain

$$R = 1 - \sqrt{8\varepsilon_0\omega/\sigma} \tag{2.59}$$

and

$$A = \sqrt{8\varepsilon_0\omega/\sigma} \tag{2.60}$$

Expressions (2.59) and (2.60) are known as Hagen–Rubens relations [see Sokolov (1967), p. 16] because their experimental data established that these two expressions hold good for many metals at room temperature in the far-infrared region ($\lambda = 25\,\mu\mathrm{m}$) for frequencies up to $3 \times 10^{13}\,\mathrm{s}^{-1}$. However, the theory of anomalous skin effect and Schulz's (1957) experimental results suggest that Eqs. (2.38), (2.59), and (2.60) are valid in the microwave and far-infrared ($\lambda \geqslant 100\,\mu\mathrm{m}$) regions. Equations (2.59) and (2.60) are not valid in the optical region of the electromagnetic spectrum, which can be explained by using electron theory: Owing to their inertia electrons cannot respond quickly enough to follow the rapidly varying field and, consequently, the electric current does not become $\sigma\mathbf{E}$ at each instant of time. As a result, the metal behaves like a material of lower conductivity at shorter wavelengths.

Equations (2.60) shows that the absorptivity decreases as the conductivity increases, but Eq. (2.36) shows that the absorption coefficient increases as the conductivity increases, which may lead to the conclusion that Eqs. (2.60) and (2.36) are contradictory. However, Eqs. (2.36) and (2.60) do not really contradict each other because $k$ and $\eta$ essentially represent the damping of light waves in metals due to the skin effect and the reflection of a large fraction of the incident beam. Only a small portion of the incident energy is absorbed and converted into heat within the metal.

## 2.6. LORENTZ MODEL FOR OPTICAL PROPERTIES OF DIELECTRICS

So far, we have discussed a macroscopic theory, that is, the purely phenomenological electromagnetic theory, for the optical properties of various materials. The electromagnetic theory of light is found to be applicable at room temperature and above, when the assumptions of the theory are valid. In many cases, the results of this theory do not agree with experimental data in various wavelength ranges. As discussed above, the

Hagen–Rubens relation is not satisfactory in the infrared and visible regions of the electromagnetic spectrum. Also, the Maxwell relations, $n^2 = \mu_r \varepsilon_r$ for dielectrics and $\hat{n}^2 = \mu_r \hat{\varepsilon}_r$ for metals, are inadequate because $\varepsilon_r$, $\mu_r$, and $\sigma$ are functions of frequency (dispersion). To explain these effects, a microscopic theory for the interactions of electromagnetic waves with materials is required. The miroscopic analysis is based on the assumption that electromagnetic waves interact only with the outer electrons (sometimes called optical electrons), and it does not take the detailed atomic structure of the material into account. In dispersion theory, the material is considered to be surrounded by a large number of electrons which interact with the incident wave. Lorentz (1909, 1952) developed a model for such interactions in dielectric materials (the classical electron theory of absorption and dispersion) under the following two assumptions:

1. The electrons of dielectric materials are bound by quasi-elastic forces which are proportional to their displacement from the equilibrium position.
2. The electrons have a natural frequency of oscillation, which is a consequence of the first assumption.

With these assumptions, the motion of an electron due to a linearly polarized electromagnetic wave can be written as

$$m\frac{d^2\mathbf{r}}{dt^2} + m\Gamma\frac{d\mathbf{r}}{dt} + K\mathbf{r} = e\mathbf{E}_0 e^{-i\omega t} \qquad (2.61)$$

It should be noted [Wooten (1972)] that the choice of $\mathbf{E} = \mathbf{E}_0 e^{i\omega t}$ leads to $\hat{n} = n - ik$, and the choice of $\mathbf{E} = \mathbf{E}_0 e^{-i\omega t}$ leads to $\hat{n} = n + ik$. In Eqs. (2.61), $\mathbf{E}_0$ is constant, $e = -|e|$, and the first, second, third, and fourth terms represent the product of mass and acceleration of the electron, damping force that accounts for radiative losses of energy due to various scattering processes in a solid, restoring force, and the external force due to the incident wave, respectively. Here $\Gamma$ represents damping due to the resistance of the material. The natural frequency of a harmonic oscillator corresponds to the frequency of oscillations in the absence of any damping and external forces. Thus, Eq. (2.61) can be written as

$$m\frac{d^2\mathbf{r}}{dt^2} + K\mathbf{r} = 0 \qquad (2.62)$$

to determine the natural frequency of the electrons of dielectrics. The solution of Eq. (2.62) is given by

$$\mathbf{r} = \mathbf{A}_1 \exp[i\sqrt{(K/m)}\,t] + \mathbf{B}_1 \exp[-i\sqrt{(K/m)}\,t] \qquad (2.63)$$

where $A_1$ and $B_1$ are constants of integration. Equation (2.63) indicates that the natural frequency

$$\omega_0 = \sqrt{K/m} \tag{2.64}$$

Substituting Eq. (2.64) into Eq. (2.61), we can write the solution of Eq. (2.61) as

$$\mathbf{r} = \mathbf{A}_2 \exp\left[-\frac{1}{2}(\Gamma + \sqrt{\Gamma^2 - 4\omega_0^2})t\right] + \mathbf{B}_2 \exp\left[-\frac{1}{2}(\Gamma - \sqrt{\Gamma^2 - 4\omega_0^2})t\right]$$
$$+ \frac{e}{m} \frac{(\omega_0^2 - \omega^2) + i\Gamma\omega}{(\omega_0^2 - \omega^2)^2 + \Gamma^2\omega^2} \mathbf{E}_0 \exp(-i\omega t) \tag{2.65}$$

where $A_2$ and $B_2$ are constants of integration. After a sufficiently long time, the transient motion, that is, the first two terms on the right-hand side of Eq. (2.65), will decay, and only the electromagnetic field $\mathbf{E} = \mathbf{E}_0 e^{i\omega t}$ of the incident wave will influence the oscillation of the electrons. In this situation, the polarization or dipole moment of an electron can be written as

$$\hat{p} = re = \frac{e^2}{m} \frac{(\omega_0^2 - \omega^2) + i\Gamma\omega}{(\omega_0^2 - \omega^2)^2 + \Gamma^2\omega^2} \mathbf{E} \tag{2.66}$$

It should be noted that the microscopic polarization $\hat{p}$ is complex, and is given by

$$\hat{p} = \hat{\alpha}(\omega)\mathbf{E} \tag{2.67}$$

where $\hat{\alpha}(\omega)$ is the frequency-dependent microscopic or atomic polarizability, and is also complex. The macroscopic polarization $\hat{P}$ can be written as

$$\hat{P} = N\hat{p} = \frac{Ne^2}{m} \frac{(\omega_0^2 - \omega^2) + i\Gamma\omega}{(\omega_0^2 - \omega^2)^2 + \Gamma^2\omega^2} \mathbf{E} \tag{2.68}$$

To determine the complex macroscopic polarization, we generalize the expression $\mathbf{D} = \varepsilon_0 \mathbf{E} + \mathbf{P}$ and the constitutive relation (2.6) to include complex variables, which leads to the following relations:

$$\hat{D} = \hat{\varepsilon}_r \varepsilon_0 \mathbf{E} = \varepsilon_0 \mathbf{E} + \hat{P} \tag{2.69}$$

Using Eqs. (2.68) and (2.69), we can write

$$\hat{\varepsilon}_r = 1 + \frac{Ne^2}{\varepsilon_0 m} \frac{(\omega_0^2 - \omega^2) + i\Gamma\omega}{(\omega_0^2 - \omega^2)^2 + \Gamma^2\omega^2} \qquad (2.70)$$

Using Eqs. (2.24) and (2.70), we can write

$$\sigma = \frac{Ne^2}{m} \frac{\Gamma\omega^2}{(\omega_0^2 - \omega^2)^2 + \Gamma^2\omega^2} \qquad (2.71)$$

and

$$\varepsilon_r = 1 + \frac{Ne^2}{\varepsilon_0 m} \frac{\omega_0^2 - \omega^2}{(\omega_0^2 - \omega^2)^2 + \Gamma^2\omega^2} \qquad (2.72)$$

Setting $\mu_r = 1$ in Eqs. (2.27) and (2.28), and using them in Eqs. (2.71) and (2.72), we obtain

$$n^2 - k^2 = \varepsilon_r = 1 + \frac{Ne^2}{\varepsilon_0 m} \frac{\omega_0^2 - \omega^2}{(\omega_0^2 - \omega^2)^2 + \Gamma^2\omega^2} \qquad (2.73)$$

and

$$2nk = \frac{\sigma}{\varepsilon_0 \omega} = \frac{Ne^2}{\varepsilon_0 m} \frac{\Gamma\omega}{(\omega_0^2 - \omega^2)^2 + \Gamma^2\omega^2} \qquad (2.74)$$

Figure 2.2 shows qualitatively the variations of $n$, $k$, and $R$ with $\omega$, where $n$, $k$, and $R$ attain maximum values near the resonance frequency $\omega_0$. The damping constant $\mu$ determines the width of the resonance. It should be noted that resonance causes the absorption index as well as the reflectivity to become very high. Although we have determined various optical properties for materials having only one natural frequency, many real materials can have many resonances, and for such materials, Eq. (2.70) takes the following form:

$$\hat{\varepsilon}_r = 1 + \frac{e^2}{\varepsilon_0 m} \sum_j N_j \frac{(\omega_j^2 - \omega^2) + i\Gamma_j\omega}{(\omega_j^2 - \omega^2)^2 + \Gamma_j^2\omega^2} \qquad (2.75)$$

$$\sum_j N_j = N \qquad (2.76)$$

where $N_j$ is the number of electrons per unit volume that have natural frequency $\omega_j$ and damping constant $\Gamma_j$.

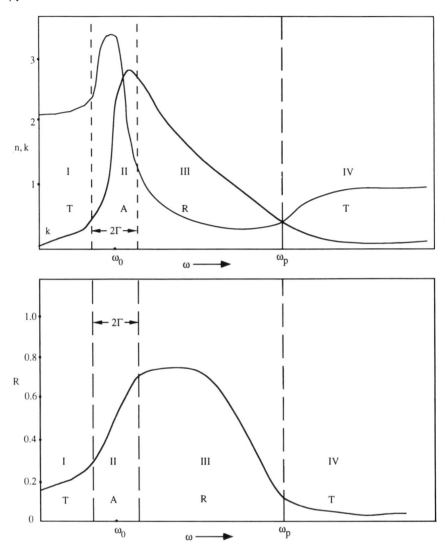

Figure 2.2. Frequency dependence of the refractive index and the Fresnel reflectivity.

Nonmetals (insulators or semiconductors) have only bound electrons in the absence of any excited states, and are transparent except near the natural frequencies (resonances). The resonance corresponding to the energy of interband transitions, that is, the transitions of valence-band electrons to the conduction band, can be utilized in laser–materials processing. The energy of each incident photon must be higher than or equal to the band-gap energy to cause interband transitions. The optical properties of materials can be affected by the free carriers, electrons and holes, which are generated in pairs during interband transitions. Due to free carriers, many semiconductors exhibit metal-like reflectivity in the visible region of the electromagnetic spectrum. The band-gap energies of semiconductors are in the visible and infrared regions of the spectrum, whereas, insulators have band gaps in the ultraviolet region.

Besides interacting with the electromagnetic waves through such interband transitions, many nonmetals can interest with the incident radiation through optical phonons in the near-infrared part of the spectrum. The interactions between the optical phonons, lattice vibrations, and the incident radiation can be analyzed by using Eq. (2.61) with appropriate values of $m$, $\mu$, $K$, and $e$ characterizing the lattice vibrations instead of the electronic vibrations.

Based on the nature of the interactions between the incident radiation and nonmetals, the electromagnetic spectrum can be divided into four regions as shown in Fig. 2.2:

REGION I: In this region, $\omega \ll \omega_0$ and, consequently, Eqs. (2.73) and (2.74) imply that $n^2 - k^2 > 1$ and $2nk \approx 0$. Thus, $n > 1$ and $k \approx 0$ and, therefore, the nonmetals are transparent and have a small reflectivity and no absorption in this region.

REGION II: This region, where $\omega_0 - \mu \leqslant \omega \leqslant \omega_0 + \mu$, exhibits strong absorption and considerable reflectivity, which means that the high values of $n$ and $k$ enhance the reflectivity of the material in this region, but the radiation, which is not reflected, is absorbed strongly in the material.

REGION III: In this region $\omega \gg \omega_0$, which means that the photon energy is much higher than the binding energy of the electrons of nonmetals and, consequently, these electrons interact with the incident radiation as if they were free electrons. The insulators exhibit metal-like reflectance in this region. However, good insulators exhibit such reflectance in the ultraviolet region, which is not observable visually. On the other hand, semiconductors, such as Ge and Si, have band-gap energy in the infrared region, and exhibit metal-like reflectance in the visible part of the spectrum. For this reason, Ge and Si appear to be metallic and KCl is transparent to the eye.

REGION IV: In this region $\omega \gg \omega_0 \gg \Gamma$, which implies $n^2 - k^2 < 1$ according to Eq. (2.73). This region begins at a frequency $\omega_p$, which is called the plasma frequency, at which $\varepsilon_r = n^2 - k^2 = 0$. Applying this condition to Eq.

(2.73), we obtain

$$\omega \phi_p^2 = Ne^2/\varepsilon_0 m \tag{2.77}$$

Nonmetals become transparent in this region.

### ANOMALOUS DISPERSION

The variations of $\varepsilon_r$ or $n^2 - k^2$ and $2nk$ with $\omega$ are shown in Fig. 2.3; $\varepsilon_r$ usually increases as $\omega$ increases, and this phenomenon is called normal dispersion. However, in the vicinity of the resonance frequency $\omega_0$, the value of $\varepsilon_r$ decreases as $\omega$ increases, and this is called anomalous dispersion. The width of the anomalous dispersion region, which is $\Gamma$ as shown in Fig. 2.3, can be determined from the following expression, which is obtained by taking the derivative of Eq. (2.73) with respect to $\omega$, equating the derivative to zero, and conidering that the anomalous dispersion region is very narrow so that $\omega_0 \approx \omega_m$, where $\omega_m$ is the frequency at which $\varepsilon_r$ has a maximum or

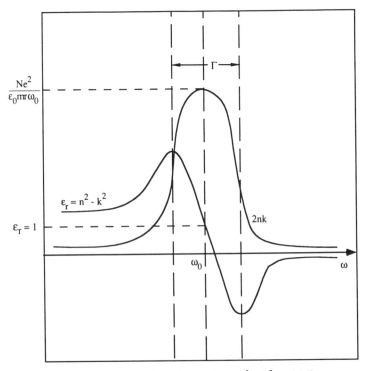

Figure 2.3. Frequency dependence of $\varepsilon_r$ or $n^2 - k^2$, and $2nk$.

minimum value:

$$\omega_0 - \omega_m = \pm \Gamma/2 \qquad (2.78)$$

$|\omega_0 - \omega_m|$ represents the half-width of the anomalous dispersion region whose full width is given by $\Gamma$. From Eq. (2.74), the maximum value of $2nk$ is given by

$$(2nk)_{\max} = Ne^2/\varepsilon_0 m \Gamma \omega \qquad (2.79)$$

Using Eqs. (2.74) and (2.79) and assuming that $\omega_0 \approx \omega_1$, the full width, that is, $2|\omega_0 - \omega_1|$, for the $2nk$-curve at half-maximum can be shown to be $\Gamma$, where $\omega_1$ is the frequency at which $2nk$ is equal to one-half of its maximum value. It should be noted that the results of Section 2.6 are appliable to homogeneous materials. In nonmetals, the optical properties are affected by the scattering of light at the grain boundaries or inclusions if the inhomogeneity exists on the scale of one wavelength or more, because such scatterings increase the light path inside the material. For this reason, many materials that are supposed to be transparent, for example, ceramics, exhibit strong absorption of radiation.

## 2.7. DRUDE–ZENER THEORY FOR OPTICAL PROPERTIES OF METALS

Drude and Zener developed a free-electron model [Lorentz (1952), Drude (1959), Zener (1953)] for metals to explain various optical properties of metals. They assumed that metals have ions, which form the crystal lattice of the metal, and a free-electron gas which moves between the ions. Normally, these electrons and ions are in thermal equilibrium. When a constant electric field is applied to the metal, the electrons move in the direction of the applied field to produce an electric current. If there were no collisions between the electrons and ions (lattice), the electrons would not lose any energy, but continue to gain energy owing to their interactions with the applied field; this would result in the energy of the electrons becoming infinite, giving rise to infinite conductivity, that is, zero resistance for metals. However, metals exhibit finite conductivity, which means that the free electrons of metals collide with the ions and transfer part of the energy that they acquire from the applied field, to the lattice. In this way, the metal is heated up. Using the Drude theory, we can write the equation of motion of a free electron in a metal as

$$m\frac{d^2\mathbf{r}}{dt^2} + m\Gamma_m\frac{d\mathbf{r}}{dt} = e\mathbf{E}_0 \qquad (2.80)$$

where $\Gamma_m$ is the damping constant for metals, which represents the scattering of electrons, which leads to the loss of electron energy and, consequently to the electrical resistivity of metals. It essentially represents a resistance to the motion of electrons, due to the collisions between the electrons and ions in metals. $\mathbf{E}_0$ is the constant electric field applied to the metal, and $e = -|e|$. The quantity $\Gamma_m$ can be related to the DC conductivity $\sigma_0$ and the mean free time between collisions (that is, the relaxation time $\tau$) as follows:

To obtain an expression for the average velocity from Eq. (2.80), the average acceleration $\bar{\ddot{\mathbf{r}}}$ and average velocity $\bar{\dot{\mathbf{r}}}$ are defined as

$$\bar{\ddot{\mathbf{r}}} = \frac{1}{\tau}\int_0^\tau \ddot{\mathbf{r}}\,dt \quad \text{and} \quad \bar{\dot{\mathbf{r}}} = \frac{1}{\tau}\int_0^\tau \dot{\mathbf{r}}\,dt$$

With these definitions and the condition that $\bar{\ddot{\mathbf{r}}} = 0$, Eq. (2.80) yields

$$\bar{\dot{\mathbf{r}}} = \frac{e}{m\Gamma_m}\mathbf{E}_0 \qquad (2.81)$$

The current density $\mathbf{j}$ is given by

$$\mathbf{j} = \sigma_0\mathbf{E}_0 = Ne\bar{\dot{\mathbf{r}}} \qquad (2.82)$$

From Eqs. (2.81) and (2.82), $\Gamma_m$ can be written as

$$\Gamma_m = Ne^2/m\sigma_0 \qquad (2.83)$$

The average velocity of the electron is given by [see Wooten (1972), p. 87]

$$\bar{\dot{\mathbf{r}}} = \frac{e\tau}{m}\mathbf{E}_0 \qquad (2.84)$$

Comparing Eqs. (2.81) and (2.84), we obtain

$$\Gamma_m = 1/\tau \qquad (2.85)$$

Equation (2.80) is similar to Eq. (2.61) except that $K = 0$ and $\Gamma = \Gamma_m$ in equation. (2.80). Therefore, setting $\omega_0 = 0$ and $\Gamma = \Gamma_m = 1/\tau$ in Eqs. (2.71),

(2.73) and (2.74), we obtain the following expressions for metals:

$$\sigma = \frac{Ne^2}{m}\frac{\tau}{1+\omega^2\tau^2} = \sigma_0\frac{1}{1+\omega^2\tau^2} \quad (2.86)$$

$$\varepsilon_r = n^2 - k^2 = 1 - \frac{Ne^2}{\varepsilon_0 m}\frac{\tau^2}{1+\omega^2\tau^2} = 1 - \frac{\sigma_0\tau}{\varepsilon_0}\frac{1}{1+\omega^2\tau^2} \quad (2.87)$$

$$2nk = \frac{Ne^2}{\varepsilon_0 m\omega}\frac{\tau}{1+\omega^2\tau^2} = \frac{\sigma_0}{\varepsilon_0\omega}\frac{1}{1+\omega^2\tau^2} \quad (2.88)$$

where the DC conductivity $\sigma_0 = Ne^2\tau/m$ as given by Eq. (2.83). To determine a characteristic frequency for metals, we obtain the following expression for the electron plasma frequency by taking $\varepsilon_r = 0$ and $\omega\tau \gg 1$ in Eq. (2.87):

$$\omega_p = \sqrt{\sigma_0/\varepsilon_0\tau} = \sqrt{Ne^2/\varepsilon_0 m} \quad (2.89)$$

where $\omega_p$ is in the vacuum ultraviolet region for most metals.

In the case of nonmetals, the frequency regime was divided into four regions based on the characteristic frequency $\omega_0$ to explain the transmission, absorption, and reflection phenomena. For metals, the frequency regime is divided into two regions based on the electron plasma frequency. When $\omega < \omega_p$, the Fresnel reflectivity and absorption coefficients are large, and when $\omega > \omega_p$ they are small.

### 2.7.1. Real Metals

The free-electron theory is inadequate for many metals where the interband transition of electrons is dominant. The electrons of metals usually undergo two types of transitions: intraband and interband transitions. In the intraband transition, the electrons are optically excited from a state that is below the Fermi energy to another state that is above it within the same band. The intraband transitions can occur only in metals, do not require any threshold energy, and are described by the Drude–Zener theory. For such transitions to occur, the material must have partially filled bands so that the electrons can go from a filled state that is below the Fermi energy to an empty state within the same band. Owing to the lack of partially filled bands, the electrons of insulators cannot undergo intraband transitions, and the insulators have nonconducting properties.

In interband transitions, the electrons are excited from one band to another. This process requires a threshold energy that is similar to the band-gap energy required to excite an electron in an insulator. Hence, the Lorentz model can be used to describe the interband transitions in metals.

## 2.7.2. Perfect Metals

In perfect metals, the mean free path of electrons, the mean time between collisions of electrons (relaxation time) $\tau$, and the conductivity are infinite. The electrons do not transfer any energy to the lattice of a perfect metal and, therefore, the lattice cannot absorb the energy of the incident electromagnetic radiation. As $\tau$ approaches infinity, Eqs. (2.87) and (2.88) take the following form for perfect metals:

$$\varepsilon_r = n^2 - k^2 = 1 - Ne^2/\varepsilon_0 m\omega^2 \quad (2.90)$$

$$2nk = 0 \quad (2.91)$$

Depending on the wavelength of the incident radiation, we can divide the frequency regime into two regions to describe the optical properties of perfect metals:

SHORT-WAVELENGTH REGION: For short-wavelength radiations, $\omega$ is very large, and $Ne^2/(\varepsilon_0 m\omega^2) < 1$. Thus, from Eq. (2.90), we obtain

$$n^2 - k^2 < 1 \quad (2.92)$$

Since $n$ and $k$ are real numbers, Eqs. (2.91) and (2.92) suggest that $k = 0$ and $n < 1$. The latter inequality may seem to be unusual because, according to the definition of $n$ ($n = c/v$), it should be always greater than or equal to unity. However, it should be noted that the phase velocity $c/n$, which represents the speed with which each of the cophasal surfaces or wave surfaces advances, can be greater than $c$ in certain cases [see Born and Wolf (1980), p. 18, 621]. As signals cannot travel faster than $c$, according to the theory of relativity, the phase velocity does not represent a velocity of signal propagation. In the short-wavelength region, a perfect metal is transparent for normally incident radiation, and the radiation undergoes total reflection at the metal surface when the angle of incidence exceeds a critical value.

LONG-WAVELENGTH REGION: In this region, $\omega$ is small, and $Ne^2/\varepsilon_0 m\omega^2 > 1$. Thus, from Eq. (2.90), we obtain

$$n^2 - k^2 < 0 \quad (2.93)$$

Equations (2.91) and (2.93) lead to $k = 0$ and imaginary $n$. In this region, the incident wave undergoes total reflection for all angles of incidence.

It should be noted that when the mean free path of the electrons, the relaxation time, and the damping constant are finite, the electrons absorb the incident energy, hold the absorbed energy for a finite interval of time $\tau$, and then transfer their energy to the lattice. Under such circumstances, that is, for finite $\tau$, even when $Ne^2/\varepsilon_0 m\omega^2 > 1$, the reflectivity of the material is less than unity because a portion of the energy of the incident wave is absorbed in a surface layer of the metal. Also, for finite $\tau$, when $Ne^2/\varepsilon_0 m\omega^2 < 1$, a thin surface layer of the metal attenuates the incident wave, and the rest of the material behaves like an opaque object.

## 2.8. EFFECTS OF TEMPERATURE ON THE ABSORPTIVITY OF METALS

The absorption of radiation by metals is found to increase as the temperature of the material is increased. Higher absorption makes the temperature of the material higher, which, in turn, induces further absorption, and thus a positive feedback loop is set up. Due to this effect, "thermal runaway" can occur in metals, especially in the infrared region, where the absorptivity is small. The effect of temperature on absorptivity can be explained on the basis of the following three phenomena.

1. Free-electron effects: As the temperature increases, the electrons move faster and the time between electrons–lattice collisions decreases. Owing to this phenomenon, the electrons transfer energy to the lattice more rapidly at higher temperatures than at lower temperatures.
2. Interband transitions: In hot metals, the electrons have a considerable amount of energy, and can absorb the incident radiation to undergo interband transitions, which leads to an enhancement of the absorptivity of the material.
3. Surface effects: The chemistry of the surface of hot metals can be affected by chemical reactions between the surrounding gas and the hot metal surface. A change in surface chemistry can alter the absorptivity of the material.

Accounting for these three effects, the absorptivity of metals can be written as

$$A = A_e + A_i + A_s$$

where $A_e$, $A_i$, and $A_s$ refer to the absorptivities due to free electrons, interband transitions, and surface effects, respectively. There are no general theories dealing with the effects of temperature on $A_i$ and $A_s$, but $A_e$ can be determined by using the free-electron theory of metals. From the Fresnel relation given by Eq. (2.56), we obtain

$$A_e = 1 - R = \frac{4n}{(n+1)^2 + k^2} \tag{2.94}$$

which can be simplified for the following two cases.

CASE 1: $\omega\tau \ll 1 \ll \omega_p\tau$ — This criterion corresponds to the far-infrared region. In terms of plasma frequency, Eqs. (2.87) and (2.88) can be written as

$$n^2 - k^2 = 1 - \frac{\omega_p^2 \tau^2}{1 + \omega^2 \tau^2} \approx -\omega_p^2 \tau^2 \tag{2.95}$$

$$2nk = \frac{\omega_p^2 \tau}{\omega} \frac{1}{1 + \omega^2 \tau^2} \approx \frac{\omega_p^2 \tau}{\omega} \tag{2.96}$$

Solving Eqs. (2.95) and (2.96) and noting that $n^2$ should be positive, we obtain

$$n^2 = \frac{\omega_p^2 \tau^2}{2}[-1 + \sqrt{1 + 1/(\omega\tau)^2}]$$

which leads to

$$n \approx \omega_p \sqrt{\tau/\omega} \tag{2.97}$$

Equations (2.96) and (2.97) yield

$$k \approx \omega_p \sqrt{\tau/2\omega} \approx n \tag{2.98}$$

Noting that $n \gg 1$, we obtain the following expression for $A_e$ from Eqs. (2.94), (2.97), (2.98), and (2.89):

$$A_e = 2/n = \sqrt{8\varepsilon_0 \omega / \sigma_0(T)} \tag{2.99}$$

CASE 2: $\omega_p \tau \gg \omega \tau \gg 1$ — This criterion corresponds to the near-infrared and visible regions. In terms of plasma frequency, Eqs. (2.87) and (2.88) can be written as

$$n^2 - k^2 = 1 - \frac{\omega_p^2 \tau^2}{1 + \omega^2 \tau^2} \approx -\frac{\omega_p^2}{\omega^2} \tag{2.100}$$

$$2nk = \frac{\omega_p^2 \tau}{\omega} \frac{1}{1+\omega^2\tau^2} \approx \frac{\omega_p^2}{\omega^3 \tau} \qquad (2.101)$$

Solving Eqs. (2.100) and (2.101) and noting that $n^2$ should be positive, we obtain

$$\begin{aligned} n^2 &= \frac{1}{2}\frac{\omega_p^2}{\omega^2}\left(-1 + \sqrt{1 + \frac{1}{1+\omega^2\tau^2}}\right) \\ &= \frac{1}{2}\frac{\omega_p^2}{\omega^2}\left[-1 + \left(1 + \frac{1}{2}\frac{1}{1+\omega^2\tau^2} + \cdots\right)\right] \\ &= \frac{1}{4}\frac{\omega_p^2}{\omega^4 \tau^2} \end{aligned}$$

which leads to

$$n = \tfrac{1}{2}(\omega_p/\omega^2\tau) \qquad (2.102)$$

Equations (2.101) and (2.102) yield

$$k = \omega_p/\omega \qquad (2.103)$$

Noting that $k \gg n$, we obtain the following expression for $A_e$ from Eqs. (2.94), (2.102), (2.103), and (2.89).

$$A_e \approx \frac{4n}{k^2} = \frac{2\varepsilon_0 \omega_p}{\sigma_0(T)} \qquad (2.104)$$

The variation of $\sigma_0(T)$ with temperature is given by the Wiedemann–Franze law [Kittel (1984)]:

$$\sigma_0(T) = \frac{3}{\pi^2}\left(\frac{e}{k_B}\right)^2 \frac{\kappa}{T} \qquad (2.105)$$

for solid metals at not-too-low temperatures. $\sigma_0(T) \propto 1/T$ for solid metals above room temperature [Bass (1982)], and for liquid metals at not-too-high temperatures [Grosse (1966)]. However, the proportionality constant of liquid metals is usually different from that of solid metals. The electrical conductivity of most metals decreases after melting.

The absorptivities given by Eqs. (2.99) and (2.104) should be corrected to account for the anomalous skin effect. In the usual electrical circuit theory, the current of high frequency is found to be concentrated near the surface of the conductor. This phenomenon, known as the skin effect, can be explained by using Maxwell's equations and Ohm's law. On the basis of his experimental results, Pippard (1947) showed that at low temperatures, the skin resistance is independent of the DC resistance, which departs from the normal theory of skin effect. This phenomenon is known as the anomalous skin effect, which arises at low temperatures and at certain frequencies, because the current density at a given point in the metal cannot be determined by the electric field alone, that is, Ohm's law is not applicable. When the mean free path of electrons is equal to or much longer than the penetration depth of the field into the metal, the electrons will experience different field strengths while traveling through a distance of one mean free path in one mean free time $\tau$; therefore, an electron's velocity will depend on the field strength along its entire path. For this reason, the relation $\mathbf{j} = \sigma \mathbf{E}$, where $\sigma$ is taken to be constant for all parts of the metal, must be replaced by the relation $\mathbf{j} = f(\mathbf{E}, z)$, where $z$ is the impedance of the material. In classical theory, the current density is determined by considering that the electric field is uniform, which is valid only when the field does not essentially vary over one mean free path of the conduction electrons, that is, when the mean free path of the electrons is much shorter than the penetration depth of the field into the metal. Accounting for this anomalous effect, we can write Eqs. (2.99) and (2.104), respectively, as [see Sokolov (1967), p. 215]:

$$A_e = \frac{3}{4}\frac{v_e}{c} + \sqrt{\frac{8\varepsilon_0 \omega}{\sigma_0(T)}}$$

and

$$A_e = \frac{3}{4}\frac{v_e}{c} + \frac{2\varepsilon_0 \omega}{\sigma_0(T)}$$

## 2.9. EFFECTS OF LASER INTENSITY ON ABSORPTION

The expressions for optical properties that have been derived in the previous sections of this chapter are applicable when the incident electromagnetic radiation is so weak that it essentially does not alter the states of the electrons and atoms of the material [Allmen (1987)]. High-power lasers can affect the states of the electrons and atoms of the interacting materials: therefore, the interactions of high-power lasers with materials involve dynamic processes which cannot be described by a static dielectric function. The optical properties of materials are affected by the intensity of the incident laser beam due to the following phenomena, which can occur during laser–materials interactions [Allmen (1987)].

THERMAL EFFECT: The heat generated during laser–materials interactions changes the density and electronic structure of the material, leading to thermal self-focusing in transparent materials and thermal runaway in semiconductors and metals.

OPTICAL EFFECT: The photons of the laser beam can induce interband transitions, which produce free carriers and increase the absorption coefficient of the material.

ELECTRIC FIELD EFFECT: The electric field of a high-intensity laser beam induces nonlinear optical phenomena [Newell and Maloney (1992)], such as multiphoton absorption and self-focusing. Also, the distortion of electron orbitals and molecules becomes nonlinear in the presence of a strong electric field.

GEOMETRICAL EFFECT: When the material is heated up during laser–materials interactions, the roughness and morphology of the surface of the material alter, which affects the absorptivity of the material.

In the classical theory of electromagnetic radiation, the polarization and free-electron current density are considered to depend linearly on the electric field strength, which does not hold good in the presence of a high-intensity laser beam owing to nonlinear effects. Nonlinear optical phenomena are very important because many of the optical devices that have been fabricated, such as optically bistable elements, frequency converters, and tunable parametric devices are based on the principles of nonlinear optics. The linear absorption given by Beer's law,

$$I = I_0 e^{-\eta z} \tag{2.106}$$

is also affected by the following nonlinear optical phenomena.

## 2.9.1. Two-Photon Absorption

Multiphoton absorption, especially two-photon absorption, is found to occur in semiconductors, where the valence electrons are excited to the conduction band due to the absorption of two photons, when the band-gap energy

$$E_g \leq h\nu_1 + h\nu_2$$

In the case of a monochromatic beam of light, such as a laser beam, $\nu_1 = \nu_2$; in that case, the probability of two-photon absorption, which is proportional to the probability of the availability of two photons, is proportional to the square of the laser intensity. To account for optical nonlinear effects, the polarization vector is written as [Milonni and Eberly (1988)]

$$\mathbf{P}_l(\omega_3) = \varepsilon_0 \sum_{m=1}^{3} \chi_{lm}(\omega_3) E_m(\omega_3)$$
$$+ \varepsilon_0 \sum_{m=1}^{3} \sum_{n=1}^{3} \chi_{lmn}(-\omega_3, \omega_1, \omega_2) E_m(\omega_2) E_n(\omega_2) + \cdots \quad (2.107)$$

In the notation for nonlinear susceptibility $\chi_{lmn}(-\omega_3, \omega_1, \omega_2)$, a minus sign is introduced in front of $\omega_3$ just for convenience rather than for any specific reason. In Eq. (2.107), the first term on the right-hand side represents the linear effect of the electric field on the polarization, and the second- and higher-order terms represent the nonlinear optical effects. In the linear theory, which the frequency $\omega$ of the incident radiation is equal to the transition frequency $\omega_0$, the linear susceptibility $\chi(\omega)$ increases, and the imaginary part of the susceptibility determines the absorption of radiation. Similarly, the imaginary part of the nonlinear susceptibility is associated with the absorption of radiation in materials due to nonlinear effects. In the presence of a radiation of frequency $\omega$, the third-order polarization is given by

$$\mathbf{P}_{III}(\omega) = \tfrac{3}{4}\varepsilon_0 \chi_{III}(-\omega, \omega, -\omega, \omega)|\mathbf{E}(\omega)|^2 \mathbf{E}(\omega)$$

and Beer's law is modified to the following form:

$$I(z) = \frac{\eta I(0) e^{-\eta z}}{\eta + \eta_2 I(0)(1 - e^{-\eta z})} \quad (2.108)$$

where $\eta$ is the usual absorption coefficient, and $\eta_2$ is the two-photon absorption coefficient, which is given by

$$\eta_2 = \frac{3\omega \chi_i}{2\varepsilon_0 c^2 n^2(\omega)} \quad (2.109)$$

where $\chi_i$ is the imaginary part of $\chi_{III}(-\omega, \omega, -\omega, \omega)$. Usually, only two-photon absorption processes are observed. This is because three-photon or higher-order multiphoton absorption processes require a very high optical field, but two-photon absorption can also occur in such fields, and then the absorption mechanism can change, reducing the possibility of higher-order multiphoton absorption.

### 2.9.2. Free-Carrier Absorption

When a free carrier is produced by exciting an electron with a photon, across the band gap of a semiconductor or insulator, to a new absorbing state, the free carrier can further absorb photons leading to enhanced absorption. It should be noted that the free-carrier absorption process is similar to the two-photon absorption process, except that the time for two successive absorptions of photons can be as long as the lifetime of the free carrier in its excited state, which is from 0.1 to 1 $\mu$sec. For this reason, if the laser pulse is shorter than the free-carrier lifetime, then (a) two-photon absorption depends on the laser intensity, and (b) free-carrier absorption depends on the laser fluence and the number of free carriers produced.

## 2.10. THIN FILM OPTICS

In LCVD processes, the optical properties of the substrate surface change as the film deposition progresses. At first, the atoms of the bare surface of the substrate interact with the incident laser beam, and part of the laser energy is converted into thermal energy. As the film begins to deposit on the substrate surface, the atoms of the film interact with the laser beam and begin to limit this conversion. The film thickness also plays an important role in determining the reflection, transmission, absorption, and interference of the laser beam. Proper understanding of the absorption of laser energy by the substrate and the film is very important for pyrolytic LCVD because the absorptivity determines the surface temperature, which affects the chemical reaction and the film deposition. For this reason, the reflection and transmission phenomena are discussed below.

### 2.10.1. Fresnel Coefficients

Fresnel coefficients provide a convenient means of expressing the transmittivity and reflectivity of multiple layers. In this section, we will derive the Fresnel reflection and transmission coefficients for nonconducting ($\sigma = 0$) mediums, which can be applied to conducting ($\sigma \neq 0$) mediums by replacing the refractive index $n$ by the complex refractive index, $\hat{n} = n + ik$. The amount of light reflected and transmitted at a boundary separating two mediums can be determined by solving Maxwell's equations with appropriate boundary conditions, that is, that the tangential components of both electric and magnetic fields are continuous at the boundary.

For an isotropic and nonconducting ($\sigma = 0$) medium, Maxwell's equation for the electric field is given by Eq. (2.18), which is rewritten below:

$$\frac{\varepsilon_r \mu_r}{c^2} \frac{\partial^2 \mathbf{E}}{\partial t^2} = \nabla^2 \mathbf{E} \tag{2.110}$$

Setting $\sigma = 0$ in Eq. (2.11), taking the curl of Eq. (2.11), noting that $\nabla \times (\nabla \times \mathbf{H}) = \nabla \times (\nabla \cdot \mathbf{H}) - \nabla^2 \mathbf{H}$, and using Eqs. (2.9) and (2.10), we obtain the following wave equation for the magnetic field:

$$\frac{\varepsilon_r \mu_r}{c^2} \frac{\partial^2 \mathbf{H}}{\partial t^2} = \nabla^2 \mathbf{H} \tag{2.111}$$

To solve Eqs. (2.110) and (2.111), we consider a plane wave incident on a plane boundary separating two semi-infinite mediums of different optical properties as shown in Fig. 2.4. The $z = 0$ plane, that is, the x0y plane

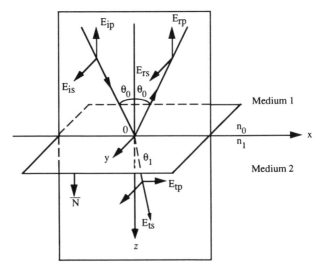

Figure 2.4. Reflection and transmission of a plane wave at a plane interface.

separates mediums 1 and 2 in Fig. 2.4, and the x0z plane indicates the plane of incidence of the radiation. The plane of incidence represents a plane that contains the normal **N** to the boundary and the direction of propagation of the incident wave. The refractive indices of mediums 1 and 2 are $n_0$ and $n_1$, respectively. A plane wave with electric field $\mathbf{E}_i$ is incident on medium 2 from medium 1 at an angle of incidence $\theta$ at the origin $O$, the reflected wave $\mathbf{E}_r$ propagates back through medium 1, and the transmitted wave $\mathbf{E}_t$ propagates through medium 2 at an angle of refraction $\theta_1$.

In Fig. 2.4, $E_{ip}$, $E_{rp}$, and $E_{tp}$ represent the components of the incident, reflected, and transmitted electric fields, respectively, that are parallel to the plane of incidence. The parallel component is sometimes referred to as the TM (transverse magnetic) mode or p-polarization, which comes from the German word "parallel." On the other hand, $E_{is}$, $E_{rs}$, and $E_{ts}$ represent the components of the incident, reflected, and transmitted electric fields, respectively, that are perpendicular to the plane of incidence. The perpendicular component is also referred to as the TE (transverse electric) mode or s-polarization, which comes from the German word "senkrecht" meaning perpendicular. In general,

$$E_\| = E_p = \mathrm{TM} = \text{Electric field parallel to the plane of incidence}$$
$$E_\perp = E_s = \mathrm{TE} = \text{Electric field perpendicular to the plane of incidence}$$

The components of the magnetic fields are not shown in Fig. 2.4, because they can be visualized by noting that **H** is perpendicular to **E**. The phases of the electromagnetic radiation can be written as [Heavens (1955)]

$$\exp i\left(\frac{2\pi n_0 x \sin\theta_0}{\lambda} + \frac{2\pi n_0 z \cos\theta_0}{\lambda} - \omega t\right) \quad \text{for the incident wave}$$

$$\exp i\left(\frac{2\pi n_0 x \sin\theta_0}{\lambda} - \frac{2\pi n_0 z \cos\theta_0}{\lambda} - \omega t\right) \quad \text{for the reflected wave}$$

and

$$\exp i\left(\frac{2\pi n_1 x \sin\theta_1}{\lambda} + \frac{2\pi n_1 z \cos\theta_1}{\lambda} - \omega t\right) \quad \text{for the transmitted wave}$$

At the point of incidence, that is, at $x = y = z = 0$, the total components of the electric and magnetic fields along the $x$ and $y$ directions are given by

FOR MEDIUM 1:

$$E_{0x} = (E_{ip} + E_{rp})\cos\theta_0$$

$$E_{0y} = E_{is} + E_{rs}$$

$$H_{0x} = \frac{n_0}{\mu_{r_1}\mu_0 c}(-E_{is} + E_{rs})\cos\theta_0$$

$$H_{0y} = \frac{n_0}{\mu_{r_1}\mu_0 c}(E_{ip} - E_{rp})$$

and

FOR MEDIUM 2:

$$E_{1x} = E_{tp}\cos\theta_1$$

$$E_{1y} = E_{ts}$$

$$H_{1x} = -\frac{n_1}{\mu_{r_2}\mu_0 c}E_{ts}\cos\theta_1$$

$$H_{2y} = \frac{n_1}{\mu_{r_2}\mu_0 c}E_{tp}$$

The relative permeability is almost equal to unity for wavelengths ranging from the long-wave infrared region to the visible and ultraviolet region [see Sokolov (1967), p. 15]. Thus, we can take $\mu_{r_1} \approx \mu_{r_2} \approx 1$. Satisfying the boundary conditions by using the above equations, we obtain the following ratios of the amplitudes of the transmitted and reflected waves to those of the incident wave:

$$\frac{E_{rp}}{E_{ip}} = \frac{n_0\cos\theta_1 - n_1\cos\theta_0}{n_0\cos\theta_1 + n_1\cos\theta_0} = r_{1p} \qquad (2.112)$$

$$\frac{E_{tp}}{E_{ip}} = \frac{2n_0\cos\theta_0}{n_0\cos\theta_1 + n_1\cos\theta_0} = t_{1p} \qquad (2.113)$$

$$\frac{E_{rs}}{E_{1s}} = \frac{n_0\cos\theta_0 - n_1\cos\theta_1}{n_0\cos\theta_0 + n_1\cos\theta_1} = r_{1s} \qquad (2.114)$$

$$\frac{E_{ts}}{E_{is}} = \frac{2n_0\cos\theta_0}{n_0\cos\theta_0 + n_1\cos\theta_1} = t_{1s} \qquad (2.115)$$

where $r_{1_p}$ and $r_{1_s}$ are the Fresnel reflection coefficients and $t_{1_p}$ and $t_{1_s}$ are the Fresnel transmission coefficients, which can be used to express the reflection and transmission phenomena in multilayer materials in a convenient way.

## 2.10.2. Reflectivity and Transmissivity of a Multilayer Nonabsorbing ($k = 0$) Medium

Consider a nonabsorbing medium with $n$ layers of films of different materials. The refractive index and thickness of the $m$th layer is $n_m$ and $d_m$, respectively. The $j$th component ($j = s$ or $p$) of the incident and reflected waves and the angle of incidence at the boundary of the $m$th and ($m + 1$)th layer are denoted by $E_{ijm}$, $E_{rjm}$, and $\theta_m$, respectively, as shown in Fig. 2.5. By following the procedure used in Section 2.10.1, that is, by determining the $x$ and $y$ components of the **E** and **H** vectors in the $m$th and ($m - 1$)th layer, and satisfying the boundary conditions, it can be shown [Heavens (1955)] that

$$\begin{pmatrix} E_{ijm-1} \\ E_{rjm-1} \end{pmatrix} = \frac{1}{t_{jm}} \begin{bmatrix} e^{i\delta_{m-1}} & r_{jm}e^{i\delta_{m-1}} \\ r_{jm}e^{-i\delta_{m-1}} & e^{i\delta_{m-1}} \end{bmatrix} \begin{pmatrix} E_{ijm} \\ E_{rjm} \end{pmatrix} \qquad (2.116)$$

where $t_{jm}$ and $r_{jm}$ are the Fresnel coefficients as given by Eqs. (2.112)–(2.115) for $j = p$ and $s$. Equation (2.116) is applicable to both the $p$- and $s$-polarized component of the electric field **E** for $j = p$ and $j = s$, respectively, and, for this reason, the subscript $j$ will be dropped in the following discussion. The expression for $\delta_m$ is

$$\delta_m = \frac{2\pi n_m \cos \theta_m}{\lambda} \sum_{i=1}^{m-1} d_i$$

Since the reflectivity and transmissivity are given by

$$R = \frac{(E_{r0})(E_{r0})^*}{(E_{i0})(E_{i0})^*} \quad \text{and} \quad T = \frac{(E_{i_{n+1}})(E_{i_{n+1}})^*}{(E_{i0})(E_{i0})^*}$$

we obtain the following expression from Eq. (2.116):

$$\begin{pmatrix} R_{i0} \\ E_{r0} \end{pmatrix} = \frac{(C_1)(C_2)\cdots(C_{n+1})}{t_1 t_2 \cdots t_{n+1}} \begin{pmatrix} E_{i_{n+1}} \\ E_{r_{n+1}} \end{pmatrix} \qquad (2.117)$$

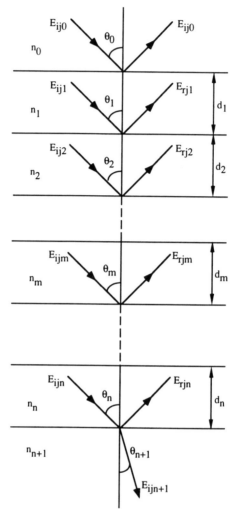

Figure 2.5. Reflection and transmission phenomena in a multilayer nonabsorbing ($k = 0$) medium.

where the matrix $(C_m)$ is given by

$$(C_m) = \begin{bmatrix} e^{i\delta_{m-1}} & r_m e^{i\delta_{m-1}} \\ r_m e^{-i\delta_{m-1}} & e^{-i\delta_{m-1}} \end{bmatrix} \quad (2.118)$$

and the matrix product $(C_1)(C_2)\cdots(C_{n+1})$ can be written as

$$(C_1)(C_2)\cdots(C_{n+1}) = \begin{pmatrix} a & b \\ q & d \end{pmatrix} \quad (2.119)$$

Since there is no reflection in the $(n+1)$th medium, $E_{r_{n+1}} = 0$, and we obtain the following relations from Eqs. (2.117) and (2.119):

$$R_i = \frac{E_{r0}}{R_{i0}} = \frac{q}{a} \quad \text{and} \quad T_i = \frac{E_{i_{n+1}}}{E_{i0}} = \frac{t_1 t_2 \cdots t_{n+1}}{a}$$

and, therefore,

$$R = \frac{qq^*}{aa^*} \tag{2.120}$$

$$T = \frac{n_{n+1}}{n_0} \frac{(t_1 t_2 \cdots t_{n+1})(t_1^* t_2^* \cdots t_{n+1}^*)}{aa^*} \tag{2.121}$$

For a single film, $n = 1$ and, since $\delta_0 = 0$, Eq. (2.119) becomes

$$(C_1)(C_2) = \begin{pmatrix} 1 & r_1 \\ r_1 & 1 \end{pmatrix} \begin{pmatrix} e^{i\delta_1} & r_2 e^{i\delta_1} \\ r_2 e^{-i\delta_1} & e^{-i\delta_1} \end{pmatrix} = \begin{pmatrix} a & b \\ q & d \end{pmatrix}$$

which yields

$$a = e^{i\delta_1} + r_1 r_2 e^{-i\delta_1}$$

$$q = r_1 e^{i\delta_1} + r_2 e^{-i\delta_1}$$

Since $r_1$, $r_2$, $t_1$, $t_2$, and $\delta_1$ are real for a transparent film on a transparent substrate,

$$R = \frac{qq^*}{aa^*} = \frac{(r_1 e^{i\delta_1} + r_2 e^{-i\delta_1})(r_1 e^{-i\delta_1} + r_2 e^{i\delta_1})}{(e^{i\delta_1} + r_1 r_2 e^{-i\delta_1})(e^{-i\delta_1} + r_1 r_2 e^{i\delta_1})} = \frac{r_1^2 + 2r_1 r_2 \cos 2\delta_1 + r_2^2}{1 + 2r_1 r_2 \cos 2\delta_1 + r_1^2 r_2^2} \tag{2.122}$$

$$T = \frac{n_2}{n_0} \frac{(t_1 t_2)(t_1^* t_2^*)}{aa^*} = \frac{n_2}{n_0} \frac{t_1^2 t_2^2}{1 + 2r_1 r_2 \cos 2\delta_1 + r_1^2 r_2^2} \tag{2.123}$$

## 2.10.3. Reflectivity and Transmissivity of a Multilayer Absorbing ($k \neq 0$) Medium

For an absorbing medium, the analysis of the previous section can be used by redefining the Fresnel coefficients and refractive index using complex variables. The refractive index for the $m$th layer is taken to be

$\hat{n}_m = n_m + ik_m$, and the Fresnel coefficients at the $m$th interface are defined as

$$r_m = g_m - ih_m \quad \text{and} \quad t_m = 1 - g_m + ih_m$$

For the case of normal incidence, it can be shown [Heavens (1955)] that

$$g_m = \frac{n_{m-1}^2 + k_{m-1}^2 - n_m^2 - k_m^2}{(n_{m-1} + n_m)^2 + (k_{m-1} + k_m)^2} \quad (2.124)$$

$$h_m = \frac{2(n_{m-1}k_m - n_m k_{m-1})}{(n_{m-1} + n_m)^2 + (k_{m-1} + k_m)^2} \quad (2.125)$$

For an absorbing medium,

$$e^{i\delta_{m-1}} = e^{i2\pi/\lambda}(n_{m-1} + ik_{m-1})d_{m-1} = e^{-\alpha_{m-1}} e^{i\gamma_{m-1}}$$

where $\alpha_{m-1} = (2\pi/\lambda)k_{m-1}d_{m-1}$ and $\gamma_{m-1} = (2\pi/\lambda)n_{m-1}d_{m-1}$.
The matrix $(C_m)$ of Eq. (2.118) is now given by

$$(C_m) = \begin{bmatrix} e^{i\delta_{m-1}} & r_m e^{i\delta_{m-1}} \\ r_m e^{-i\delta_{m-1}} & e^{-i\delta_{m-1}} \end{bmatrix} \equiv \begin{bmatrix} \mathbf{p}_m + i\mathbf{q}_m & \mathbf{r}_m + i\mathbf{s}_m \\ \mathbf{t}_m + i\mathbf{u}_m & \mathbf{v}_m + i\mathbf{w}_m \end{bmatrix} \quad (2.126)$$

where

$$\mathbf{p}_m = \beta_{m-1} \cos \gamma_{m-1}$$
$$\mathbf{q}_m = \beta_{m-1} \sin \gamma_{m-1}$$
$$\mathbf{r}_m = \beta_{m-1}(g_m \cos \gamma_{m-1} - h_m \sin \gamma_{m-1})$$
$$\mathbf{s}_m = \beta_{m-1}(h_m \cos \gamma_{m-1} - g_m \sin \gamma_{m-1})$$
$$\mathbf{t}_m = \bar{\beta}_{m-1}(g_m \cos \gamma_{m-1} + h_m \sin \gamma_{m-1})$$
$$\mathbf{u} = \bar{\beta}_{m-1}(h_m \cos \gamma_{m-1} - g_m \sin \gamma_{m-1})$$
$$\mathbf{v}_m = \bar{\beta}_{m-1} \cos \gamma_{m-1}$$
$$\mathbf{w}_m = -\bar{\beta}_{m-1} \sin \gamma_{m-1}$$
$$\beta_{m-1} = e^{-\alpha_{m-1}} \quad \text{and} \quad \bar{\beta}_{m-1} = e^{\alpha_{m-1}}$$

Using expression (2.126), and carrying out the matrix product $(C_1)(C_2) \cdots (C_{n+1})$, we can determine the values of $a$ and $q$ from Eq. (2.119); then $R$ and $T$ can be calculated from Eqs. (2.120) and (2.121), respectively, keeping in mind that $n_{n+1}$ and $n_0$ should be replaced with complex refractive indices in Eq. (2.121) for an absorbing medium.

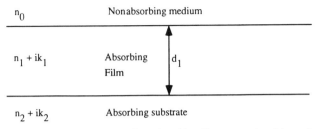

Figure 2.6. Geometric configuration of an absorbing film on an absorbing substrate.

For a single layer of an absorbing film on an absorbing substrate, $n = 1$, and the reflectivity is given by [Heavens (1955)]

$$R = \frac{(g_1^2 + h_1^2)e^{-2\alpha_1} + (g_2^2 + h_2^2)e^{2\alpha_1} + A_1 \cos 2\gamma_1 + B_1 \sin 2\gamma_1}{e^{-2\alpha_1} + (g_1^2 + h_1^2)(g_2^2 + h_2^2)e^{2\alpha_1} + C_1 \cos 2\gamma_1 + D_1 \sin 2\gamma_1} \quad (2.127)$$

for the geometric configuration shown in Fig. 2.6. In Eq. (2.127), the various variables are as follows:

$$g_1 = \frac{n_0^2 - n_1^2 - k_1^2}{(n_0 + n_1)^2 + k_1^2}$$

$$g_2 = \frac{n_1^2 - n_2^2 + k_1^2 - k_2^2}{(n_1 + n_2)^2 + (k_1 + k_2)^2}$$

$$h_1 = \frac{2n_0 k_1}{(n_0 + n_1)^2 + k_1^2}$$

$$h_2 = \frac{2(n_1 k_2 - n_2 k_1)}{(n_1 + n_2)^2 + (k_1 + k_2)^2}$$

$$A_1 = 2(g_1 g_2 + h_1 h_2) \qquad B_1 = 2(g_1 h_2 - g_2 h_1)$$

$$C_1 = 2(g_1 g_2 - h_1 h_2) \qquad D_1 = 2(g_1 h_2 + g_2 h_1)$$

$$\alpha_1 = 2\pi k_1 d_1/\lambda \qquad \text{and} \qquad \gamma_1 = 2\pi n_1 d_1/\lambda$$

## 2.10.4. Applicability of Thin Film Optics

The Fresnel coefficients, which have been used in the previous sections to determine the reflectivity and transmissivity of various media, have been derived in Section 2.10.1 under the assumptions that (1) the incident electromagnetic radiation is a plane wave, and (2) the interface on which the radiation impinges is a plane. Assumption (1) holds good for many applications, because even the spherical wave from a point source can be treated locally as a collimated plane wave. However, assumption (2) is valid only for optically smooth surfaces. A surface is considered to be optically smooth if the root mean square (rms) value of the roughness of the surface is much less than the wavelength of the incident radiation. Most unpolished surfaces can be considered optically smooth for wavelengths in the far-infrared region and beyond. For wavelengths in the visible region, optically smooth surfaces can sometimes be produced with graded sandpaper treatment and diamond paste polishing. For the utraviolet region, optically smooth surfaces are produced by depositing a thin film in an ultrahigh vacuum. Unpolished surfaces have asperities which induce multiple reflections of the radiations having wavelengths in the visible and near-infrared regions; for this reason, the actual reflectivity of an unpolished surface is less than that of a smooth surface in these two regions of the electromagnetic spectrum.

### 2.10.5. Example — Calculation of Optical Properties at High Temperatures

From among various optical properties of materials, the most important one is the optical absorptivity of materials. Such data for many materials can be found in Wolfe and Zissis (1978). Sometimes complex refractive indices are reported in the literature which can be used to determine the absorptivity under suitable conditions by using Fresnel's relation. Optical constants (refractive indices) for various noble and transition metals at room temperature can be found in Johnson and Christy (1972, 1974); and Lenham and Treherne (1966) as a function of wavelength of the incident electromagnetic radiation. However, the optical constants of materials depend not only on the wavelength of the incident radiation but also on the temperature of the materials. Wolfe and Zissis (1978) provide the temperature variation of optical constants of some materials. In pyrolytic LCVD processes, the temperature of the laser-irradiated spot is very high and hence the absorptivity data of the substrate materials at high temperature is important. Since optical constants of materials are usually reported for the room temperature condition, we will show how high-temperature data can be determined from the available room-temperature data.

We will consider a multilayer medium which is made up of a SS 304 substrate, a Ti film on the surface of SS 304 and a vapor phase containing

TiBr$_4$, and the chemical reaction products. These three layers are arranged as shown in Fig. 2.6. Since LCVD is usually carried out at low pressures, we will assume that the optical properties of the vapor phase are the same as those of a vacuum. From the room-temperature data, we will first determine the optical constants ($n$ and $k$, the real and imaginary parts of the refractive index, respectively) of Ti and SS 304 at 1173 K as the decomposition temperature of the chemical reaction [Funaki *et al.* (1961)], 4TiBr → 3Ti + TiBr$_4$, which produces the Ti film, is greater than 1173 K. Using the high-temperature values of $n$ and $k$, the reflectivity of the composite material (made up of SS 304 and Ti film) is determined by using Eq. (2.127).

The $n$ and $k$ values of Ti and SS 304 at room temperature are obtained from the experimental data of Lenham and Treherne (1966), where $2nk/\lambda$ and $k^2 - n^2$ values of various transition metals have been reported. Here $\lambda$ is the wavelength of the incident electromagnetic radiation in $\mu$m. To compute the $n$ and $k$ values of SS 304, we consider that the composition of SS 304 is 71.5% Fe, 19% Cr, and 9.5% Ni by weight and that

$$n_{SS} = 0.715 n_{Fe} + 0.095 n_{Ni} + 0.19 n_{Cr}$$

and

$$k_{SS} = 0.715 k_{Fe} + 0.095 k_{Ni} + 0.19 k_{Cr}$$

Here, $n_i$ and $k_i$ are the real and the imaginary parts of the refractive indices of the $i$th material, where $i$ stands for SS 304 (SS), Fe, Ni, and Cr. Table 2.2 is a result of this consideration and the data of Lenham and Treherne (1966); for a CO$_2$ laser, the wavelength $\lambda = 10.6\,\mu$m.

Table 2.2. Optical Properties of Various Materials

| Materials | Values at room temperature (300 K) | | | |
| --- | --- | --- | --- | --- |
| | $2nk/\lambda$ | $k^2 - n^2$ | $n$ | $k$ |
| Ti | 25 | 83 | 9.8665 | 13.4293 |
| Fe | 50 | 833 | | |
| Ni | 60 | 1333 | | |
| Cr | 54 | 583 | | |
| SS 304 | — | — | 9.1417 | 30.1586 |

Also, we have the following data from Goldsmith et al. (1961),

$$\rho_{Ti,300} = 44\ \mu\Omega\text{-cm}, \qquad \rho_{Ti,1173} = 160\ \mu\Omega\text{-cm}$$
$$\rho_{SS,300} = 74.5\ \mu\Omega\text{-cm}, \qquad \rho_{SS,1173} = 122.5\ \mu\Omega\text{-cm}$$

where $\rho_{i,j}$ is the DC resistivity of the $i$th material at the $j$th temperature. It should be noted that Goldsmith et al. (1961) provide the resistivity data for SS 303, which we are using for SS 304.

To obtain the $n$ and $k$ values at higher temperature, we use the following empirical relation from Arata and Miyamoto (1978):

$$A = 1 - R = 112.2 \times 10^{-4}\sqrt{\rho} \tag{2.128}$$

where $R$ is the reflectivity and $\rho$ is the DC resistivity in $\mu\Omega$-cm. It should be noted that Eq. (2.128) is same as the Hagen–Rubens relation (2.60), which is applicable in the far-infrared ($\lambda \geqslant 100\ \mu m$) and microwave regions [see Sokolov (1967), p. 16, 17, 142]. However, Arata and Miyamoto (1978) have shown experimentally that Eq. (2.128) is valid for a 10.6 $\mu m$ $CO_2$ laser for many metals, including Ag, Au, W, Mo, Ni, Cr, Cu, Al, Fe, Ta, Ti, and Zr.

Taking the derivative of Eq. (2.128) with respect to temperature $T$, we obtain

$$\frac{dR}{dT} = \frac{112.2 \times 10^{-4}}{2\sqrt{\rho}}\frac{d\rho}{dT} \tag{2.129}$$

Equation (2.129) will be used to determine $dR/dT$ at any temperature by assuming that the resistivities of Ti and SS 304 vary linearly with $T$.

For normally incident beams, we can use the Fresnel relation (2.56). Taking the derivative of expression (2.56) with respect to $T$, we get the following equation for $dn/dT$ and $dk/dT$.

$$\frac{dn}{dT} + A_2\frac{dk}{dT} = A_3 \tag{2.130}$$

here

$$A_2 = A'_1/A'_0 \text{ and } A_3 = A'_2/A'_0$$

where

$$A'_0 = (n_0 - 1)[(n_0 + 1)^2 + k_0^2] - (n_0 + 1)[(n_0 - 1)^2 + k_0^2]$$
$$A'_1 = 4n_0 k_0$$
$$A'_2 = \tfrac{1}{2}[(n_0 + 1)^2 + k_0^2]^1 \left(\frac{dR}{dT}\right)_0$$

The subscript 0 is used in this section to refer to the values of the variables $n$, $k$, and $dR/dT$ at room temperature. From Eqs. (2.56) and (2.128), we obtain

$$\frac{(n-1)^2 + k^2}{(n+1)^2 + k^2} = 1 - 112.2 \times 10^{-4}\sqrt{\rho} \qquad (2.131)$$

which is applicable at any temperature. We assume that $n$ and $k$ vary linearly with temperature, and therefore

$$n = n_0 + (T - T_0)\frac{dn}{dT} \qquad (2.132)$$

and

$$k = k_0 + (T - T_0)\frac{dk}{dT} \qquad (2.133)$$

Substituting Eqs. (2.132) and (2.133) into Eq. (2.131), we obtain

$$\left(\frac{dn}{dT}\right)^2 + \left(\frac{dk}{dT}\right)^2 + A_4 \frac{dn}{dT} + A_5 \frac{dk}{dT} = A_6 \qquad (2.134)$$

Here

$$A_4 = A'_4/A'_3, \qquad A_5 = A'_5/A'_3 \quad \text{and} \quad A_6 = A'_6/A'_3$$

where

$$A'_3 = 112.2 \times 10^{-4}(T - T_0)^2\sqrt{\rho}$$
$$A'_4 = 224.4 \times 10^{-4}(n_0 - 1)(T - T_0)\sqrt{\rho}$$
$$A'_5 = 224.4 \times 10^{-4} k_0 (T - T_0)\sqrt{\rho}$$
$$A'_6 = (1 - 112.2 \times 10^{-4}\sqrt{\rho})[(n_0 + 1)^2 + k_0^2](n_0 - 1)^2 - k_0^2$$

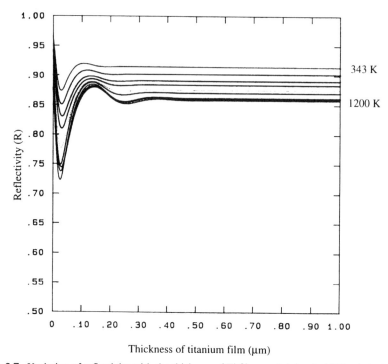

Figure 2.7. Variation of reflectivity with the thickness of Ti film on stainlss steel 304 substrate.

Equations (2.130) and (2.134) are solved by substituting the value of $T$ at which $n$ and $k$ values are required. Knowing $dn/dT$ and $dk/dT$ from the solutions of Eqs. (2.130) and (2.134), $n$ and $k$ are determined from Eqs. (2.132) and (2.133) for SS 304 and Ti, and then the reflectivity of the composite material, which is made up of SS 304 and Ti, is obtained by using Eq. (2.127) for a wide range of the Ti film thickness, as shown in Fig. 2.7. The reflectivity of the composite medium is found to be 0.86351 for Ti film 0.4-$\mu$m thick and 0.86350 for Ti film of thickness 0.8, 1, 10, 100, 200, 500, 700, 900, and 1000 $\mu$m. Thus, Ti films of thickness 0.4 $\mu$m and above do not affect the reflectivity of the composite medium of SS 304 and Ti at 1173 K. However, Fig. 2.7 shows that Ti films thinner than 0.4 $\mu$m do affect the reflectivity of the composite medium. By using the $n$ and $k$ values determined at the high temperature (1173 K) by the above procedure in Eq. (2.127), the reflectivity of SS 304 is found to be 0.8758. The absorptivity of SS 304 obtained from this value of reflectivity agrees very well with the experimental data of Duley et al. (1979).

## 2.11. CHEMICAL KINETICS

Chemical kinetics [Laidler (1987), Levenspiel (1975)] is very important for an understanding of the reactions that occur during pyrolytic LCVD processes, as it affords insight into the mechanism of reactions and the formation and breakage of chemical bonds and allows determination of the energy and stability of such bonds. It also provides a method for analyzing the heat and mass transfer processes during chemical reactions. For these reasons, chemical kinetics and the reaction rate constants of various reactions are very important for modeling LCVD processes. Various basic concepts of chemical kinetics are discussed below:

### 2.11.1. Chemical Reaction and Reaction Rate

If $a$ moles of a chemical reactant A react with $b$ moles of another chemical reactant B to produce $m$ and $w$ moles of the products M and W, respectively, the chemical reaction is represented by

$$aA + bB \underset{k_r}{\overset{k_f}{\rightleftharpoons}} mM + wW, \qquad \Delta H \qquad (2.135)$$

The heat of reaction $\Delta H$ is given by

$$\Delta H = mH_M + wH_W - aH_A - bH_B$$

$\Delta H < 0$ and $\Delta H > 0$ represent the exothermic and endothermic reactions, respectively. The rate of reaction of the $i$th species $r_i$, where $i = A, B, M,$ or $W$, is defined as the change in the moles of the $i$th species due to chemical reaction per unit volume per unit time and is given by

$$r_i = \frac{1}{V}\left(\frac{dN_i}{dt}\right)_{\text{by reaction}}$$

It should be noted that the number of moles of the $i$th species can change at any point in the reaction volume due to chemical reactions, diffusion, and convection, but only the rate of change due to chemical reaction determines the reaction rate. If $i$ is a reaction product, $r_i$ is positive and represents the rate of production of the $i$th species. On the other hand, if $i$ is a reactant, $r_i$ is negative and $-r_i$ is the rate of consumption of the $i$th species. In general, $r_i$ and $-r_i$ represent the production and consumption rates, respectively. Although the above definition of $r_i$ is based on the volume of the reaction zone, the reaction rate can be defined in terms of other dimensions as discussed below:

1. Based on unit mass of solid in fluid–solid systems,

$$r_{i,ms} = \frac{\text{Change in the moles of the } i\text{th species by reaction}}{(\text{Mass of solid})(\text{Time})}$$

$$= \frac{1}{M_s}\left(\frac{dN_i}{dt}\right)_{\text{by reaction}}$$

2. Based on unit interfacial surface area in two-phase systems, such as gas–liquid, gas–solid, or liquid–solid

$$r_{i,sa} = \frac{\text{Change in the moles of the } i\text{th species by reaction}}{(\text{Surface area of the interface})(\text{Time})}$$

$$= \frac{1}{S}\left(\frac{dN_i}{dt}\right)_{\text{by reaction}}$$

This definition of the reaction rate is useful for modeling pyrolytic LCVD processes.

3. Based on unit volume of solid in gas–solid systems,

$$r_{i,vs} = \frac{\text{Change in the moles of the } i\text{th species by reaction}}{(\text{Volume of solid})(\text{Time})}$$

$$= \frac{1}{V_s}\left(\frac{dN_i}{dt}\right)_{\text{by reaction}}$$

4. Based on unit volume of reactor,

$$r_{i,vr} = \frac{\text{Change in the moles of the } i\text{th species by reaction}}{(\text{Volume of reactor})(\text{Time})}$$

$$= \frac{1}{V_r}\left(\frac{dN_i}{dt}\right)_{\text{by reaction}}$$

### 2.11.2. Types of Reactions

Chemical reactions can be classified into several types based on the directionality and/or the mechanism of the reaction, the value of $\Delta H$, and the geometrical location of the reaction zone:

CLASSIFICATION BY THE REACTION DIRECTIONALITY: If only the forward reaction occurs, that is, if the products do not react to form the reactants, the reaction is said to be irreversible. On the other hand, if the reaction proceeds in both forward and reverse directions, that is, if the products react to form the reactants, the reaction is said to be reversible.

CLASSIFICATION BY REACTION MECHANISMS: If a reaction occurs in a single step without any reaction intermediates, the reaction is categorized as elementary. For a long time, the reaction

$$H_2 + I_2 \rightarrow 2HI$$

was considered to be an elementary one, but it is now considered to occur through a mechanism involving the following intermediate products or elementary steps:

$$I_2 \rightleftharpoons 2I$$
$$I + H_2 \rightarrow HI + H$$
$$H + I_2 \rightarrow HI + I$$

CLASSIFICATION BY THE VALUE OF $\Delta H$: If $\Delta H > 0$, the reaction is endothermic, and if $\Delta H < 0$, the reaction is exothermic.

CLASSIFICATION BY THE GEOMETRICAL LOCATION OF THE REACTION ZONE: A reaction is called homogeneous if it occurs in one phase alone. If a reaction occurs in the presence of at least two phases, it is called heterogeneous. In pyrolytic LCVD processes, the chemical reactions are usually considered to be heterogeneous.

## 2.11.3. Stoichiometric Coefficient, Molecularity, Order of a Reaction, and Reaction Rate Constant

The stoichiometric coefficient of a species represents the number of moles of the species in a balanced chemical equation, and it is positive for products and negative for reactants. In the balanced chemical equation (2.135), $-a$ and $-b$ are the stoichiometric coefficients for the reactants, and $m$ and $w$ are the stoichiometric coefficients for the products.

The stoichiometric coefficient refers to any type of reaction, but molecularity is relevant only for elementary reactions. The molecularity of a species in an elementary reaction is the number of moles of the species involved in the reaction, and is found to be one or two, and occasionally three.

From chemical reaction (2.135), the rate of production of $M$ or $W$ and the rate of consumption of $A$ or $B$ are given by

$$\frac{1}{m}r_M = \frac{1}{w}r_W = -\frac{1}{a}r_A = -\frac{1}{b}r_B = kC_A^a C_B^b \qquad (2.136)$$

where $k$ is the reaction rate constant. The order of the reaction is indicated by the powers to which the concentrations are raised. Therefore, reaction (2.135) is: $a$th order with respect to the species $A$, $b$th order with respect to the species $B$, and the overall reaction order $= n = a + b$. It should be noted that the order of a reaction is not necessarily related to the molecularity and the stoichiometric coefficients. The rate expression (2.136) is determined empirically; therefore, the order of a reaction need not be an integer, but can have a fractional value. The molecularity of a reaction is always an integer, because it refers to the reaction mechanism, and is applicable to an elementary reaction.

The dimension of the rate constant $k$ is

$$(\text{Concentration})^{1-n}(\text{Time})^{-1}$$

when the overall order of the reaction is $n$, and it is simply $(\text{Time})^{-1}$ for a first-order reaction. The concept of "reaction order" is used only for those reactions whose rate expressions are of the form of Eq. (2.136). The nonelementary reaction

$$H_2 + Br_2 \rightarrow 2HBr$$

has a rate expression

$$r_{HBr} = \frac{k_1[H_2][Br_2]^{1/2}}{k_2 + [HBr]/Br_2}$$

and the term reaction order is not used for such nonelementary reactions. In the above rate expression, $k_1$ and $k_2$ are constants which are referred to as kinetic parameters, and $[X]$ denotes the concentration of $X$, where $X = H_2$, $Br_2$, or $HBr$. For many nonelementary reactions, only a single reaction is observed, although it involves a series of elementary reactions or intermediate products, because the amount or the lifetime of the intermediate products is so small that the elementary steps are not easily detectable.

Usually, the rate expressions of elementary reactions correspond to the form given by Eq. (2.136).

## 2.11.4. Thermodynamics and the Equilibrium Constant for Elementary Reactions

The thermodynamics of a chemically reacting system provides two important results, which are the heat of reaction or the change in enthalpy and the change in the standard free energy (standard Gibbs function). The heat of reaction allows us to determine the amount of energy absorbed or liberated during the reaction. The change in the standard free energy (standard Gibbs function) is related to the equilibrium constant, which can be written for Eq. (2.135) as

$$\Delta G^0 = mG_M^0 + wG_W^0 - aG_A^0 - bG_B^0 = -RT \ln K$$

The maximum possible yield of the reaction products can be determined from the value of $K$.

The equilibrium condition of a chemical reaction can be characterized in any one of the following three ways:

THERMODYNAMICS: From the thermodynamics point of view, a system is in equilibrium with its surroundings at a given pressure and temperature if the Gibbs free energy of the system has the lowest possible value. When the system is away from the equilibrium condition, we have

$$(\Delta G)_{P,T} > 0$$

where $(\Delta G)_{P,T}$ is the Gibbs free energy of the nonequilibrium system minus the Gibb's free energy of the equilibrium system at a given pressure and temperature.

STATISTICAL MECHANICS: In statistical mechanics, the state that is attained by the system with the highest probability is called the equilibrium state. The equilibrium system contains the largest number of equally likely molecular configurations that are macroscopically identical.

CHEMICAL KINETICS: From the chemical kinetics point of view, a system is at equilibrium if the forward and reverse reactions occur at equal rates. An equilibrium system represents a dynamic steady state condition at which the rates of changes of the reactant and product molecules are equal, instead of a static condition at which everything is at rest.

The energy diagram for the chemical reaction (2.135) and the activation energies $E_f$ and $E_r$ for the forward and reverse reactions, respectively, are shown in Fig. 2.8. Considering that Eq. (2.135) represents an elementary reaction, the rate of production of $M$ by the forward reaction is

$$r_f = k_f C_A^a C_B^b$$

and the rate of consumption of $M$ by the reverse reaction is

$$-r_r = k_r C_M^m C_W^w$$

Since the rates of production and consumption are equal at the equilibrium condition, we have

$$\frac{k_f}{k_r} = \frac{C_M^m C_W^w}{C_A^a C_B^b} \quad \text{at equilibrium}$$

From the definition of the equilibrium constant, we have

$$K = \frac{C_M^m C_W^w}{C_A^a C_B^b} \quad \text{at equilibrium}$$

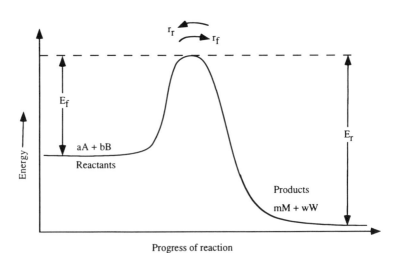

Figure 2.8. Energy diagram for a chemical reaction.

and, therefore, for the elementary reaction (2.135), we can write

$$K = \frac{k_f}{k_r} = \frac{C_M^m C_W^w}{C_A^a C_B^b} \quad \text{at equilibrium}$$

The relationships among the equilibrium constant, reaction rate constants, and concentrations are not so simple for nonelementary reactions, as shown by Denbigh (1955).

### 2.11.5. Reaction Quotient $Q$

The equilibrium constant $K$ relates the concentrations of the reactants with those of the products when the system is at equilibrium. The reaction quotient $Q$ is used to define a relationship between the nonequilibrium concentrations of the reactants and products. If the following reaction

$$a'A + b'B \rightleftharpoons m'M + w'W$$

is at the nonequilibrium condition, $Q$ is given by

$$Q = \frac{C_M^{m'} C_W^{w'}}{C_A^{a'} C_B^{b'}}$$

The value of $Q$ indicates how the equilibrium state will be attained: (a) $Q = K$ indicates that the system is at equilibrium; (b) $Q < K$ indicates that the forward reaction will occur to achieve the equilibrium state, that is, the concentrations of the reactants will decrease and those of the reactants will increase in order to establish equilibrium; (c) $Q > K$ indicates that the reverse reaction will occur to achieve the equilibrium state, that is, the concentrations of the products will decrease and those of the reactants will increase in order to establish equilibrium.

### 2.11.6. Le Chatelier's Principle

Le Chatelier's principle states that if a thermodynamic variable, such as the temperature, pressure, or concentration of an equilibrium system is changed, the chemical reaction or physical transformation of the system will proceed in such a way that the products will facilitate in accommodating the effect of the change in the thermodynamic variable. Essentially, an equilibrium system is transformed into a new equilibrium state due to a change in its temperature, pressure, or concentration in order to minimize the effect of the change on the original system.

Le Chatelier's principle can be used to explain why the melting temperature of ice decreases when the pressure on the ice is increased. The effect of increasing the pressure on ice is to reduce its volume, but the volume of water is smaller than the volume of the same mass of ice. Therefore, ice will melt so that the product, that is, water, can help the system in accommodating the effect of increasing the pressure. In this way, the effect of increasing pressure on ice, that is, the reduction in the volume of the ice is minimized.

We consider the reaction

$$N_2 + O_2 \rightleftharpoons 2NO, \quad \Delta H > 0$$

which is endothermic. Since the reactants and the product have the same number of moles, according to Le Chatelier's principle a change in pressure has no effect on the position of the equilibrium. An increase in temperature increases the enthalpy of the system, but the enthalpy of the product is higher than the total enthalpy of the reactants since $\Delta H > 0$. Therefore, an increase in temperature, which increases both the forward and reverse reaction rates, will favor the formation of the product in endothermic reactions.

We consider the reaction

$$N_2 + 3H_2 \rightleftharpoons 2NH_3, \quad \Delta H < 0$$

which is exothermic. Since the number of moles of the product $NH_3$ is less than the total number of moles of the reactants, an increase in the total pressure of the system, whose effect is to reduce the volume, will favor the formation of $NH_3$. The enthalpy of the product $NH_3$ is less than the total enthalpy of the reactants, $N_2$ and $H_2$ since $\Delta H < 0$. Thus, an increase in temperature, which favors the reaction that absorbs heat, will lead to the decomposition of the product molecules into the reactant molecules in exothermic reactions.

In pyrolytic LCVD, the substrate is heated with a laser beam, and chemical reactions take place at the hot spot. If the temperature of the hot spot increases during the film deposition process, the film deposition rate will increase for endothermic reactions and decrease for exothermic reactions.

## 2.11.7. Effects of Temperature on the Reaction Rate Constant

The rate of many reactions, especially elementary reactions, can be expressed as a product of two terms, one of which depends on the concentrations of the reactants and the other on the temperature. The dependence of the reaction rate on the concentrations is determined from the stoichiometric equation of the chemical reaction. The temperature-dependent term, which is known as the reaction rate constant, is determined from experimental data, or by using a suitable theory of chemical reactions. Various theories, such as collision theory and transition-state theory, have been proposed, and both classical and quantum mechanical calculations have been carried out for the reaction dynamics to obtain the reaction potential surface in order to determine the reaction rate constant [Johnston (1966)]. Some of these theories are briefly reviewed below.

### 2.11.7.1. Arrhenius' Law

Arrhenius pointed out that the effect of temperature on the reaction rate constant can be expressed by the following expression:

$$k = k_0 e^{-E/RT} \tag{2.137}$$

which has been found to be a very good approximation for many reactions, where $k_0$ and $E$ are constants which are determined from experimental data or theoretical calculations as discussed below.

### 2.11.7.2. Thermodynamics

For an elementary reversible reaction, such as Eq. (2.135), the effect of temperature on the equilibrium constant $K$ is given by the following Van't Hoff equation

$$\frac{d(\ln K)}{dT} = \frac{\Delta H}{RT^2}$$

Noting that $K = k_f/k_r$ and $\Delta H = E_f - E_r$, we can write the above equation as

$$\frac{d(\ln k_f)}{dT} - \frac{E_f}{RT^2} = \frac{d(\ln k_r)}{dT} - \frac{E_r}{RT^2} \tag{2.138}$$

From this equation, it is concluded that

$$\frac{d(\ln k_f)}{dT} = \frac{E_f}{RT^2} \quad \text{and} \quad \frac{d(\ln k_r)}{dT} = \frac{E_r}{RT^2}$$

although these results do not follow mathematically from Eq. (2.138). In general, we write

$$\frac{d(\ln K)}{dT} = \frac{E}{RT^2} \tag{2.139}$$

which can be shown to be the same as Eq. (2.137).

### 2.11.7.3. Collision Theory of Gases

Consider the bimolecular collisions of like molecules $A$ in the gas phase. From the kinetic theory of gases, the number of collisions of molecules in a unit volume of gas per unit time is given by

$$Z_{AA} = \sigma_A^2 N^2 \sqrt{\frac{4\pi RT}{M_A}} C_A^2 \tag{2.140}$$

For bimolecular collisions of unlike molecules $A$ and $B$ in a gas phase, the number of collisions per unit volume per unit time is given by kinetic theory as

$$Z_{AB} = \left(\frac{\sigma_A + \sigma_B}{2}\right)^2 N^2 \sqrt{8\pi RT \left(\frac{1}{M_A} + \frac{1}{M_B}\right)} C_A C_B \tag{2.141}$$

Equations (2.140) and (2.141) represent the reaction rates of unimolecular and bimolecular reactions, respectively, if each collision leads to the formation of the products. The experimentally observed reaction rates are much smaller than the values predicted by Eqs. (2.140) and (2.141). We thus conclude that not all collisions induce chemical reactions, and that only those molecules that have energies higher than a certain minimum energy $E$ undergo reactions. From the Maxwell law of distribution of molecular energies, the probability of a molecule having energies higher than $E$ can be approximated by $e^{-E/RT}$ when $E \gg RT$. This approximation holds good for energetic collisions. The number of moles of the molecules $A$ consumed per unit volume per unit time to form the products of the collisions between the molecules $A$ and $B$ is given by

$$-r_A = -\frac{1}{V}\frac{dN_A}{dt} = kC_A C_B \qquad (2.142)$$

Dividing Eq. (2.141) by $N$ to obtain the number of moles of the molecules $A$ colliding with the molecules $B$ per unit volume per unit time, we can write

$$kC_A C_B = \frac{1}{N} Z_{AB} e^{-E/RT}$$

which leads to

$$k = k'_0 \sqrt{T} e^{-E/RT} \qquad (2.143)$$

where

$$k'_0 = \left(\frac{\sigma_A + \sigma_B}{2}\right)^2 N \sqrt{8\pi R \left(\frac{1}{M_A} + \frac{1}{M_B}\right)}$$

### 2.11.7.4. Collision Rate with a Surface

In pyrolytic LCVD, the reactant molecules collide with the substrate surface where the chemical reaction takes place. From the kinetic theory of gases, the number of such collisions per unit area of the substrate surface per unit time is given by [see Johnston (1966), pp. 137, 138]

$$Z_{As} = \frac{N}{4}\left(\frac{8RT}{\pi M_A}\right)^{1/2} C_A \qquad (2.144)$$

It should be noted that Eq. (2.144) is valid for elastic collisions, but chemical reactions involve inelastic collisions.

The rate of consumption of the molecules $A$ in terms of the moles of $A$ per unit area of the substrate surface is given by

$$-r_A = -\frac{1}{S}\frac{dN_A}{dt} = kC_A \qquad (2.145)$$

Dividing Eq. (2.144) by $N$ to obtain the number of moles of the molecules $A$ colliding with the substrate per unit area of the substrate surface per unit time, and by following the argument used for bimolecular collisions, we can write

$$kC_A = \frac{1}{N} Z_{A_s} e^{-E/RT}$$

which leads to

$$k = k'_0 \sqrt{T}\, e^{-E/RT} \qquad (2.146)$$

where $k'_0 = (R/2\pi M_A)^{1/2}$.

### 2.11.7.5. Activation Energy

The collision theory described above allows us to determine the Arrhenius constant $k_0$. The activation energy $E$ can be determined by carrying out molecular dynamics and quantum mechanical calculations [Johnston (1966)]. However, as such calculations provide fairly acurate values of $E$ only for simple reactions, $E$ needs to be determined from experimental data.

For highly endothermic reactions, that is, when $\Delta H$ is 20 kcal/mol or higher, $E \approx \Delta H$. $E$ decreases as $\Delta H$ becomes smaller and negative, and is approximately zero for highly exothermic reactions.

Taking the second derivative of $k$ with respect to $T$ in Eq. (2.137) and setting $d^2k/dT^2 = 0$, we find that the temperature at the inflection point is $T^* = E/2R$, and at that temperature $k = k_0/e^2$. Almost all chemical reactions occur at temperatures below the inflection temperature, (which satisfies the condition $E \gg RT$ used in collision theory), usually, $5\,\text{kcal/mol} \leqslant E \leqslant 50\,\text{kcal/mol}$, which means $1250\,\text{K} \leqslant T^* \leqslant 12{,}500\,\text{K}$.

### 2.11.7.6. Limitations of the Theories for Determining $k_0$ and $E$

Equations (2.143) and (2.146) show that

$$k = k'_0 \sqrt{T}\, e^{-E/RT}$$

as opposed to the Arrhenius expression (2.137). From this equation, we find that $k$ depends on temperature through the exponential term and the $\sqrt{T}$ term. Since $e^{-E/RT}$ varies strongly with $T$, the effect of the $\sqrt{T}$ term on $k$ is essentially insignificant. For this reason, the Arrhenius expression (2.137) is used for almost all calculations.

The experimentally determined reaction rates are usually found to be of the same order of magnitude or lower than those predicted by collision theory. Collision theory allows us to estimate the upper bound of the reaction rate, but for complex reactions, especially for catalytic reactions, the observed reaction rates are sometimes much higher than the predicted values.

For simple elementary reactions, if enough thermodynamic data are available, transition-state theory provides a better estimate for the reaction rates. When sufficient thermodynamic data are available, transition-state theory predicts the reaction rates better than collision theory.

Molecular dynamics and quantum mechanical calculations allow us to determine the activation energy $E$ for simple reactions.

### 2.11.8. Experimental Determination of the Pre-Exponential Factor $k_0$, Activation Energy $E$, and the Order of a Reaction

The values of $k_0$, $E$, and the reaction order are very useful not only for mathematical modeling of LCVD processes, but also for understanding the reaction mechanisms. There are two main techniques, the differential and the integral methods, to evaluate these three constants for a reaction.

#### 2.11.8.1. Differential Method

To explain various steps involved in this method, we consider the following reaction

$$aA \to wW \quad (2.147)$$

STEP 1: Assume a mechanism and write a rate equation for the chemical reaction under consideration. For chemical reaction (2.147), we assume that the order of the reaction is "$a$" with respect to the species $A$, and we write the rate of disappearance of the reactant $A$ as

$$-r_A = -dC_A/dt = kC_A^a \quad (2.148)$$

STEP 2: Obtain the values of the concentration $C_A$ at various times $t$ from experiment as the chemical reaction (2.147) progresses, and plot $C_A$ vs. $t$.

STEP 3: Join these data points with a smooth curve.
STEP 4: Select a few concentration values, such as $C_{A_i}$ for $i = 1, 2, 3, \ldots$, and determine the slopes of the smooth curve at these values. The slope $dC_{A_i}/dt$ represents the rate of reaction at the concentration $C_{A_i}$.
STEP 5: Plot $\log(-dC_{A_i}/dt)$ vs. $\log C_{A_i}$. This plot must yield a straight line. The slope of this straight line represents the value of the reaction order "$a$," and the intercept of this straight line with the log $(-dC_{A_i}/dt)$ axis gives the value of $\log k$. If the plot does not yield a straight line, assume another mechanism for the reaction in Step 1 and repeat the above steps. In this way, we can determine the values of "$a$" and $k$ at the temperature $T$ at which the reaction was carried out in Step 2. It should be noted that the reaction order "$a$" does not depend on temperature, but $k$ varies with temperature.
STEP 6: Carry out the chemical reaction at various temperatures, such as $T_i$ for $i = 1, 2, 3, \ldots$, and determine the rate constant at each of these temperatures. Let $k_i$ be the rate constant at temperature $T_i$.
STEP 7: Assuming that Eq. (2.137) holds good, plot $\ln k_i$ vs. $-1/RT_i$, which must yield a straight line whose slope is $E$ and intercept with the $\ln k_i$-axis is $\ln k_0$. We now have the values of $k_0$, $E$, and the reaction order.

### 2.11.8.2. Integral Method

We will explain this technique for chemical reaction (2.147):

STEP 1: Follow Step 1 of the differential method.
STEP 2: Obtain the values of $C_A$ at various times $t$ from experiment as the chemical reaction (2.147) progresses.
STEP 3: Integrate Eq. (2.148) to obtain

$$I(C_A) = -\int_{C_{A0}}^{C_A} \frac{d\bar{C}_A}{\bar{C}_A^a} = \frac{1}{a-1}\left(\frac{1}{C_A^{a-1}} - \frac{1}{C_{A0}^{a-1}}\right) = \int_0^t k\,dt = kt$$

where $C_{A0}$ is the initial concentration of the reactant $A$, that is, $C_A = C_{A0}$ at time $t = 0$.

STEP 4: Assume a value of the reaction order "$a$", evaluate $I(C_A)$ for the values of $C_A$ obtained in Step 2, and plot $I(C_A)$ vs. $t$, which must yield a straight line passing through the origin. The slope of this straight line is $k$. In this way, we determine the values of $k$ and "$a$". If the plot is not a straight line passing through the origin, repeat Steps 1, 3, and 4.

STEP 5: Follow Steps 6 and 7 of the differential method to determine $k_0$ and $E$.

#### 2.11.8.3. Comparison of the Differential and the Integral Methods

The differential method is very reliable for investigating the mechanism of a reaction when there is not much information available about the reaction. It is useful for complicated reactions, but requires a large amount of data.

The integral method is easy to use and is useful for chemical reactions having simple rate expressions. It is also useful when the experimental data concerning the variations of concentration with time are so irregular that the derivatives needed in the differential method cannot be determined reliably.

### 2.11.9. Example—Estimation of $k_0$ and Equations for Transport Coefficients in a Gaseous Medium

The optical properties, such as the absorptivities of the substrate and the film, and the chemical kinetics data, such as the pre-exponential factor, activation energy, and reaction order, are important to understand LCVD processes. Similarly, the transport properties, such as the diffusion coefficient, thermal conductivity, and viscosity, in the gas phase and at the gas–substrate interface are important to analyze and control the shape, thickness, and homogeneity of the film. The absorptivities of the substrate and the film allow us to determine how much of the laser energy is converted into thermal energy. The chemical kinetics data are used to calculate the rate at which the atoms of the film material are produced. The thermal conductivity, diffusion coefficient, and viscosity allow us to determine the distributions of the thermal energy, the concentrations and velocities of various species, respectively.

#### 2.11.9.1. Estimation of $k_0$

Since the chemical reaction that produces the film material is considered to occur at the substrate surface, we will use Eq. (2.146) to estimate $k_0$. Although $k_0$ is taken to be independent of temperature, Eq. (2.146) shows that the pre-exponential factor depends on temperature and, for this reason,

$k_0$ is estimated from the following expression:

$$k_0 = \left(\frac{R}{2\pi}\right)^{1/2} \sqrt{\frac{T}{M_A}} = 3.64 \times 10^3 \sqrt{\frac{T}{M_A}} \quad [\text{cm/s}] \quad (2.149)$$

where the temperature $T$ is in Kelvin, and the molecular weight $M_A$ of species $A$ is in g/mol.

During the pyrolytic LCVD of Ti from $TiBr_4$, the gas phase constitutes the carrier gas, the reactant $TiBr_4$, and various reactant products such as $TiBr_2$ and TiBr. The thermal decomposition temperature of TiBr is about 1173 K. At this temperature, we will estimate the values of $k_0$ for $TiBr_4$, $TiBr_2$, and TiBr.

For $TiBr_4$

$$k_0 = 3.64 \times 10^3 \sqrt{\frac{1173}{367.564}} = 6.5 \times 10^3 \text{ cm/s}$$

For TiBr

$$k_0 = 3.64 \times 10^3 \sqrt{\frac{1173}{207.732}} = 8.65 \times 10^3 \text{ cm/s}$$

For TiBr

$$k_0 = 3.64 \times 10^3 \sqrt{\frac{1173}{127.816}} = 11.03 \times 10^3 \text{ cm/s}$$

### 2.11.9.2. Formulas for Transport Coefficients in the Gas Phase [Bird et al. (1962)]

*2.11.9.1a. Viscosity*

$$\mu_A = 2.6693 \times 10^{-5} \sqrt{M_A T / \sigma_A^2 \Omega}$$

where $\mu$ is the viscosity of species $A$ (g cm$^{-1}$ s$^{-1}$), $T$ is the temperature of the gaseous medium (K), $M_A$ is the molecular weight of species $A$ (g/mol), $\sigma_A$ is the diameter of species $A$ (Å), and $\sigma_A$ is the interaction parameter. $\Omega = 1$ for rigid sphere model. It should be noted that the above expression for the viscosity $\mu$ was derived [Bird et al. (1962)] for monatomic gases, but it gives very good results for polyatomic gases as well.

When the gas phase is a mixture of several species, the viscosity of the mixture is given by

$$\mu_m = \sum_{i=1}^{n} x_i \mu_i \Big/ \sum_{j=1}^{n} x_j \phi_{ij}$$

where $\mu_m$ is the viscosity of the gaseous mixture, $n$ is the number of chemical species in the mixture, $x_k$ is the mole fraction of species $k$ for $k = i$ or $j$, $\mu_k$ is the viscosity of species $k$ for $k = i$ or $j$, and

$$\phi_{ij} = \frac{1}{\sqrt{8}}\left(1 + \frac{M_i}{M_j}\right)^{-1/2}\left[1 + \left(\frac{\mu_i}{\mu_j}\right)^{1/2}\left(\frac{M_j}{M_i}\right)^{1/4}\right]^2$$

with $M_k$ being the molecular weight of species $k$ for $k = i$ or $j$.

*2.11.9.4b. Thermal Conductivity.* For polyatomic gases, Eucken's formula [Bird et al. (1962)] provides the following expression for thermal conductivity:

$$k_A = \left(C_p + \frac{5}{4}\frac{R}{M_A}\right)\mu_A$$

where $k_A$ is the thermal conductivity of species $A$, $C_p$ is the heat capacity per unit mass per unit temperature change, and $M_A$ and $\mu_A$ are the same as those defined for viscosity. For a mixture, we have

$$k_m = \sum_{i=1}^{n}\left(x_i k_i \Big/ \sum_{j=1}^{n} x_j \phi_{ij}\right)$$

where $k_m$ is the thermal conductivity of the gaseous mixture, $k_i$ is the thermal conductivity of species $i$, and $x_i$, $x_j$, and $\phi_{ij}$ are the same as those defined for viscosity.

*2.11.9.4c. Diffusion Coefficient.* The diffusion coefficient [Bird et al. (1962)] of the species $i$ with respect to the species $j$ is given by

$$D_{ij} = 2.2646 \times 10^{-5} \frac{[T(1/M_i + 1/M_j)]^{1/2}}{C_i \sigma_{ij}^2 \Omega_{ij}}$$

where $D_{ij}$ is the diffusion coefficient of the species $i$ with respect to the species $j$ (cm$^2$/s) and $\sigma_{ij} = \frac{1}{2}(\sigma_i + \sigma_j)$, with $\sigma_i$ being the diameter of the species $i$ (Å),

$\sigma_j$ the diameter of the species $j$(Å), $C_i$ the concentration of the $i$th species (g-mol cm³), and $\Omega_{ij}$ the potential field parameter. $\Omega_{ij} = 1$ for rigid sphere model and $T$, $M_i$, and $M_j$ are the same as those defined for viscosity.

In a mixture of gases, the diffusion coefficient of the $i$th species with respect to the mixture is given by

$$D_{im} = \left[ (1 - x_i) \bigg/ \sum_{\substack{j=1 \\ i \neq j}}^{n} x_j/D_{ij} \right]$$

where $D_{im}$ is the diffusion coefficient of the $i$th species with respect to the mixture, $D_{ij}$ is the diffusion coefficient of the $i$th species with respect to the $j$th species, and $x_i$ and $x_j$ are the same as those defined for viscosity.

## NOMENCLATURE

| | |
|---|---|
| $a, b, m, w$ | Number of moles of the $i$th species involved in the chemical reaction, where $i = A$, $B$, $M$, and $W$, respectively. In the present discussion, "$a$" is assumed to be the reaction order with respect to the species $A$ |
| $A$ | Absorptivity (in Sections 2.5–2.11) |
| $A, B$ | Chemical species (reactants) |
| $A_1$ | Amplitude for the field $E_y$ |
| $A_e$ | Absorptivity due to free electron |
| $A_i$ | Amplitude of $E_{yi}$ |
| $A_r$ | Amplitude of $E_{yr}$ |
| $A_t$ | Amplitude of $E_{yt}$ |
| $A_i^*$ | Complex conjugate of $A_i$ |
| $A_r^*$ | Complex conjugate of $A_r$ |
| **B** | Magnetic induction |
| $B_1$ | Amplitude for the field $H_z$ |
| $c$ | Speed of light in vacuum $= 1/\sqrt{\mu_0 \varepsilon_0}$ |
| $C_i$ | Concentration of the $i$th species in moles per unit volume for $i = A$, $B$, $M$, or $W$ |
| $d$ | Penetration depth, where $I = I_0/e$ |
| $\hat{D}$ | Complex electric displacement |
| **D** | Electric displacement |
| $e$ | Charge of an electron (note that $e$ is negative, that is, $e = -\|e\|$, for an electron) |
| $\|e\|$ | Magnitude of an electronic charge |
| $\mathbf{e}_0$ | Electric field when $t = 0$ and $\mathbf{r} = 0$ |
| $E$ | Activation energy (in Section 2.11) |

| | |
|---|---|
| **E** | Electric field |
| $\mathbf{E}_0$ | Electric field at time $t = 0$ |
| $E_f$ | Activation energy for the forward reaction |
| $E_r$ | Activation energy for the reverse reaction |
| $E_y$ | $y$-component of **E** |
| $E_{yi}$ | $y$-component of **E** for the incident wave |
| $E_{yr}$ | $y$-component of **E** for the reflected wave |
| $E_{yt}$ | $y$-component of **E** for the transmitted wave |
| $G_i$ | Standard free energy (standard Gibbs function) of the $i$th species for $i = A, B, M,$ or $W$ |
| $\Delta G^0$ | Change in the standard free energy (standard Gibbs function) |
| **H** | Magnetic field |
| $H_i$ | Enthalpy of the $i$th species for $i = A, B, M,$ and $W$ |
| $H_z$ | $z$-component of **H** |
| $H_{zi}$ | $z$-component of **H** for the incident wave |
| $H_{zr}$ | $z$-component of **H** for the reflected wave |
| $H_{zt}$ | $z$-component of **H** for the transmitted wave |
| $\Delta H$ | Change in enthalpy, heat of reaction |
| $\Delta H$ | Heat of reaction [see Eq. (2.135)] |
| $I$ | Energy density at any point whose position vector is **r** |
| $I_0$ | Energy density at $\mathbf{r} = 0$ |
| $J$ | Current density |
| $k$ | Absorption index |
| $k$ | Reaction rate constant (in Section 2.11) |
| $\hat{k}$ | Wave number $= 2\pi/\lambda$ |
| $k_0$ | Frequency factor, or pre-exponential factor, or Arrhenius constant (in Section 2.11) |
| $k_B$ | Boltzmann constant |
| $k_f$ | Reaction rate constant for reverse chemical reaction |
| $k_r$ | Reaction rate constant for reverse chemical reaction |
| $K$ | Equilibrium constant (in Section 2.11) |
| $K$ | Restoring force per unit distance (in Section 2.6) |
| $m$ | Mass of an electron |
| $M_i$ | Molecular weight of the $i$th molecule for $i = A$ or $B$ |
| $M_s$ | Mass of solid |
| $M, W$ | Chemical species (products of a chemical reaction) |
| $\hat{n}$ | Complex refractive index |
| $n$ | Overall order of a reaction (in Section 2.11). |
| $n$ | Refractive index |
| $\hat{n}^*$ | Complex conjugate of $\hat{n}$ |
| $N$ | Avogadro's number ($6.023 \times 10^{23}$ molecules/mol) (in Section 2.11) |

| | |
|---|---|
| $N$ | Number of electrons per unit volume (in Sections 2.1–2.10) |
| $N_i$ | Number of moles of the $i$th species for $i = A, B, M$, or $W$ |
| $\hat{p}$ | Microscopic polarization, or dipole moment of an electron, with real and imaginary components |
| $\hat{P}$ | Macroscopic polarization, or dipole moment of electrons per unit volume, with real and imaginary components |
| $\mathbf{P}$ | Macroscopic polarization, or dipole moment of electrons per unit volume |
| $\mathbf{P}_{III}(\omega)$ | Third-order polarization |
| $\mathbf{P}_l(\omega_3)$ | Nonlinear polarization |
| $\mathbf{r}$ | Position vector |
| $r_i$ | Rate of reaction of the $i$th species |
| $R$ | Reflectivity (in Sections 2.5–2.11) |
| $R$ | Universal gas constant |
| $\mathbf{s}$ | Unit vector in the direaction of propagation of the beam |
| $S$ | Surface area of the interface |
| $t$ | Time variable |
| $T$ | Temperature (K) (in Sections 2.8, 2.10.5, 2.11.3). |
| $T$ | Transmittivity (in Sections 2.5, 2.6, 2.10.2) |
| $v$ | Speed of light in a medium |
| $v_e$ | Speed of the conduction electrons at the Fermi surface |
| $V$ | Volume of the space where reaction takes place |
| $V_r$ | Volume of reactor |
| $V_s$ | Volume of solid |
| $z$ | Impedance of the material (in Section 2.8) |
| $Z_{AA}$ | Number of collisions of $A$ with $A$ in the gas phase per unit volume per unit time |
| $Z_{AB}$ | Number of collisions of $A$ with $B$ in the gas phase per unit volume per unit time. |
| $Z_{As}$ | Number of collisions of $A$ with the substrate per unit area of the substrate surface per unit time |

GREEK SYMBOLS

| | |
|---|---|
| $\hat{\alpha}(\omega)$ | Frequency-dependent microscopic or atomic polarizability |
| $\Gamma$ | Damping constant for nonmetals |
| $\Gamma_m$ | Damping constant for metals |
| $\varepsilon$ | Capacitance or permittivity of the dielectric |
| $\varepsilon_0$ | Capacitance or permittivity of free space $= (36\pi \times 10^9)^{-1}$ F/m |
| $\hat{\varepsilon}_r$ | Complex dielectric constant of a dielectric medium |
| $\varepsilon_r$ | Relative capacitance or relative permittivity or dielectric constant $= \varepsilon/\varepsilon_0$ (obtained from Rojansky (1971) |
| $\eta$ | Absorption coefficient |

| | |
|---|---|
| $\eta_2$ | Two-photon absorption coefficient |
| $\kappa$ | Thermal conductivity |
| $\lambda$ | Wavelength of light in a medium |
| $\lambda_0$ | Wavelength of light in vacuum |
| $\mu$ | Inductivity or magnetic permeability of the medium |
| $\mu_0$ | Inductivity or magnetic permeability of free space $= (4\pi \times 10^{-7})\,\text{H/m}$ |
| $\mu_r$ | Relative permeability $= \mu/\mu_0$ |
| $\nu_i$ | Frequency of a photon ($i = 1, 2$) |
| $\rho$ | Charge density (in Section 2.4) |
| $\rho$ | Direct current (DC) resistivity in $\mu\Omega$-cm (in Section 2.10.5) |
| $\rho_0$ | Charge density at time $t = 0$ |
| $\sigma$ | Electrical conductivity |
| $\sigma_0$ | Direct current (DC) conductivity |
| $\sigma_i$ | Diameter of the $i$th molecule for $i = A$ or $B$ (in Sections 2.11.1–2.11.3). |
| $\tau$ | Mean free time between collisions (relaxation time) |
| $\chi_i$ | Imaginary part of $\chi_{\text{III}}(-\omega, \omega, -\omega, \omega)$ |
| $\chi_{\text{III}}(-\omega, \omega, -\omega, \omega)$ | Nonlinear susceptibility for third-order polarization |
| $\chi(\omega)$ | Linear susceptibility |
| $\chi_{lm}$ | Linear susceptibility tensor |
| $\chi_{lmn}$ | Nonlinear susceptibility tensor |
| $\omega$ | Angular frequency of the monochromatic radiation ($\omega = 2\pi c/\lambda_0 = 2\pi v/\lambda$) |
| $\omega_0$ | Natural frequency of the electrons of dielectrics |
| $\omega_i$ | Angular frequency of the incident radiation ($i = 1, 2, 3$) |
| $\omega_p$ | Plasma frequency |

## REFERENCES

Allmen, M. V. (1987), *Laser-Beam Interactions with Materials*, Springer-Verlag, New York, pp. 18–19.

Arata, Y., and Miyamoto, I. (1978), *Technocrat* **11**, 33.

Bass, J. (1982), in: *Landolt-Börnstein*, New Ser., Vol. 15a, Hellwege, K., and Olsen, J. L., eds., Springer-Verlag, Berlin, pp. 5–137.

Bass, M. (1991), Laser-Materials Interactions, *Encyclopedia of Lasers and Optical Technology*, Meyers, R. A., ed., Academic, New York, pp. 181–197.

Bäuerle, D. (1986), *Chemical Processing with Lasers*, Springer-Verlag, Berlin.

Bird, R. B., Stewart, W. E., and Lightfoot, E. N. (1962), *Transport Phenomena*, Wiley, New York, 2nd Printing, pp. 23, 24, 257, 258, 511, and 571.

Born, M. and Wolf, E. (1980), *Principles of Optics*, Pergamon Press, New York, 6th Ed., Reprinted 1987, pp. 611–718.
Boyd, I. W. (1987), *Laser Processing of Thin Films and Microstructures*, Springer-Verlag, Berlin.
Denbigh, K. G. (1955), *The Principles of Chemical Equilibrium*, Cambridge University Press, Cambridge, 442 pp.
Drude, P. K. L. (1959), *Theory of Optics*, Dover, New York.
Duley, W. W., Simple, D. J., Morency, J. P., and Gravel, M., (1979), *Optics and Laser Technology*, **Dec.**, 313.
Funaki, K., Uchimura, K., and Kuniya, Y. (1961), *Kogyo Kagaku Zasshi* **64**, 1914.
Goldsmith, A., Waterman, T. E., and Hirschhorn, H. J. (1961), *Handbook of Thermophysical Properties of Solid Materials*, Revised Edition, Vol. II: Alloys, MacMillan, New York, pp. 671, 171.
Grosse, A. V. (1966), *Rev. Hautes Temp. Refract.* **3**, 115–146.
Heavens, O. S. (1955), *Optical Properties of Thin Solid Films*, Academic, New York, pp. 46–95.
Johnson, P. B., and Christy, R. W. (1972), *Appl. Opt.* **11**, 643.
Johnson, P. B., and Christy, R. W. (1974), *Phys. Rev.* **B9**, 5056.
Johnston, H. S. (1966), *Gas Phase Reaction Rate Theory*, Ronald, New York.
Kittel, C. (1984), *Introduction to Solid State Physics*, Wiley Eastern Ltd., Calcutta, 5th Ed., Calcutta, 4th Wiley Eastern Reprint, p. 178.
Laidler, K. J. (1987), *Chemical Kinetics*, Harper and Row, New York, 3rd Ed.
Lenham, A. P., and Treherne, D. M. (1966), *J. Opt. Soc. Am.* **56**, 1137.
Levenspiel, O. (1975), *Chemical Reaction Engineering*, Wiley Eastern Limited, New Delhi, 2nd Ed., 2nd Wiley Eastern Reprint.
Lorentz, H. A. (1909), *Theory of Electrons and Its Applications to the Phenomena of Light and Radiant Heat*, G. E. Stechert, New York.
Lorentz, H. A. (1952), *The Theory of Electrons*, Dover, New York.
Milonni, P. W., and Eberly, J. H. (1988), *Lasers*, Wiley, New York, pp. 661–682.
Newell, A. C., and Maloney, J. V. (1992), *Nonlinear Optics*, Addison-Wesley, New York.
Pippard, A. B. (1947), *Proc. Roy. Soc. London* **A191**, 385.
Rojansky, V. (1971), Electromagnetic Fields and Waves, Prentice Hall, Englewood Cliffs, pp. 232, 233, 309.
Schulz, L. G. (1957), *Advan. Phys.* **6**, 102.
Sokolov, A. V. (1967), *Optical Properties of Metals*, Translated by Chomet, S., Elsevier, New York, English edition.
Wolfe, W. L., and Zissis, G. J., eds. (1978), *The Infrared Handbook*, Infrared Information and Analysis Center, Environmental Research Institute of Michigan, pp. 7.1–7.76.
Wooten, F. (1972), *Optical Properties of Solids*, Academic, New York, pp. 42–67.
Zener, C. (1953), *Nature* **132**, 968.

# THREE

# PHOTOLYTIC LCVD

## 3.1. INTRODUCTION

Laser is an electromagnetic radiation, which is a form of energy and does not require any medium to travel through space. The properties of electromagnetic radiation can be characterized by its wave-like or particle-like (corpuscular) behavior. The waves of electromagnetic radiation are characterized by wavelength or frequency. The wave theory of electromagnetic radiation explains various observed phenomena, such as the diffraction, interference, and polarization of light [Calvert and Pitts (1966)]. If an opaque object is placed between a point source and a screen, alternate bands of light are formed within the geometrical shadow of the object, which is known as diffraction. This phenomenon can be explained by using Huygen's principle, which states that each point on a wave front of light acts as a point source and emits a secondary wavelet in all directions. According to the wave theory, unpolarized light, which we encounter most commonly, consists of a large number of waves with the electric and magnetic vectors of each wave oscillating in fixed planes perpendicular to the direction of the radiation propagation, but the orientations of the planes of the various waves are random with respect to one another. In plane polarized light, which can also consist of many waves, the electric vectors of all the waves lie in a single plane and the magnetic vectors of all the waves lie in another plane perpendicular to the electric field. From the point of view of the wave theory, laser is an electromagnetic wave of a single frequency, that is, a monochromatic beam of radiation.

Although the concept of oscillating electric and magnetic fields explains many of the optical phenomena, we cannot consider that light travels as an infinitely long wave train and that the light which extends from an emitter to a receiver is a fraction of such an infinitely long wave train. It should be noted that a wave train refers to a collection of waves over an appreciable distance along the direction of the wave propagation, where the profile of a wave is given by a simple sine curve with constant or slowly varying amplitude, and the displacement elsewhere is zero [Ditchburn (1957)]. The radiation is considered to travel in discrete units called quanta or photons. The quanta were named photons by G. N. Lewis in 1926. The photon or quantum of radiation is a quantized form of electromagnetic wave. The relationship between the energy of a photon and the wavelength of an electromagnetic wave is given by the following Planck equation:

$$E = h\nu = hc/\lambda$$

The wave theory of light does not explain several phenomena, such as the absorption and emission of light and the effects (photoelectric effect and Compton effect) induced by the absorbed light. Hertz (1887) provided experimental support for Maxwell's electromagnetic equations by demonstrating the existence of electromagnetic waves. While carrying out these experiments, Hertz found that when a metal surface is irradiated with light of very short wavelength, it emits electrons, and this is known as the photoelectric effect. The effects of photons on the emitted electrons are found to be the following:

1. The number of electrons that are emitted due to the photoelectric effect is proportional to the intensity of the incident radiation.
2. The kinetic energy of the emitted electrons is proportional to the frequency and independent of the intensity of the incident radiation.

These observed phenomena cannot be explained by the classical electromagnetic wave theory. Einstein (1905a) used the quantum theory of Planck (1901) to explain photoelectric phenomena by suggesting that:

1. The energy $h\nu$ of a quantum of light which is absorbed by the metal is used to eject an electron.
2. The kinetic energy of the ejected electron is related to the absorbed energy by the expression

$$\tfrac{1}{2}m_e v^2 = h\nu - w_0 \tag{3.1}$$

where $w_0$ is the work function of the metal, that is, $w_0$ refers to the minimum energy required to remove an electron from the metal.

Millikan (1916) provided experimental support for Eq. (3.1) by obtaining the following results from his experimental data:

1. The plot of the kinetic energy of the photoelectrons vs. the frequency of the incident radiation was found to be a straight line for each metal.
2. The value of the slopes of the straight lines was found to be equal to the Planck constant for blackbody radiation, $h$.
3. The absolute values of the intercepts $w_0$ were found to be different for each metal, but the value of $w_0$ for a given metal was shown to be equal to its work function obtained independently by other investigators from thermionic emission data.

By studying the scattering of monochromatic X-rays, Compton (1923a,b) found that the scattered X-rays are made up of radiation of a wavelength that is same as that of the incident radiation plus new radiation of longer wavelength, which is known as the Compton effect, and is explained by assuming that:

1. The radiation behaves as a particle or quantum with energy $h\nu$ and momentum $h\nu/c$.
2. The quantum loses a portion of its energy during its collision with an electron of the scattering material, which leads to a decrease in the frequency and an increase in the wavelength.

As discussed above, wave theory has failed to explain the photoelectric and Compton effects, which have been explained successfully by the quantum or particle theory of light. In this way, it has been established that radiation exhibits particle character which interacts with materials obeying the laws of collisions between particles of matter. In laser-induced chemical reactions, sometimes wave theory is useful and sometimes the particle theory of radiation is convenient.

## 3.1.1. Electromagnetic Radiation Spectrum

The wavelength of laser plays an important role in affecting the dissociation of chemical species because the absorption of radiation by the chemical species depends on the wavelength of the incident radiation. On the basis of wavelength, the electromagnetic radiation spectrum is divided into several regions, such as optical radiation, X-rays, and radio waves. The word light refers to the optical radiation, which is again divided into three regions:

visible light, infrared, and ultraviolet ranges. Table 3.1 shows a portion of the electromagnetic radiation spectrum and the energies of the associated photons [Calvert and Pitts (1966), Rabek (1987), Lide (1992)]. The wavelength of thermal radiation ranges from approximately 0.1 to 100 μm [Incropera and DeWitt (1985)]. The energies, such as the enthalpy and free energy, of chemical species are usually expressed in terms of kcal/mol to study chemical reactions. For this reason, the energies of photons are given in terms of kcal/einstein in Table 3.1, where 1 einstein is equal to 1 mol of quanta or photons. The energy of one quantum or photon is given by the Planck equation $E = hc/\lambda$, and the energy of 1 mol of quanta or photons is given by $E_m = hAc/\lambda$. In terms of kcal/einstein, $E_m$ can be expressed as

$$E_m = \frac{(6.6256 \times 10^{-27} \text{ erg s/quantum})(6.03 \times 10^{23} \text{ quantum/mol of quanta})}{(\lambda \text{Å})(10^{-8} \text{ cm/Å})}$$
$$\times (2.9979 \times 10^{10} \text{ cm/s})(2.3871 \times 10^{-11} \text{ kcal/erg})$$
$$= \frac{2.8591 \times 10^5}{\lambda} \text{ [kcal/einstein]}$$

Table 3.1. Wavelength and Energy Values for Typical Electromagnetic Radiations

| Typical classification of radiation | Approximate wavelength range | Energy (kcal/einstein) | Types of chemical reaction |
|---|---|---|---|
| Radiowave | $10^3$–$10^5$ m | $2.86 \times 10^{-8}$–$2.86 \times 10^{-10}$ | |
| Shortwave radio | 1–$10^3$ m | $2.86 \times 10^{-5}$–$2.86 \times 10^{-8}$ | |
| Microwave | 1–100 cm | $2.86 \times 10^{-3}$–$2.86 \times 10^{-5}$ | |
| Extreme far-infrared | 0.3–10 mm | 0.095–$2.86 \times 10^{-3}$ | |
| Far-infrared | 1.4–300 μm | 20.42–0.095 | |
| Near-infrared | 750–1400 nm | 38.12–20.42 | |
| Red | 640–750 nm | 44.67–38.12 | |
| Yellow | 580–640 nm | 49.29–44.67 | |
| Green | 495–580 nm | 57.76–49.29 | |
| Blue | 440–495 nm | 64.98–57.76 | Photochemical |
| Violet | 380–440 nm | 75.24–64.98 | |
| Near-ultraviolet | 300–380 nm | 95.30–75.24 | |
| Far-ultraviolet | 200–300 nm | 142.96–95.30 | |
| Vacuum-ultraviolet | 300–2000 Å | 953.03–142.96 | |
| Long X-ray | 0.3–300 Å | $9.53 \times 10^5$–953.03 | |
| Short X-ray | 0.03–0.3 Å | $9.53 \times 10^6$–$9.53 \times 10^5$ | Radiation-chemical |
| Gamma ray | 0.003–0.03 Å | $9.53 \times 10^7$–$9.53 \times 10^6$ | |
| Cosmic ray | 0.0001–0.003 Å | $2.86 \times 10^9$–$9.53 \times 10^7$ | |

where $\lambda$ is in Å. It should be noted that the bond energies of most chemical species are about 40 kcal/mol or higher, which must be supplied by the electromagnetic radiation to cause dissociation of the reactant molecules through the single-photon absorption mechanism. The data of Table 3.1 suggest that such chemical changes can be produced by using light of wavelengths less than 700 nm, which corresponds to an energy of 40.8 kcal/einstein. However, light of wavelengths higher than 700 nm, such as near-infrared radiation, can produce chemical changes through the two-photon absorption mechanism, if: (a) the intensity of the radiation is very high, which can be achieved with a laser beam, and (b) the lifetime if the single-photon excited state of the reactant species is long enough for the two-photon absorption to occur.

### 3.1.2. Interactions of Radiations with Chemical Reactants

Electromagnetic radiation can be classified into several categories depending on the wavelength of the radiation, as shown in Table 3.1. Besides electromagnetic radiation, there is another type of radiation called particle radiation, such as $\alpha$-particles (He nuclei), $\beta$-particles (electrons), cathode rays (electrons), proton and deuteron beams, etc. Particle radiations usually lose energy continuously by undergoing a large number of collisions with the medium through which they travel and transferring a small amount of energy in each collision. On the other hand, the photons of electromagnetic radiation lose a lot of energy when they interact with materials [Spinks and Woods (1990)]. However, only some of the incident photons interact with the materials of the medium through which they travel, and the rest of them pass through the medium without any change of energy or direction.

When visible light or low-frequency ultraviolet radiation is absorbed by a molecule, the molecule either dissociates or attains an electronically excited state which is usually not in thermal equilibrium with its surroundings [Wilkinson (1980)]. These electronically excited molecules lose energy through the following mechanisms:

PHOTOPHYSICAL PROCESSES

1. Radiative processes: luminescence (fluorescence and phosphorescence), that is, emission of radiation.
2. Radiationless process: nonradiative intramolecular energy transfer.
3. Collisional quenching: intermolecular energy transfer, that is, transfer of energy from the electronically excited molecules to other molecules.

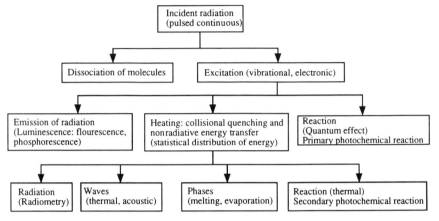

Figure 3.1. Various phenomena associated with the interactions of radiations with materials.

A PHOTOCHEMICAL PROCESS: These are chemical reactions, for example, photooxidation, photoreduction, photoassociation, photodissociation, and photoisomerization, to form new products. Photochemical phenomena can be divided into two types as follows:

1. Primary photochemical process: the transformation, that is, the production, the decay, or the chemical reaction of species which are not in thermal equilibrium with their surroundings, photolytic LCVD involves the primary photochemical processes.
2. Secondary photochemical process: reactions where the elementary reactions are not different from normal thermal elementary reactions; it should be noted that pyrolytic LVCD involves thermal reactions.

The mechanisms of the interactions of radiations with chemical reactants are shown in Fig. 3.1 [Hess (1989)].

## 3.2. DEFINITIONS OF BASIC PHOTOCHEMICAL PROCESSES

In this section, we will define the terminology and concepts that are used in studies concerning the interactions of radiation with chemical reactants.

PHOTOLYSIS: The word photolysis comes from the Greek words "photo," meaning light, and "lysis," meaning splitting. The absorption of radiation leads to the formation of excited species which subsequently produce atoms and radicals in the primary photochemical process as discussed above, and such a process is referred to as photolysis. Photochemistry deals with such phenomena. In all photochemical reactions, we encounter electronically excited singlet and/or triplet states [Rabek (1987)]. A singlet state is one which has electrons with paired spins. A triplet state has electrons with parallel spins, that is, with $S = 1$ and multiplicity $M = 2S + 1 = 3$, which corresponds to three closely spaced levels. Each excited state has a definite electronic structure, lifetime, and energy. The excited species are chemically different from the ground state species, and can behave differently.

RADIATION-CHEMICAL PROCESS: Production of ions by radiations, usually particle radiation, high-frequency ultraviolet radiation, X-ray, and $\gamma$-rays, is called a radiation-chemical process. Radiation chemistry deals with such processes.

LUMINESCENCE: When a molecule is electronically excited as a result of its interaction with radiation, it loses its excitation energy by emitting radiation, which is called luminescence.

FLUORESCENCE: Refers to the luminescence due to the transition of electrons between two states of the same multiplicity. Fluorescence is a spin-allowed radiative emission between states with $\Delta S = 0$. The wavelength of fluorescence radiation is larger than the wavelength of absorbed radiation, as illustrated in Fig. 3.2. However, as can be seen in the figure in many compounds, the shorter-wavelength region of the fluorescence spectrum overlaps the longer-wavelength region of the absorption spectrum. The absorption and fluorescence spectra are mirror images of each other if they are plotted in a frequency scale rather than a wavelength scale. In Fig. 3.2, the absorption spectrum has three peaks, at wavelengths $\lambda_1$, $\lambda_2$, and $\lambda_3$, which correspond to the excited singlet states $S_3$, $S_2$, and $S_1$, respectively. The lifetime of an excited species is about $10^{-8}$ s after which there is fluorescent decay. For polyatomic molecules in solution or in the gas phase at pressures $\geqslant 1000\,\text{Nm}^{-2}$, fluorescence usually occurs [Wilkinson (1980)] as a result of the transition of electrons only from the vibrationally relaxed lowest excited singlet state $S_1$ to the ground singlet state $S_0$.

PHOSPHORESCENCE: Refers to the luminescence due to the transition of electrons between two states of different multiplicity. Phosphorescence, which is a spin-forbidden radiative emission, is usually found to occur due to the transition of electrons only from the lowest excited triplet state to the

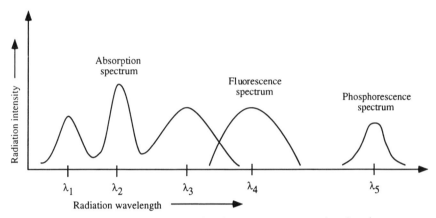

Figure 3.2. Relative positions of various spectra on a wavelength scale.

ground singlet state [Wilkinson (1980)], that is, $T_1 \to S_0$. The wavelength of the phosphorescence radiation is larger than the wavelength of the absorbed and fluorescence radiations, as illustrated in Fig. 3.2, because $T_1$ has an energy lower than $S_1$. In phosphorescent decay, the lifetime of an excited ranges from about $10^{-3}$ s to many seconds and even minutes. Due to the long lifetime, the $T_1$ state is susceptible to collisional quenching reactions which reduce the phosphorescence in fluid mediums, where the excited species collide with one another. However, phosphorescence occurs with high efficiency in rigid mediums due to the lack of collisional quenching. It can be observed in fluid mediums at low temperatures when the intermolecular collisions are less frequent.

INTERNAL CONVERSION (IC): A nonradiative transition between two states of the same multiplicity, e.g., $S_i \to S_1$ and $T_i \to T_1$. IC involves the spin-allowed changes in the electronic states by keeping the internuclear separation constant. The lifetime of the IC process is about $10^{-13}$ s, which is shorter than the period of molecular vibration.

INTERSYSTEM CROSSING (ISC): A nonradiative transition between two states of different multiplicity, e.g., $S_1 \to T_1$. ISC is a spin-forbidden process, and it occurs at $10^{-7}$ to $10^{-8}$ s after the chemical reactant is excited. Since fluorescence occurs after about $10^{-8}$ s, the frequency of ISC is of the same order of magnitude as that of fluorescent decay.

Both IC and ISC involve radiationless transitions, that is, no radiation is emitted during these processes. When an electron jumps from an upper energy state to a lower energy state in these two processes, the excess energy, which is equal to the difference in energy between these two states, appears as the vibrational energy of the species in the lower state. This vibrational energy is usually lost in solution or in the gas phase at pressures $\geqslant 1000 \, \text{Nm}^{-2}$ as a result of collisions [Wilkinson (1980)], and consequently, the temperature of the surrounding medium increases very slightly. Usually, an electronically excited state with an equilibrium distribution of vibrational levels is produced within about $10^{-12}$ s after the occurrence of a non-radiative transition.

CHEMILUMINESCENCE: Refers to the emission of radiation as the result of a chemical reaction. Chemiluminescence is usually observed in gas phase reactions. Due to the high probability of the deactivation of excited species by collisions in dense phases, it is rarely observed in liquid phase reactions. The gas phase reaction between Na vapor and Cl exhibits chemiluminescence through the following mechanism [Laidler (1987)]:

$$Na + Cl_2 \rightarrow NaCl + Cl$$

The chlorine atoms react with $Na_2$ to produce vibrationally excited NaCl', according to the following reaction:

$$Na_2 + Cl \rightarrow Na + NaCl'$$

The NaCl' molecules collide with Na atoms to generate electronically excited Na* atoms, which emit the yellow Na D radiation:

$$NaCl' + Na \rightarrow NaCl + Na^*$$

QUANTUM YIELD [Calvert and Pitts (1966), Rabek (1987), Laidler (1987)]: The quantum yield is a very important quantity for analyzing photochemical reactions because its magnitude and variation with respect to the experimental parameters provide a lot of information about the nature of the reaction. If a reactant $A$ undergoes photochemical reaction, $A + hv \rightarrow B$, to produce $B$, the quantum yield of the product $B$ is defined as

$$\phi_B = \frac{\text{Molecules of } B \text{ produced per unit volume per unit time}}{\text{Number of quanta or photons absorbed by } A \text{ per unit volume per unit time}}$$

The quantum yield can also be defined [Laidler (1987)] in terms of the reactant molecules:

$$\phi_A = \frac{\text{Number of reactant molecules transformed per unit volume per unit time}}{\text{Number of quanta or photons absorbed by the reactant molecules per unit volume per unit time}}$$

In the above expressions, the numerators are determined by conventional chemical analysis, and the denominator is obtained by dividing the total energy of the absorbed radiation of the wavelength $\lambda$ by the energy of one quantum or photon, $h\nu$. The value of the quantum yield varies over a wide range (0 to $10^6$) for different reaction systems. There are several types of quantum yields, such as the product quantum yields, quantum yields of fluorescence $\phi_F$, phosphorescence $\phi_P$, decomposition, etc. $\phi_F$ and $\phi_P$ are defined as

$$\phi_F = \frac{\text{Rate of fluorescent emission}}{\text{Rate of excitation}}$$

and

$$\phi_P = \frac{\text{Rate of phosphorescent emission}}{\text{Rate of excitation}}$$

### 3.2.1. Energy Level Diagrams

The absorption of radiation by any material and the resulting emission of radiation and radiationless transitions are described by Jablonsky or Franck–Condon energy level diagrams [Rabek (1987), Wilkinson (1980)].

#### 3.2.1.1. Jablonsky Energy Level Diagrams

Figure 3.3 shows a typical Jablonsky energy level diagram indicating various radiative and nonradiative processes for a single electron in an atom. In this figure, $S_0$ represents the ground state of the atom, that is, the ground singlet electronic state, and $S_i$ and $T_i$ represent the excited singlet and triplet electronic states, respectively, for $i = 1, 2, 3, \ldots$ As the energy of the atom increases, the electron jumps to higher and higher excited states which are progressively closer to one another, and eventually the discrete excited states merge with a continuum of energy levels as shown in Fig. 3.3. The upper limit of the discrete excited states depends on the ionization energy at which the electron leaves the atom and enters into a continuum of energy levels or on the dissociation energy at which the chemical bond between nuclei breaks.

In Fig. 3.3, the excited singlet states $S_1, S_2, S_3, \ldots$ are generated by the absorbed photon. During this process, one of the electrons jumps into a higher orbital, but the electrons remain paired as shown in Fig. 3.4. The

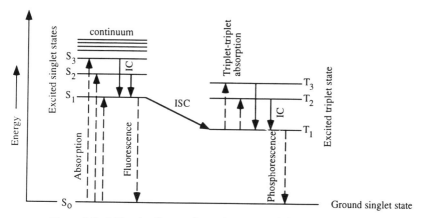

Figure 3.3. Jablonsky diagram for various types of electronic transitions.

electrons jump from the upper excited singlet states, $S_2, S_3, \ldots$, to the lowest singlet state $S_1$ at a rapid rate due to the radiationless IC process, which makes the photochemical reactions less likely to occur from the upper excited states. The photochemical reactions are found to occur from the lowest excited singlet state $S_1$, When the electron jumps from $S_1$ to $S_0$, we observe fluorescence.

The lowest excited triplet state $T_1$ is formed mainly due to the nonradiative ISC of an electron from the lowest excited singlet state $S_1$, and for this reason the electronic energy of the $T_1$ state is lower than the electronic energy of the $S_1$ state. In the triplet states, the spins of the electrons are unpaired, as shown in Fig. 3.4. It should be noted that the formation of a triplet state from the ground singlet state $S_0$ by direct absorption of a photon is a spin-forbidden transition. The higher triplet states $T_2, T_3, \ldots$ are formed from the lowest triplet state $T_1$ due to the absorption of a photon, which is called triplet–triplet absorption. When the electron jumps from $T_1$ to $S_0$, we observe phosphorescence.

Besides the energy levels shown in Fig. 3.3, there are other types of energy levels for crystalline or paracrystalline structures where the molecules are arranged in a regular array. In such materials, the electrons remain attached to the array as a whole even if they have enough energy to dissociate from specific molecules. The continuum energy level is divided into two parts: a narrow continuum of energies, called the exciton band, and a broad continuum of energies, called the conduction band, which are shown in Fig. 3.5. Electrons that have energies in the conduction band conduct electric current in the material.

It should be noted that each electronic state has many vibrational levels, and each vibrational state has many rotational levels, as shown in

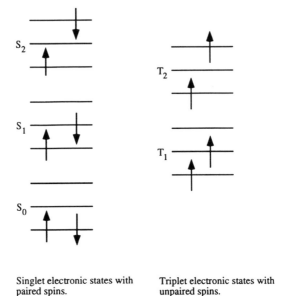

Singlet electronic states with paired spins.

Triplet electronic states with unpaired spins.

Figure 3.4. Spins of electrons in the singlet and triplet states.

Fig. 3.6. The vibrational energy levels of different electronic states can overlap one another, which usually happens when the ground vibrational levels of the respective electronic states are close to each other. As indicated in Fig. 3.6, the energy difference between the ground and excited electronic states is very high. Also, for each electronic state, the energy difference

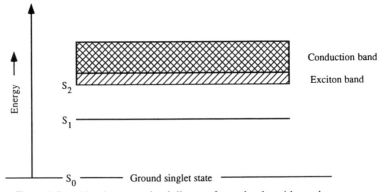

Figure 3.5. Jablonsky energy level diagram for molecules with regular array.

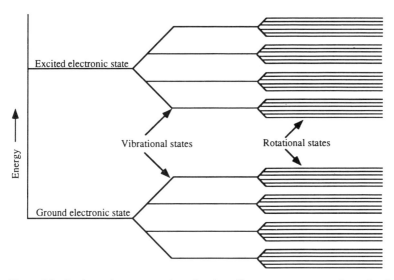

Figure 3.6. A schematic representation of various discrete energy states of a molecule.

between two adjacent levels of the vibrational state is higher than the energy difference between two adjacent levels of the rotational state. A molecule can be electronically, vibrationally, or rotationally excited by the following types of photons:

1. A photon of light or ultraviolet radiation can excite a molecule from the ground electronic state to an excited electron state.
2. An energetic infrared photon can excite a molecule to a higher vibrational level.
3. A low-energy microwave photon can excite a molecule to a higher rotational level.

### 3.2.1.2. The Franck–Condon Diagram

As discussed above, each electronic state is associated with many vibrational and rotational energy levels. On a two-dimensional plot, the Franck–Condon diagram represents the energy of a diatomic molecule along with the vibrational energy levels as a function of the distance between the two nuclei. At each vibrational level, the internuclear distance oscillates due to the vibration of the atoms of the molecule, but the energy of the molecule remains constant. The Franck–Condon diagram becomes three dimensional for triatomic molecules, and $n$-dimensional for polyatomic molecules containing $n$ atoms. Figure 3.7 shows a two-dimensional Franck–Condon diagram for two electronic states of a diatomic molecule. The vibrational levels $V_i$ and $V_i^*$, for $i = 1,2,3,\ldots,n$, represent the vibrational energy states of the molecules in the ground and excited electronic states, respectively. The Gaussian-shaped curves that are associated with the vibrational levels in Fig. 3.7 indicate the nuclear position probability, which is determined by the square of the vibrational wavefunction [Wilkinson (1980)]. The end points of the vibrational levels, such as $V_1$, $V_2$, $V_3$, $V_1^*$, $V_2^*$, etc., correspond to the extreme internuclear separations at the corresponding vibrational energy. Higher vibrational levels refer to higher total energy content, more vigorous oscillations of the atoms, and a longer internuclear distance of the molecule. However, the vibrational levels become closer to one another at higher vibrational energy states. The electronic state of the molecule is designated by the envelope of all the discrete (quantized) vibrational levels. In Fig. 3.7, the curve LBR is such an envelope which represents the ground electronic state of the molecule. In each electronic state, the left side of the envelope, such as LB, becomes vertical at high vibrational levels. On the other hand, the right side, such as BR, becomes horizontal at high vibrational energy levels to indicate that the internuclear distance of the molecule increases as its vibrational energy increases, and that the chemical bond between the nuclei eventually breaks, leading to separation of the nuclei. The lowest point of the envelope, such as B, corresponds to the state of the

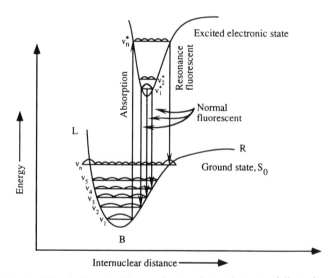

Figure 3.7. A typical Frank–Condon diagram for two electronic states of diatomic molecules.

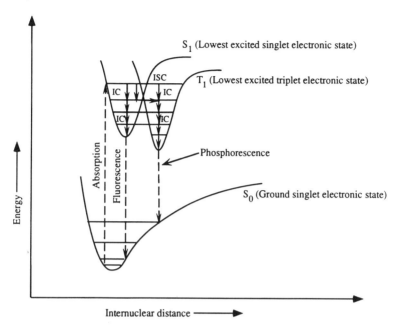

Figure 3.8. Frank–Condon diagram for various types of electronic transitions. The dashed (– →) and solid (→)' arrows represent the radiative and nonradiative processes, respectively. IC and ISC indicate the internal conversion and intersystem crossing processes, respectively.

molecule in the absence of any vibration, which is considered to be attainable at 0 K. However, the lowest vibrational level, such as $V_1$, represents the vibrational ground state of the molecule at the temperature for which the Franck–Condon diagram is plotted. The emissions of radiations from the originally excited state and the lowest vibrational level of the excited electronic state are known, respectively, as resonance emission, such as resonance fluorescence, and normal emission, such as normal fluorescence, as shown in Fig. 3.7.

Figure 3.8 is a Franck–Condon diagram for radiation absorption and radiative and radiationless transition processes that were discussed while the Jablonsky diagram shown in Fig. 3.3 was explained. As shown in Fig. 3.8, the absorption of a photon excites the molecule to an excited vibrational level of the excited singlet electronic state, such as $S_1$. The molecule experiences many cycles of oscillation during its excited lifetime, and jumps to the lower vibrational level of the excited electronic state, such as $S_1$, through the IC process, which is a rapid one with a lifetime of about $10^{-13}$ s. The IC process, which is a radiationless transition, involves the conversion of electronic energy into other forms of energy without any change in the total energy of the molecule. After a total excited lifetime of about $10^{-8}$ s, the molecule jumps from the lowest vibrational level of the excited singlet state $S_1$ to the ground singlet state $S_0$ through the fluorescence process.

Sometimes the molecule enters into the triplet state $T_1$ from the singlet state $S_1$ through the ISC process, which occurs at $10^{-7}$ to $10^{-8}$ s after the radiation is absorbed. In the triplet state, the molecule loses energy by vibrational decay through the IC process, and eventually reaches the lowest vibrational level of the triplet state $T_1$. After a total excited lifetime of about $10^{-4}$ s or more, the molecule jumps from the lowest vibrational level of the triplet state $T_1$ to the ground singlet state $S_0$ through the phosphorescence process.

It should be noted that the intensity distribution for the transitions between various vibrational levels or continua is not uniform and usually has a maximum. This phenomenon is explained by the Franck–Condon principle [Franck (1925), Condon (1928), Okabe (1978), Turro (1967)], which states that the relative position and velocity of the nuclei of a molecule remain approximately the same during the absorption or emission of light by the molecule, since the vibrational periods ($\sim 10^{-12}$ s) of the nuclei are much longer than the time necessary for the absorption or the emission of a photon, which is very short ($\sim 10^{-15}$ s). Thus, the maximum intensity is obtained for the transition in which the internuclear distance does not change. In the Franck–Condon diagram, the vertical transition corresponds to the one that gives the maximum intensity.

### 3.2.2. Selection Rules

So far, we have explained what happens after the absorption of a photon by an atom or a molecule, but we have not discussed the conditions under which a photon will be absorbed by the material. It should be noted that a photon will not necessarily be absorbed by the material even when the energy of the photon is equal to the energy difference between two states of an atom or a molecule [Wilkinson (1980)]. The absorption or emission of a photon, that is, the transition of an atom or a molecule from one energy state to another is governed by a set of selection rules that concern the relationships between the quantum numbers of the two states. Various quantum numbers and their possible values for an atom with one electron (that is, a hydrogen atom), an atom with more than one electron, and the diatomic and linear molecules are discussed below.

### 3.2.2.1. An Atom with One Electron (Hydrogen Atom)

The state of an electron in a hydrogen atom is described by the following quantum numbers: (a) The principal quantum number $n$ determines the energy of the electron in a particular orbital of the hydrogen atom and has integral values ranging from 1 to infinity, that is, $n = 1, 2, 3, \ldots, \infty$. (b) The orbital angular momentum of the electron is specified by the quantum numbers $l$ and $m_l$. It should be noted that the orbital angular momentum is a vector quantity whose magnitude is given by $\sqrt{[l(l+1)]}\hbar$, where $l = 0, 1, 2, 3, \ldots, (n-1)$, which refers to the electrons in $s$, $p$, $d$, $f$, etc., orbitals, respectively; $m_l$ determines the component of the orbital angular momentum long a particular direction, and the magnitude of this component is given by $m_l\hbar$, where $m_l = l, l-1, l-2, \ldots, -l$. (c) The spin angular momentum of the electron has the magnitude $\sqrt{[s(s+1)]}\hbar$, where $s = \frac{1}{2}$. The magnitude of the component of the spin angular momentum along a particular direction is given by $m_s\hbar$, where $m_s = +\frac{1}{2}$ or $-\frac{1}{2}$.

### 3.2.2.2. An Atom with More than One Electron

When there is more than one electron in an atom, the electrons interact among themselves leading to the coupling of various angular momenta. Such couplings are usually described by the Russell–Saunders coupling scheme, which yields the following results:

The magnitude of the total orbital angular momentum is $\sqrt{[L(L+1)]}\hbar$, and the magnitude of the total spin angular momentum is is $\sqrt{[S(S+1)]}\hbar$. For a two-electron system with $l = l_i$ and $s = s_i$ for the $i$th electron, where $i = 1$ and 2, the possible values of $L$ and $S$ are given by

$$L = l_1 + l_2, \quad l_1 + l_2 - 1, \quad l_1 + l_2 - 2, \ldots, |l_1 - l_2|$$

and

$$S = s_1 + s_2, \quad s_1 - s_2 = 1, 0, \quad \text{since} \quad s_1 = s_2 = \tfrac{1}{2}$$

The total angular momentum quantum number $J$ has the following possible values:

$$J = L + S, \quad L + S - 1, \quad L + S - 2, \ldots, |L - S|$$

The spin multiplicity $M = 2S + 1$. It should be noted that $J$ has $M$ possible values for a given pair of $L$ and $S$ values when $J \geq S$. The electronic state can be described by its angular momentum quantum numbers, and is expressed in terms of $L, S,$ and $J$ by the following term symbol:

$$\text{Term symbol} = {}^{2S+1}L_J$$

While writing the term symbol for an atom, the numerical value of $L$ is sometimes replaced by $S, P, D, F, G$, etc., which correspond to $L = 0, 1, 2, 3, 4,$ etc., respectively.

### 3.2.2.3. Diatomic and Linear Molecules

For diatomic and linear molecules, the electronic states are represented by a vector which is a component of the total electronic orbital angular momentum along the internuclear axis. The magnitude of this component is given by $\Lambda \hbar$, where $\Lambda = |M_L|$. For a given value of $L$, the numerical values of $\Lambda$ are sometimes replaced in the term symbol by $\Sigma, \Pi, \Delta, \Phi$, respectively.

Table 3.2. Quantum Numbers and the Corresponding Symbols for Various Angular Momenta

| Angular momentum | Quantum number | Symbol for the following values (0, 1, 2) of the quantum number ($\lambda, \Lambda, \Sigma,$ or $J$) | | |
|---|---|---|---|---|
| | | 0 | 1 | 2 |
| Projection of angular momentum of one electron along the internuclear axis | $\lambda$ | $\sigma$ | $\pi$ | $\delta$ |
| Projection of molecular electronic orbital angular momentum along the internuclear axis | $\Lambda$ | $\Sigma$ | $\Pi$ | $\Delta$ |
| Projection of molecular electronic spin angular momentum along the internuclear axis | $\Sigma$ | Singlet (doublet, quartet, etc., are possible) | Triplet (doublet, quartet, etc., are possible) | Quintuplet (doublet, quartet, etc., are possible) |
| Total molecular angular momentum excluding nuclear spins | $J$ | | | |

For electrons, a quantum number $\lambda = 0, 1, 2, 3, \ldots, l$ can be assigned to represent the electrons in $\sigma, \pi, \delta, \phi$, etc., molecular orbitals, respectively. This notation is given in Table 3.2 [Bauman (1962)].

The total spin quantum number for diatomic and linear molecules is denoted by $S$, and the term symbol is given by

$$\text{Term symbol} = Y^{2S+1}\Lambda_i^j$$

where $Y$ refers to the state of the molecule, and $i$ and $j$ provide information about the symmetry of the wavefunction and the sign of the reflected wavefunction, respectively. For a diatomic molecule,

$$Y = \begin{cases} X \text{ refers to the ground state of the molecule} \\ A, B, C, D, \text{etc., refers to the excited states of the same multiplicity} \\ a, b, c, d, \text{etc., refers to the excited states of different multiplicities.} \end{cases}$$

If the molecule has a center of symmetry, then

$$i = \begin{cases} g \text{ implies that the wavefunction of the state is symmetrical} \\ \quad \text{with respect to the center, where the letter } g \text{ comes} \\ \quad \text{from the German word "gerade" meaning symmetrical.} \\ u \text{ implies that the wavefunction of the state is non-} \\ \quad \text{symmetrical with respect to the center, where the letter } u \\ \quad \text{comes from the German word "ungerade" meaning} \\ \quad \text{nonsymmetrical.} \end{cases}$$

$$j = \begin{cases} + \text{ if the wavefunction remains unchanged after reflection} \\ \quad \text{from the plane passing through the internuclear axis.} \\ - \text{ if the wavefunction becomes reversed in sign after} \\ \quad \text{reflection from the plane passing through the internuclear} \\ \quad \text{axis.} \end{cases}$$

EXAMPLE: The ground state of the oxygen molecule is written as $O_2(X^3\Sigma_g^-)$, where $X$ indicates that the oxygen molecule is in the ground state, $2s + 1 = 3$ indicates that the state is a triplet state with $S = 1$, $\Sigma$ refers to the state for which $\Lambda = 0$, and $g$ implies that the wavefunction of the state is symmetrical with respect to the center of symmetry of the $O_2$ molecule, and indicates that the wavefunction becomes reversed in sign after reflection from the plane passing through the internuclear axis of the $O_2$ molecule.

### 3.2.2.4. Selection Rules for Radiative Transitions

The absorption or emission of a photon requires that the difference in the spins of the two states between which the transition occurs be zero, that is, $\Delta S = 0$. This spin selection rule is applicable to the atoms and molecules which obey the Russell–Saunders coupling scheme. Apart from the spin selection rule, the atoms must satisfy the following conditions:

$$\Delta L = \pm 1, \quad \Delta J = 0 \text{ or } \pm 1, \quad \text{but } \Delta J \neq 0 \text{ if } J = 0$$

The dynamic and linear polyatomic molecules obey the following selection rules:

$$\Delta S = 0, \quad \Delta \Lambda = 0 \text{ or } \pm 1$$

$$u \to g, \quad g \to u, \quad \Sigma^+ \to \Sigma^+, \quad \Sigma^- \to \Sigma^-$$

The transitions which obey these selection rules are called allowed transitions, and those which do not obey them are referred to as forbidden transitions. The negative selection rules are written as

$$\Delta L \neq 0, \quad u \underset{\text{forbidden}}{\longleftrightarrow} u, \quad g \underset{\text{forbidden}}{\longleftrightarrow} g, \quad \Sigma^+ \underset{\text{forbidden}}{\longleftrightarrow} \Sigma^-, \quad \Sigma^- \underset{\text{forbidden}}{\longleftrightarrow} \Sigma^+$$

It should be noted that these selection rules are based on approximate theories which are good approximations for many applications. For this reason, forbidden transitions are often observed but with low transition probabilities.

### 3.2.2.5. Selection Rules for Radiationless Transitions

Diatomic molecules obey the following selection rules for radiationless transitions:

$$\Delta S = 0, \quad \Delta \Lambda = 0 \text{ or } \pm 1, \quad \Delta J = 0$$

$$u \longrightarrow u, \quad g \longrightarrow g, \quad g \underset{\text{forbidden}}{\longleftrightarrow} u$$

$$\Sigma^+ \underset{\text{forbidden}}{\longrightarrow} \Sigma^+, \quad \Sigma^- \underset{\text{forbidden}}{\longrightarrow} \Sigma^-, \quad \Sigma^+ \longleftrightarrow \Sigma^-$$

## 3.3. THE LAWS OF PHOTOCHEMICAL REACTIONS

There are two laws of photochemical reactions:

FIRST LAW OR THE GROTTHUSS–DRAPER LAW: Grotthuss (1819) showed that radiation can cause chemical change only after it is absorbed. This fact was rediscovered by Draper (1841), and is now known as the Grotthuss–Draper law, which states that the electromagnetic radiation must first be absorbed by a material in order for any chemical change to occur in the material.

SECOND LAW: The second law relates the number of reactant molecules that undergo a photochemical reaction to the number of photons absorbed. After Einstein's studies [Einstein (1905b), (1906)] concerning the particle theory of light, Stark (1908) and independently Einstein (1912a, b, 1913) deduced that one molecule absorbs one photon in order to undergo the photochemical reaction, which is known as the second law of photochemistry, or Einstein's law of photochemical equivalence. However, with a radiation of high intensity, such as a high-intensity laser beam, some photochemical reactions are found to occur due to multiphoton absorption. It should be noted that the photochemical reaction due to multiphoton absorption is a nonlinear phenomenon, whereas, the second law of photochemistry refers to the linear phenomenon.

## 3.4. COMPARISON OF PHOTOLYTIC AND PYROLYTIC LCVD

The photolytic and pyrolytic LCVD processes involve photochemical and thermal reactions, respectively, which differ from each other in several aspects:

1. In thermal reactions, the molecules remain in their ground electronic states with the translational, vibrational, and rotational energies on the higher side of the energy distribution given by the Maxwell–Boltzmann law. The molecules undergo thermal reactions when their temperature is above a certain minimum temperature and they have the necessary activation energy. In photochemical reactions, the molecules are electronically excited due to the absorption of photons, and only those molecules which absorb photons undergo chemical changes. The photochemical reactions can occur at temperatures lower than the temperatures required for thermal reactions.
2. In pyrolytic processes, the temperature of the reactant molecule varies over a wide range, as given by the Maxwell–Boltzmann law. Due to this variation in temperature, reactions requiring low energy takes place and, consequently, electronically excited molecules are usually not encountered during thermal reactions. The products of a photolytic reaction can be different from the products of a pyrolytic process. New and thermodynamically highly unstable molecules can be produced in considerable numbers by photolysis. Thus photolysis can be used to develop new materials and to investigate the chemistry of highly unstable molecules.

3. The pyrolysis of complex molecules usually leads to the formation of free radicals and other molecular species. The yields of the free radicals are usually very low, and they are highly reactive with a very short lifetime because of the high temperatures of the pyrolytic processes. Thus, pyrolysis is unsuitable for investigating the chemistry of free radicals. However, the chemical bonds can be broken by photolysis using a suitable radiation, which can be carried out at any desired initial temperature of the reactant molecules. For this reason, photolysis is suitable for generating free radicals in order to study their chemical reactions.

## 3.5. PROCESS VARIABLES IN PHOTOLYTIC LCVD

Some of the process variables of the photolytic LCVD are similar to the process variables of the pyrolytic LCVD, although the mechanisms of these two processes are different.

INCIDENT LASER POWER DISTRIBUTION: The power of a laser beam can vary spatially from the center to the edge of the beam. Also, it can vary with time depending on the continuous wave or pulsed mode operation of the laser. The spatial and temporal distributors of the laser power allow us to determine the total energy input to the system.

LASER BEAM DIAMETER: The laser beam diameter is an important process parameter because it allows us to calculate the laser intensity, which affects the intensity of the light absorbed and, consequently, the yields of the photochemical reactions. Bunsen and Roscoe (1859) showed that the rate of photochemical reaction between hydrogen and chlorine is proportional to the intensity of the light absorbed by the system. For the hydrogen and bromine system, Bodenstein and Lutkemeyer (1924) found that the photochemical reaction rate is proportional to the square root of the intensity of the light absorbed by the system. It should be noted that the photochemical reaction rate is controlled by the intensity of the light absorbed, not by the intensity of the incident light.

LASER WAVELENGTH: According to the first law of photochemistry, the radiation must first be absorbed by the reactant molecules in order for any chemical change to occur in the molecules. However, the absorption of radiation, that is, the absorptivity, depends on the wavelength of the incident radiation. For a given reactant, the laser has to be chosen in such a way that its wavelength is suitable to induce photochemical reaction in the reactant.

OPTICAL PROPERTIES OF THE REACTANT: One of the important optical properties of the reactant is its photon absorption cross section, which affects the attenuation coefficient that appears in the Beer–Lambert law. The photon absorption cross section of the reactant determines the amount of radiation that will be absorbed by the reactant molecules and, thus influences the photochemical reaction rate.

CHEMICAL PROPERTIES OF THE REACTANT: The rate constants of various processes, such as excitation and quenching, and the chemical bond dissociation energy are important chemical properties of the reactant which govern the yield of the reaction.

THERMOPHYSICAL PROPERTIES OF THE REACTANT: Various thermophysical properties, such as density, specific heat, thermal conductivity, viscosity, and mass diffusivity, affect the transport of various species and the thermal energy inside the reaction chamber.

ELECTRIC AND MAGNETIC FIELDS: Since the chemical reactions involve the transfer of electrons from one species to another, the electric and magnetic fields can affect the pyrolytic and photolytic reactions [*Chem. Eng. News* (1966), Steiner and Wolff (1991)]. In the case of laser-induced reactions, we do not even have to apply any external electric and magnetic fields because laser is an electromagnetic radiation whose electric and magnetic fields are readily available to influence such reactions. It has been reported [*Chem. Eng. News* (1966)] than an electric field increases the rate of catalytic cationic polymerization of α-methylstyrene and *p*-methoxystyrene, isobutyl vinyl ether, and cationic polymerization of styrene. The effect has been found to be proportional to the field strength. The electric field also increases the degree of polymerization of *p*-methoxystyrene and isobutyl vinyl ether [*Chem. Eng. News* (1966)]. These effects of the electric field are explained by noting that the ends of the growing polymer chain contain ion pairs as well as free ions during ionic polymerization, and that the electric field dissociates the ion pairs to produce free ions. In this way, the electric field increases the rate and degree of polymerization because free ions can propagate much faster than ion pairs.

## 3.6. PHOTON ABSORPTION CROSS SECTION

As noted earlier, the photon absorption cross section is an important parameter for photolytic LCVD because it allows us to determine the amount of radiation absorbed by the reactant molecules. The photochemical reactions involve the interactions of photons with the reactant molecules, and such interactions are described in terms of the absorption cross sections.

### 3.6.1. Absorption Cross Section

To understand the physical significance of the absorption cross section, let us consider a cylindrical beam of photons impinging perpendicularly on an area $a$ of a thin target of thickness $d$ (Fig. 3.9). As in the case of the interactions of neutrons [Lamarsh (1966)] with matter, we can consider [Spinks and Woods (1990)] that the rate of absorption of photons by the target as the photon beam passes through it is proportional to the intensity of the photon beam and to the atomic density, area, and thickness of the target. Thus we can write

$$\dot{R} = \sigma I_0 \rho_a a d \qquad (3.2)$$

where $\sigma$ is a proportionality constant, known as the photon absorption cross section. Equation (3.2) can be rewritten as

$$\sigma = \frac{\dot{R}}{I_0(\rho_a a d)} \qquad (3.3)$$

Since $\rho_a a d$ represents the total number of atoms along the path of the photon beam in the target, Eq. (3.3) defines the absorption cross section $\sigma$ as the absorption rate per atom in the target per unit intensity of the incident beam. It should be noted that $\sigma$ has the dimensions of area.

To interpret $\sigma$ in terms of probability, we rewrite Eq. (3.3) as

$$\frac{\sigma}{a} = \frac{\dot{R}}{I_0 a} \frac{1}{\rho_a a d} \qquad (3.4)$$

Since $\dot{R}/I_0 a$ represents the probability of one photon being absorbed in the target, Eq. (3.4) defines $\sigma/a$ as the probability that one photon is absorbed per atom in the target. However, the area $a$ on which the photon beam strikes the target is fixed by the experiment and, therefore, the probability of absorption is determined by $\sigma$ alone; in this sense, $\sigma$ is referred to as the probability of absorption even though it has the dimensions of area. The cross sections are usually expressed in barns by defining 1 barn = $10^{-24}$ cm$^2$ where the dimension barn is sometimes abbreviated as b.

The absorption cross sections of several molecules that are suitable for depositing films by the photolytic LCVD technique have been given by Rothschild (1989) for various laser wavelengths, and are found to be in the range of $10^{-17}$ to $10^{-21}$ cm$^2$ per molecule.

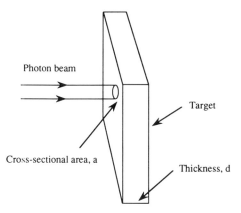

Figure 3.9. A photon beam interacting with a target.

## 3.6.2. The Beer–Lambert Law and the Attenuation Coefficient

The Beer–Lambert law states that the intensity of radiation decreases exponentially as it passes through a medium. This law is derived below under the following assumptions:

1. The cross-sectional area of a photon beam, such as a laser beam, remains unchanged as it propagates through the medium.
2. The photons of a beam are lost due to absorption only; that is, there is no loss of photons due to the scattering by the atoms in the medium.
3. The path of a photon beam remains unaltered during its propagation through the medium, as shown in Fig. 3.10; that is, if the photon beam strikes the medium parallel to the $x$-axis, it remains parallel to the $x$-axis as it passes through the medium.

We consider a photon beam of cross-sectional area $a$ that strikes the surface of a medium of thickness $d$ with intensity $I_0$; that is, the intensity of the photon beam is $I_0$ at $x = 0$. At any point $x$ in the medium, we take a thin slice of thickness $\Delta x$, and balance the number of photons across this slice to obtain the following expression:

$$I(x + \Delta x)a = I(x)a - \sigma I(x)\rho_a a \Delta x \qquad (3.5)$$

Noting that

$$\lim_{\Delta x \to 0} \frac{I(x + \Delta x) - I(x)}{\Delta x} = \frac{dI(x)}{dx}$$

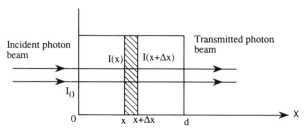

Figure 3.10. A geometric configuration for the propagation of a photon beam through a medium of thickness $d$,

Eq. (3.5) can be written as

$$\frac{dI(x)}{dx} = -\sigma I(x)\rho_a \tag{3.6}$$

Integrating Eq. (3.6) from 0 to $x$, we obtain

$$I(x) = I_0 e^{-\sigma \rho_a x} = I_0 e^{-\mu x} \tag{3.7}$$

where the attenuation coefficient, $\mu = \sigma \rho_a$. Equation (3.7) is generally known as the Beer–Lambert law.

The attenuation coefficient $\mu$ can be defined as the macroscopic absorption cross section, as in the case of the interaction of neutrons with matter [Lamarsh (1966)]. In this context, $\sigma$ can be defined as the atomic attenuation coefficient [Spinks and Woods (1990)], or the atomic absorption cross section, or the microscopic absorption cross section, as it represents the absorption cross section of an atom; $\mu$ has the dimensions of inverse length and represents the probability of a photon being absorbed per unit path length. As can be seen from Eq. (3.6), $\mu$ is essentially that fraction of incident photons that is absorbed by unit thickness of the medium.

For a molecule with a chemical formula involving $p$ atoms of X and $q$ atoms of Y, that is, $X_p Y_q$, the absorption cross section of the molecule is given by [Spinks and Woods (1990)]

$$\sigma_m = p\sigma_X + q\sigma_Y \tag{3.8}$$

Equation (3.8) can be applied to molecules containing more than two different types of atoms by adding more terms to its right-hand side. For a homogeneous mixture of type $m_1$ and $m_2$ molecules, the attenuation coefficient of the mixture is given by

$$\mu_m = \mu_{m_1} + \mu_{m_2} = \rho_{m_1}\sigma_{m_1} + \rho_{m_2}\sigma_{m_2} \tag{3.9}$$

As in the case of Eq. (3.8), Eq. (3.9) can be applied to multicomponent systems by adding more terms to its right-hand side. Equations (3.8) and (3.9) are based on the assumption that atoms $X$ and $Y$, and molecules $m_1$ and $m_2$ interact with the photons independently of each other [Lamarsh (1966)].

### 3.6.3. Two-Photon Absorption Constant

Equation (3.6) can be written as

$$-\frac{dI}{dx} \propto I\rho_a \tag{3.10}$$

where the proportionality constant $\sigma$ is the absorption cross section. Equation (3.10) indicates that the photon absorption rate is linearly proportional to the intensity of photons. When two photons are absorbed simultaneously by the atoms or molecules of the medium through which the photon beam propagates, the probability of absorption, that is, the rate of two-photon absorption, depends quadratically on the intensity of the photons [Andrews (1986)]. Thus, two-photon absorption can be expressed as

$$-\frac{dI}{dx} \propto I^2\rho_a \tag{3.11}$$

Taking the proportionality constant as $\beta$ in Eq. (3.11), and integrating it from 0 to $x$, we obtain

$$I = \frac{I_0}{1 + \beta I_0 \rho_a x} \tag{3.12}$$

which gives the intensity of the photons at any location $x$ in the medium through which the beam propagates; $\beta$ can be defined as the two-photon absorption constant. Comparing Eqs. (3.7) and (3.12), we can conclude that the usual Beer–Lambert law of Eq. (3.7) is not suitable for describing the two-photon absorption phenomena [Andrews (1986)].

### 3.6.4. Dissociation Cross Section

A molecule can be dissociated by single-photon or multiphoton absorption in which the reactant molecules are excited to higher energy states due to the vibrational and rotational transitions [Kompa and Smith (1979), Price (1972)]. When a molecule is excited, it can attain an electronically excited unstable or stable state from which it can undergo dissociation, depending on the energy of the absorbed photon and the dissociation energy and the absorption and emission characteristics of the molecule as discussed below.

ELECTRONICALLY EXCITED UNSTABLE STATE: The lifetime of this state is from $10^{-13}$ to $10^{-14}$ s and, during this time the excited molecules usually dissociate.

ELECTRONICALLY EXCITED STABLE STATE: The molecules in this state usually dissociate under the following conditions: (1) The electronically excited stable molecules undergo dissociation when $hv \geq E_d^*$. (2) There is spontaneous predissociation, that is the initially excited stable molecule, whose dissociation energy is $E_d^*$, decays into an unstable or a less stable state whose dissociation energy is less than $E_d^*$. Consequently, the molecule can dissociate from this unstable or less stable state when $hv < E_d^*$. This occurs more frequently in polyatomic molecules than in diatomic molecules. The duration for predissociation usually ranges from $10^{-6}$ to $10^{-12}$ s. (3) Sometimes the excited stable molecule decays by the IC mechanism into higher vibrational levels (that is, levels with energies higher than $E_d$) of the electronic ground state. Due to this type of transition to the electronic ground state, the molecule can undergo dissociation when $hv < E_d^*$.

For low laser intensities, the average number of dissociated molecules is proportional to the laser fluence, and is given by

$$\bar{N}_d = \bar{\sigma}_d \frac{N\phi}{hv} = \bar{\sigma}_d \frac{NI\tau}{hv} \tag{3.13}$$

When the interaction of photons with the reactant molecules is linear, the number of excited or dissociated molecules is found to be independent of whether pulsed or continuous wave laser is used [Bäuerle (1986)]. The effective dissociation cross section is given by

$$\bar{\sigma}_d = \eta \sigma_d \tag{3.14}$$

PHOTOLYTIC LCVD

where the dissociation yield $\eta$ depends on the type of reactant and carrier gas molecules and the pressure inside the reaction chamber. Under collisionless conditions which can be obtained at low pressures inside the reaction chamber $n = 1$. The dissociation cross section $\sigma_d = \sigma_m$ when $hv > E_d^*$ and

Table 3.3. The Absorption and the Effective Dissociation Cross Sections for Various Precursor Molecules

| Types of precursors | Chemical formula | Wavelength $\lambda$ (nm) | Absorption cross section $\sigma_m$ ($\times 10^{-18}$ cm$^2$) | Effective dissociation cross section $\bar{\sigma}_d$ ($\times 10^{-18}$ cm$^2$) |
|---|---|---|---|---|
| Methyls | Al$_2$(CH$_3$)$_6$ | 193 | 20 | |
| | | 257 | 0.002 | |
| | Cd(CH$_3$)$_2$ | 257 | 2 | |
| | Ga(CH$_3$)$_3$ | 193 | 5.4 | |
| | | 257 | 0.09 | |
| Carbonyls | Cr(CO)$_6$ | 193 | 12 | |
| | | 249 | 33 | |
| | | 308 | 5.2 | |
| | Fe(CO)$_5$ | 193 | 240 | |
| | | 248 | 27 | |
| | | 355 | 1.3 | |
| | Mo(CO)$_6$ | 193 | 60 | |
| | | 249 | 44 | |
| | | 308 | 11 | |
| | | 350–360 | 0.5 | |
| | Ni(CO)$_4$ | 248 | 30 | |
| | | 308 | 2.4 | |
| | W(CO)$_6$ | 193 | 12 | |
| | | 249 | 4.5 | |
| | | 308 | 2.4 | |
| | | 350–360 | 0.5 | |
| Hydrides | AsH$_3$ | 193 | 18 | |
| | B$_2$H$_6$ | 193 | 0.2 | |
| | GeH$_4$ | 193 | 0.0035 | |
| | PH$_3$ | 193 | 13 | |
| | SiH$_4$ | 193 | 0.0012 | |
| Halides | InI | 193 | | <7 |
| | NF$_3$ | 193 | 0.0053 | |
| | TlBr | 193 | | 22 |
| | TlI | 193 | | 24 |
| | | 248 | | 2.6 |

the fluorescence is negligible [Bäuerle (1986)]. The values of the absorption and the effective dissociation cross sections, $\sigma_m$ and $\bar{\sigma}_d$, respectively, are given in Table 3.3 for various reactant molecules [Bäuerle (1986)]. The bond energy of most of the molecules is so high that medium- to far-ultraviolet lasers are required to induce molecular dissociation by single-photon absorption. Although dissociation by multiphoton absorption imposes fewer restrictions on the selection of lasers, the number of the excited or dissociated molecules depends nonlinearly on the photon flux and the reaction conditions.

## 3.7. CHEMICAL BOND ENERGY

The bond energy of a molecule is an important parameter for mathematical modeling of photolytic LVCD because (1) it relates the yields of the reaction to the energy of the laser beam; (2) it allows us to select the laser wavelength for a given reactant in order to carry out the photochemical reaction. During photodissociation, the weakest bond breaks in most of the simple molecules; rarely do two or more bonds break simultaneously. The bond dissociation energy of a molecule is the energy required to dissociate a molecule from its lowest ground state level into the products that are in their respective lowest levels of the ground electronic states.

When a molecule dissociates from its lowest level of ground electronic state, the products are usually formed in their respective lowest levels of this state, and then the bond dissociation energy is equal to the dissociation energy of the reactant molecule. However, in many cases, the ground state reactant molecules dissociate into excited products that are in the ground electronic state with higher vibrational energy. For example, the dissociation energy of the ground state $N_2O$ is higher than the bond dissociation energy [Okabe (1978)]. There are several methods [Okabe (1978)] for determining the bond dissociation energies, and two of these are discussed below.

### 3.7.1. Calculation of Bond Dissociation Energies from Thermochemical Data

To illustrate the procedure for determining the bond dissociation energy, let us consider a diatomic molecule $XY$, which dissociates into $X$ and $Y$ species according to the following chemical reaction:

$$XY(g) \rightleftharpoons X(g) + Y(g)$$

In thermochemistry, the bond dissociation energy is equal to the increase in enthalpy for the gas phase reaction in the standard state at 0 K; that is,

$$D_0(X-Y) = \Delta H_0^\circ = \Delta H_{f\,o}^\circ(X, g) + \Delta H_{f\,o}^\circ(Y, g) - \Delta H_{f\,o}^\circ(XY, g) \quad (3.15)$$

where the superscript °, refers to the thermodynamic standard state of 1 atm pressure, and the subscript zero refers to the 0 K temperature. For many species, the standard enthalpy or heat of formation, $\Delta H_{f\,o}^\circ$, can be found in Lewis et al. (1961), Stull et al. (1969), Stull and Prophet (1971), Wagman et al. (1968), and Trotman-Dickenson and Milne (1967). Thus the bond dissociation energy of a molecule can be calculated by using Eq. (3.15) if the standard enthalpies of formation data are available. The value of $D_0(X-Y)$ can also be obtained from the equilibrium constant which, for the above chemical reaction, can be written as

$$K_P = \frac{P_X P_Y}{P_{XY}} \quad (3.16)$$

where the pressures $P_i$, for $i = X$, $Y$, or $XY$, are in atmospheres at the chemical reaction temperature $T$; $K_p$ is related to the standard Gibbs free energy change $\Delta G_T^\circ$ by the following equation:

$$\Delta G_T^\circ = -RT \ln K_P \quad (3.17)$$

There are two methods called the second and third law methods [Okabe (1978), Stull and Prophet (1971)] for determining the bond dissociation energy from $K_P$.

### 3.7.1.1. The Second Law Method

From the basic thermodynamic relations, we have the following expressions:

$$dG = dH - TdS - SdT \quad (3.18)$$

$$TdS = dE + PdV \quad (3.19)$$

$$dH = dE + PdV + VdP \quad (3.20)$$

where $G$, $H$, $T$, $S$, $E$, $P$, and $V$ refer to the Gibbs free energy, enthalpy, temperature, entropy, internal energy, pressure, and volume of the species, respectively.

Subtracting Eq. (3.19) from Eq. (3.20), we obtain

$$dH - TdS = VdP \qquad (3.21)$$

Taking $G$ as a function of $T$ and $P$, that is, $G = G(T, P)$, we can write the total differential of $G$ as

$$dG = \frac{\partial G}{\partial T} dT + \frac{\partial G}{\partial P} dP$$

which leads to

$$\frac{dG}{dT} = \frac{\partial G}{\partial T} \quad \text{at constant pressure}$$

Using this relationship and Eqs. (3.18) and (3.21), we obtain the following expression at constant pressure:

$$S = -\left(\frac{\partial G}{\partial T}\right)_P \qquad (3.22)$$

In Eq. (3.22), $(\partial G/\partial T)_P$ represents the change in $G$ with respect to $T$ along the isobars, that is, the constant pressure lines shown in Fig. 3.11. Let us consider that the isobar AB (see Fig. 3.11) represents the initial state of a system, which will be denoted by the subscript 1, and that the isobar CD represents the final state of the system, which will be denoted by the subscript 2. For this discussion, we assume that the isobars AB and CD correspond to the same pressure $P$. When the system is transformed from its initial state to its final state, the change in entropy can be written as

$$-\Delta S = -(S_2 - S_1) = \left(\frac{\partial G_2}{\partial T}\right)_P - \left(\frac{\partial G_1}{\partial T}\right)_P = \left[\frac{\partial (G_2 - G_1)}{\partial T}\right]_P = \left[\frac{\partial (\Delta G)}{\partial T}\right]_P$$

$$(3.23)$$

It should be noted that the English letter $d$ is used in Eqs. (3.18) through (3.21) to signify that the system changes without any chemical reactions or phase changes; that is, the system is transformed within its initial or final states (see Fig. 3.11). The greek letter $\Delta$ is used to indicate that the system is transformed from its initial to its final state due to a chemical reaction or phase change. In Eq. (3.23), $\Delta S$ represents the change in entropy due to the transformation of the system from its initial state to a final state, which can

# PHOTOLYTIC LCVD

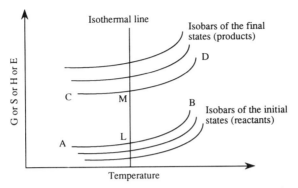

Figure 3.11. A geometric representation of various thermodynamic states.

be accomplished in many different ways and, consequently, $\Delta S$ will have several values. We are interested in the transformation at a constant temperature, that is, along the isothermal line LM. Thus, for a given temperature $T$, Eq. (3.23) can be written as

$$-\Delta S_T = \left[\frac{\partial(\Delta G_T)}{\partial T}\right]_P \tag{3.24}$$

Similarly, the change in the Gibbs free energy due to the transformation of the system from its initial to its final state at constant pressure along the isothermal line LM is given by

$$\begin{aligned}
\Delta G_{T,P} &= (G_2 - G_1)_{T,P} \\
&= (H_2 - TS_2)_{T,P} - (H_1 - TS_1)_{T,P} \\
&= (H_2 - H_1)_{T,P} - T(S_2 - S_1)_{T,P} \\
&= \Delta H_{T,P} - T\Delta S_{T,P}
\end{aligned} \tag{3.25}$$

Equations (3.24 and 3.25) yield

$$\begin{aligned}
\Delta H_{T,P} &= \Delta G_{T,P} - T\left[\frac{\partial(\Delta G_T)}{\partial T}\right]_P \\
&= -T^2\left[\frac{1}{T}\frac{\partial(\Delta G_T)}{\partial T} - \frac{1}{T^2}\Delta G_T\right]_P \\
&= -T^2\left[\frac{\partial}{\partial T}\left(\frac{\Delta G_T}{T}\right)\right]_P
\end{aligned} \tag{3.26}$$

We can derive Eq. (3.26) in a different way by noting that the Gibbs free energy $G$ is given by

$$G = H - TS \tag{3.27}$$

Combining Eqs. (3.22) and (3.27), we obtain

$$H = G - T\left(\frac{\partial G}{\partial T}\right)_P$$

$$= -T^2\left[\frac{1}{T}\left(\frac{\partial G}{\partial T}\right)_P - \frac{G}{T^2}\right]$$

$$= -T^2\left[\frac{\partial}{\partial T}\left(\frac{G}{T}\right)\right]_P \tag{3.28}$$

If $H_1$ and $H_2$ are the total enthalpies of the initial state (the reactants) and the final state (the products), respectively, at the same temperature and pressure, then

$$\Delta H_{T,P} = (H_2 - H_1)_{T,P}$$

$$= -T^2\left[\frac{\partial}{\partial T}\left(\frac{G_2}{T}\right)\right]_{T,P} + T^2\left[\frac{\partial}{\partial T}\left(\frac{G_1}{T}\right)\right]_{T,P}$$

$$= -T^2\left[\frac{\partial}{\partial T}\left(\frac{G_2 - G_1}{T}\right)\right]_{T,P}$$

$$= -T^2\left[\frac{\partial}{\partial T}\left(\frac{\Delta G_T}{T}\right)\right]_P$$

which is identical to Eq. (3.26).

Since the thermodynamic standard state is considered to be at a constant pressure of 1 atm, which is denoted by the superscript °, we omit the subscript $P$ in Eq. (3.26) and write this equation as

$$\Delta H_T^\circ = -T^2 \frac{\partial}{\partial T}\left(\frac{\Delta G_T^\circ}{T}\right) \tag{3.29}$$

Using Eq. (3.17) in Eq. (3.29), we obtain

$$\Delta H_T^\circ = -R \frac{d(\ln K_P)}{d(1/T)} \tag{3.30}$$

The bond dissociation energy can be determined from Eq. (3.30) by following the procedure outlined below:

STEP 1: Determine $K_P$ over a wide range of temperatures, and plot $\ln K_P$ vs. $1/T$.
STEP 2: Calculate the slopes of the curve obtained in Step 1 at various temperatures. These slopes are essentially equal to $d(\ln K_P)/d(1/T)$, which can be used in Eq. (3.30) to determine $\Delta H_T^\circ$ at various temperatures.
STEP 3: Plot $\Delta H_T^\circ$ vs. $T$, and extrapolate the curve to $T = 0\ K_P$ to determine $\Delta H_0^\circ$, that is, the bond dissociation energy.

It is usually difficult to determine $K_P$ over a wide range of temperatures, so the third law method is used whenever the Gibbs energy functions are available.

### 3.7.1.2. The Third Law Method

To explain this method, let us first define the Gibbs energy function [Stull and Prophet (1971)] at the standard state as

$$\bar{G}_T^\circ = \frac{G_T^\circ - H_{T_r}^\circ}{T} \tag{3.31}$$

Using Eq. (3.31) in Eq. (3.17), we obtain

$$\frac{\Delta H_{T_r}^\circ}{T} = -R \ln K_P - \Delta \bar{G}_T^\circ \tag{3.32}$$

which is used to calculate the bond dissociation energy, as indicated in the following steps:

STEP 1: Determine $K_P$ at any temperature, say $T$.
STEP 2: Obtain the value of $\Delta \bar{G}_T^\circ$ from Stull and Prophet (1971), Wagman et al. (1968), and Trotman-Dickenson and Milne (1967) or other sources at the temperature $T$. Use Eq. (3.32) to calculate $\Delta H_{T_r}^\circ$.
STEP 3: Obtain the value of $\Delta H_0^\circ - \Delta H_{T_r}^\circ$ from Stull and Prophet (1971), Wagman et al. (1968), and Trotman-Dickenson and Milne (1967), or other sources. Add this value to the value of $\Delta H_{T_r}^\circ$ calculated in Step 2, which will give the value of $\Delta H_0^\circ$, that is, the bond dissociation energy.

## 3.7.2. Calculation of Bond Dissociation Energies from Spectroscopic Data

The spectroscopic technique can be used to analyze the elementary reaction during the photodissociation or photoionization of a molecule in order to obtain its bond dissociation energy [Okabe (1978)]. The concentration of the excited species or the ion (the primary product for the purposes of the subsequent discussion) produced during photodissociation or photoionization, respectively, is monitored by varying the wavelength of the probe laser or the probe radiation in the wavelength regime, such as the vacuum-ultraviolet region, where the molecule absorbs appreciably. During photodissociation of the molecule $XY$, if the primary product $Y^*$ fluoresces, the elementary reactions for the process can be written as

$$XY + h\nu \rightarrow X + Y^*$$
$$Y^* \rightarrow Y + h\nu$$

and the bond dissociation energy $D_0(X-Y)$ is given by

$$D_0(X-Y) = h\nu_f - E^* \qquad (3.33)$$

On the other hand, during photoionization of the molecule $XY$, if the primary product is an ion $X^+$, the elementary reaction for the process can be written as

$$XY + h\nu \rightarrow X^+ + Y + e$$

and the bond dissociation energy, $D_0(X-Y)$, is given by

$$D_0(X-Y) = h\nu_i - \phi_i \qquad (3.34)$$

Equations (3.33 and 3.34) are based on the assumption that the species $XY$, $X$, and $Y^*$ and the species $XY$, $X^+$, $Y$, and $e$ have no internal and kinetic energies at the threshold of the photodissociation and photoionization, respectively. The values of $E^*$ and $\phi_i$ have been measured [Pearse and Gaydon (1963), Herzberg (1945, 1950, 1966), Rosen (1970)] for many species by using absorption spectroscopy and the photoionization technique. A problem will be solved to illustrate the procedure for determining the bond dissociation energy from the spectroscopic data.

Photodissociation and photoionization sometimes involve a small potential barrier which leads to a higher value of $D_0(X-Y)$. The bond dissociation energies of FCN, ICN, and $C_2N_2$ obtained by photoionization are slightly higher than those obtained by photodissociation, which indicates that photoionization of these molecules involves small potential barriers, while the value of $D_0$(Br–CN) obtained by photodissociation is slightly higher than that obtained by photoionization, which indicates that photodissociaton of BrCN involves a small potential barrier [Okabe (1974, 1978)]. In many cases, the bond dissociation energies obtained by the electron impact method are found to be higher than those obtained by the photon impact method [Davis and Okabe (1968), Okabe and Dibeler (1973)].

### 3.7.3. Mean Metal–Carbon Dissociation Energy

Organometallic compounds are widely used [Ehrlich et al. (1982, 1980), Allen and Bass (1979), Solanki et al. (1981, 1983); Solanki and Collins (1983), Deutsch et al. (1979)] for LCVD of thin films. Ehrlich et al. (1980, 1982) and Deutsch et al. (1979) deposited thin metal films less than 0.7 μm wide by carrying out photolytic decomposition of Cd, Sb, Al, Zn, and Ga alkyls of the form $M(CH_3)_n$ using either a frequency-doubled Ar ion laser (257 nm) or an ArF excimer laser (193 nm). Solanki and Collins (1983) and Solanki et al. (1983) deposited Zn and Al by the photolysis of dimethylzinc (DMZn) and trimethylaluminum (TMAl), respectively. Also, they were able to deposit ZnO [Solanki and Collins (1983)], and $Al_2O_3$ [Solanki et al. (1983)], by photodissociation of the corresponding metal alkyls in the presence of $N_2O$ or $NO_2$ gas. Solanki et al. (1981) deposited Cr, Mb, and W by the photolytic decomposition of the carbonyls $Cr(CO)_6$, $Mo(CO)_6$, $W(CO)_6$, respectively. Allen and Bass (1979) deposited Ni and $TiO_2$ films from $Ni(CO)_4$ and $TiCl_4 + H_2 + CO_2$, respectively, by the pyrolytic LCVD method. Although, organometallic compounds are used extensively for LCVD, very little is known about their dissociation dynamics.

Thompson and Linnett (1936) have shown that the metal alkyls and nonmetallic alkyls of the form $M(CH_3)_n$ or $M(C_2H_5)_n$ have dissociative absorption continua in the near- to medium-ultraviolet regime. Long (1961) has determined the mean metal–carbon dissociation energies for many volatile methyls from the heats of formation and the heats of atomization. These dissociation energies are listed in Table 3.4 along with the data given in Bamford and Tipper (1972), p. 253. These thermochemically derived dissociation energies are related to the intrinsic bond energies by the following expression:

$$E_i(M-C) = D_m(M-C) + R_{CH_3} + V_M/n \quad (3.35)$$

Table 3.4. Dissociation Energies of the Metal–Carbon Bonds of Various Methyls[a]

| Methyl $M(CH_3)_n$ | $D_m$ (kcal/mol) | | $D_1$ (kcal/mol) | |
|---|---|---|---|---|
| | After Long (1961) | After Bamford and Tipper (1972) | After Long (1961) | After Bamford and Tipper (1972) |
| $ZnMe_2$ | 41.5 | 43 | 47.2 | 51 |
| $CdMe_2$ | 32.8 | 34 | 43.5 | 53 |
| $HgMe_2$ | 29.5 | 30–31 | 51.3 | 58 |
| $BMe_3$ | 87.5 | | | |
| $AlMe_3$ | 62.9 | 66 | 78 | 65 |
| $GaMe_3$ | 57.5 | 59 | | 60 |
| $InMe_3$ | | 37 | | 47 |
| $TlMe_3$ | | | | 27 |
| $CMe_4$ | 85.3 | | 98.8 | |
| $GeMe_4$ | | 64 | | |
| $SiMe_4$ | 71.1 | 68 | 78.8 | 72 |
| $SnMe_4$ | 53.3 | 64.5 | 82.4 | |
| $PbMe_4$ | 35.9 | 38 | >36.9 | 48.6 |
| $NMe_3$ | 71.5 | | 100.4 | |
| $PMe_3$ | 65.3 | 67 | | |
| $AsMe_3$ | 51.5 | 58 | 54.6 | 62.8 |
| $SbMe_3$ | 49.7 | 51 | 57 | 57 |
| $BiMe_3$ | 33.7 | 35 | 44.03 | 44 |
| $OMe_3$ | 84.4 | | 98 | |
| $SMe_3$ | 70.3 | | | |

[a] $D_m$ is the mean metal–carbon bond dissociation energy, $D_1$ is the dissociation energy of the first metal–carbon bond, and Me stands for the methyl radical $CH_3$.

to account for the reorganization of the methyl radicals and the transition of the central atom M to its ground state. In expression (3.35), $E_i(M–C)$ is the intrinsic metal–carbon bond energy of the alkyls $M(CH_3)_n$, $D_m(M–C)$ is the mean metal–carbon dissociation energy of $M(CH_3)_n$, $R_{CH_3}$ is the methyl group reorganization energy, $V_M$ is the excitation energy of combination of the central atom M with the methyl radicals, and $n$ is the number of the methyl radicals in the precursor molecule $M(CH_3)_n$. Usually, $R_{CH_3}$ is nearly constant, but $V_M$ varies considerably with the electronic orbital structure of M.

Long (1961) and Jonah et al. (1971) have shown that the dissociation energy of the first metal–carbon bond, $D_1$, is markedly different from the mean metal–carbon dissociation energy $D_m$. The energy required to remove the first methyl radical is found [Long (1961)] to be higher than the mean dissociation energy, which can be explained by considering that the metal atom does not return to its ground state after the rupture of a single bond. For this reason, $D_1$ is approximately related to $D_m$ by the following relationship [Long (1961)]:

$$D_1 = D_m + V_M/n - R' \tag{3.36}$$

where $R'$ is the reorganization energy of the fragment $M(CH_3)_{n-1}$ produced in the reaction

$$M(CH_3)_n \rightarrow M(CH_3)_{n-1} + CH_3$$

In general, the photolytic reactions of organometallic compounds may be considered to be

$$M(CH_3)_n + mh\nu \rightarrow M + nCH_3$$

where $m$ is an integer less than or equal to $n$. Usually, the methyl radicals emerge in their ground state, and the dissociation is anisotropic.

## 3.8. EXAMPLE PROBLEMS

In this section, we will solve several problems to show how various concepts of photochemical reactions can be utilized to analyze the photolytic LCVD processes. It is assumed that single-photon absorption phenomena are present in all of the problems of this section.

### Problem 1

Calculate the maximum laser wavelengths required to dissociate diatomic molecules of dissociation energies 200 and 400 kJ/mol.

SOLUTION: The maximum wavelength of a reaction corresponds to the minimum energy of a photon. Since the dissociation is assumed to occur via single-photon absorption, we want the energy of 1 mol of photons, that is, 1 einstein, to be equal to 200 and 400 kJ/mol for the first and second diatomic molecules, respectively.

We know that $E = h\nu$ for one photon or quantum, and

$$1 \text{ einstein} = Ahc/\lambda$$

Thus

$$Ahc/\lambda = D_0(X-Y)$$

Therefore,

$$\lambda = \frac{Ahc}{D_0(X-Y)}$$

$$= \frac{(6.03 \times 10 \,\text{quantum/mol})(6.6256 \times 10^{-34} \,\text{J s/quantum})(2.9979 \times 10^{10} \,\text{cm/s})}{D_0(X-Y)}$$

$$= \frac{11.977 \,\text{J cm/mol}}{D_0(X-Y)}$$

When $D_0(X-Y) = 200 \,\text{kJ/mol}$

$$\lambda = \frac{11.977 \,\text{J cm/mol}}{200 \times 10^3 \,\text{J/mol}} \times (10^7 \,\text{nm/cm}) = 598.85 \,\text{nm}$$

When $D_0(X-Y) = 400 \,\text{kJ/mol}$

$$\lambda = \frac{11.977 \,\text{J cm/mol}}{400 \times 10^3 \,\text{J/mol}} \times (10^7 \,\text{nm/cm}) = 299.425 \,\text{nm}$$

**Problem 2**

Chromium hexacarbonyl is found [Breckenridge and Stewart (1986)] to undergo the following reaction

$$Cr(CO)_6 + h\nu \rightarrow Cr(CO)_5 + CO$$

in gas phase laser (355 nm) photolysis. When 1 J of energy is absorbed, 1.364 mg of $Cr(CO)_6$ is decomposed. Determine the number of molecules of $Cr(CO)_6$ that will be decomposed by one photon of the above laser.

SOLUTION: The energy of one photon or quantum is given by

$$E_1 = \frac{hc}{\lambda} = \frac{(6.6256 \times 10^{-34} \,\text{J s/quantum})(2.9979 \times 10^{10} \,\text{cm/s})}{(355 \,\text{nm})(10^{-7} \,\text{cm/nm})}$$

$$= 5.595 \times 10^{-19} \,\text{J/quantum}$$

Let $E_0$ be the energy in joules that is required to decompose $m_p$ grams of $Cr(CO)_6$. The energy $E_p$ required to dissociate one molecule of the precursor $Cr(CO)_6$ is given by

$$E_p = \frac{E_0 M_p}{A m_p}$$

where $M_p$ is the molecular weight of the precursor $Cr(CO)_6$, and is equal to $51.996 + 6(12 + 5.9994) = 219.9924$; $E_p$ can be calculated as

$$E_p = \frac{(1\text{ J})(219.9924\text{ g/mol})}{(1.364 \times 10^{-3}\text{ g})(6.03 \times 10^{23}\text{ molecules/mol})}$$
$$= 2.6747 \times 10^{-19}\text{ J/molecule}$$

Thus, the number of molecules of $Cr(CO)_6$ that will be decomposed by one photon of the 355-nm laser $= E_1/E_p$, which is found to be

$$\frac{E_1}{E_p} = \frac{5.595 \times 10^{-19}\text{ J/quantum}}{2.6747 \times 10^{-19}\text{ J/molecule}}$$
$$= 2.09\text{ molecules/quantum}$$
$$= 2\text{ molecules/quantum}$$

Breckenridge and Stewart (1986) have pointed out that 37 and 77 kcal/mol are required to remove one and two CO ligands, respectively, from $Cr(CO)_6$ and, therefore, one photon of the 355-nm laser has sufficient energy to remove two CO ligands with one photon. However, they have argued that the production of a meaningful amount of $Cr(CO)_4$ is not expected with single-photon excitation. For this reason, in the above example, each photon of the 355-nm laser is assumed to dissociate more than one molecule of $Cr(CO)_6$ by removing one CO ligand from each, rather than dissociating one molecule of $Cr(CO)_6$ by removing two CO ligands.

## Problem 3

Calculate the time required to decompose 0.1 mol of a substance with a 20-W excimer laser of wavelength 248 nm. Assume that the substance absorbs all the photons of the excimer laser, and that the quantum yield, which is defined in terms of the reactant molecules for this problem, is unity.

SOLUTION: The energy of one photon or quantum is given by

$$E_1 = \frac{hc}{\lambda} = \frac{(6.6256 \times 10^{-34}\text{ J s/quantum})(2.9979 \times 10^{10}\text{ cm/s})}{(248\text{ nm})(10^{-7}\text{ cm/nm})}$$
$$= 8.009 \times 10^{-19}\text{ J/quantum}$$

If $P$ denotes the laser power in watts, the number of photons emitted per second is given by

$$N_p = \frac{P}{E_1} = \frac{20\,\text{J}\,\text{s}}{8.009 \times 10^{-19}\,\text{J/quantum}} = 2.497 \times 10^{19}\,\text{quantum/s}$$

According to the definition of the quantum yield in terms of the reactant molecules, we have

$$\phi_A = \frac{\text{Molecules of the reactant decomposed per unit volume per unit time}}{\text{Quanta of light absorbed by the reactant per unit volume per unit time}}$$

Since the quantum yield is unity for this problem, each photon dissociates one molecule of the substance. Thus, the number of molecules decomposed per second is equal to $N_m = N_p \phi_A$, and the time required to decompose 0.1 mol of the substance is given by

$$\frac{0.1 A}{N_m} = \frac{0.1 A}{N_p \phi} = \frac{(0.1\,\text{mol})(6.03 \times 10^{23}\,\text{molecules/mol})}{(2.497 \times 10^{19}\,\text{quantum/s})(1\,\text{molecule/quantum})}$$

$$= 2414.8979\,\text{s} = 40.25\,\text{min}$$

**Problem 4**

Perlman and Rollefson (1941) found that the equilibrium constant for the reaction

$$I_2 \rightleftharpoons 2I$$

is $K_P = P_I^2/P_{I_2} = 0.1692$ atm at 1274 K. Determine the bond dissociation energy, $D_0(\text{I--I})$ for $I_2$. From Stull and Prophet (1971), we have the thermochemical data shown in Table 3.5, where the reference temperature $T_r = 298$ K.

SOLUTION: We will determine $\bar{G}_T^\circ$ at 1274 K from the data in Table 3.5 by linear interpolation. For I, we have

$$\bar{G}_T^\circ = \frac{-46.671 + 46.368}{1300 - 1200}(1274 - 1200) - 46.368$$

$$= -46.592\,\text{cal/mol K} \quad \text{at } T = 1274\,\text{K}$$

Table 3.5. Thermochemical Data [after Stull and Prophet (1971)]

| Species | Temperature (K) | $\bar{G}_T^\circ = (G_T^\circ - H_{298}^\circ)/T$ (cal mol$^{-1}$ K$^{-1}$) | $H_0^\circ - H_{298}^\circ$ (kcal mol$^{-1}$) |
|---|---|---|---|
| I | 0 |  | −1.481 |
|  | 1200 | −46.368 |  |
|  | 1300 | −46.671 |  |
| I$_2$ | 0 |  | −2.418 |
|  | 1200 | −68.010 |  |
|  | 1300 | −68.559 |  |

Similarly for I$_2$, we have

$$\bar{G}_T^\circ = \frac{-68.559 + 68.010}{1300 - 1200}(1274 - 1200) - 68.010$$

$$= -68.416 \text{ cal/mol K} \quad \text{at} \quad T = 1274 \text{ K}$$

Thus, $\Delta\bar{G}_{1274}^\circ = 2(\bar{G}_{1274}^0 \text{ for I}) - (\bar{G}_{1274}^0 \text{ for I}_2) = -2 \times 46.592 + 68.416 = -24.768$ cal/mol K. From Eq. (3.32), we have

$$\Delta H_{298}^\circ/1274 = -R\ln(0.1692) + 24.768$$

Noting that $R = 1.987165$ cal/mol K, we find that $\Delta H_{298}^\circ = 36.052$ kcal/mol. We also have

$$\Delta H_0^\circ - \Delta H_{298}^\circ = 2(H_0^\circ - H_{298}^\circ \text{ for I}) - (H_0^\circ - H_{298}^\circ \text{ for I}_2)$$

$$= -2 \times 1.481 + 2.418$$

$$= 0.544 \text{ kcal/mol}$$

Thus, the bond dissociation energy, $D_0(\text{I–I})$, is found to be

$$D_0(\text{I–I}) = \Delta H_0^\circ = \Delta H_{298}^\circ + (\Delta H_0^\circ - \Delta H_{298}^\circ) = 35.508 \text{ kcal/mol}$$

**Problem 5**

Okabe and Dibeler (1973) studied the photodissociation of $C_2HCN$, which occurs according to the reaction

$$C_2HCN + h\nu_f \rightarrow C_2H + CN^*(B^2\Sigma)$$

and found that the threshold wavelength for this photodissociation is 131.8 nm. Davis and Okabe (1968) have shown that the CN emission spectrum obtained during the transition from the $B^2\Sigma^+$ state to the $X^2\Sigma^+$ state has a strong peak at the 388-nm wavelength. Determine the bond dissociation energy of $C_2HCN$, that is, $D_0(C_2H-CN)$, from these spectral data.

SOLUTION: The minimum photon energy required to carry out the above photodissociation is given by

$$h\nu_f = hc/\lambda = \frac{(6.6256 \times 10^{-34}\,\text{J s/quantum})(2.9979 \times 10^{10}\,\text{cm/s})}{(131.8\,\text{nm})(10^{-7}\,\text{cm/nm})}$$

$$= 1.507 \times 10^{-18}\,\text{J/quantum} = 1.507 \times 10^{-18} \times 6.2418 \times 10^{18}\,\text{eV}$$

$$= 9.406\,\text{eV}$$

The electronic energy of CN* is given by

$$E^* = \frac{hc}{388 \times 10^{-7}\,\text{cm}} = 5.119 \times 10^{-19}\,\text{J/quantum}$$

$$= 5.119 \times 10^{-19} \times 6.2418 \times 10^{18}\,\text{eV} = 3.195\,\text{eV}$$

Thus,

$$D_0(C_2H-CN) = h\nu_f - E^* = 9.406 - 3.195\,\text{eV} = 6.211\,\text{eV}$$

## 3.9. REACTION STEPS, RATES, AND CORRELATION RULES IN PHOTODISSOCIATION

We have already discussed various photophysical and photochemical processes that we encounter when a photon is absorbed by a molecule. Three distinct situations are expected to arise, depending on whether the incident photon energy is less than, equal to, or more than the bond dissociation energy of the molecule. These processes, their rates of occurrence, and the spin correlation (conservation) rules will be discussed in this section.

### 3.9.1. Photophysical Processes in the Absence of Collisional Losses or Quenching of Excitation Energy

When the incident photon energy is less than the bond dissociation energy of the molecule, the molecule attains an electronically excited state, and then it decays to its ground state by undergoing various radiative and nonradiative transitions. If the pressure of the system is very low, there will be no loss of excitation energy from collisions among the molecules. Table 3.6 shows various photophysical processes (steps) and photophysical reactions involving the electronically excited states in the absence of any

Table 3.6. Photophysical Processes and Reactions Involving Electronically Excited States and the Rates of Disappearance of Various States

| Photophysical step or process | Photophysical reaction[a] | Lifetime or reciprocal of the phototophysical reaction rate constant (s) | Rate of disappearnce of various states[b] |
|---|---|---|---|
| Excitation (photon absorption) | $S_0 + h\nu \to S_1$ | $10^{-15}$ | $I'_A = k_{S_1}[S_0]$ |
| Internal conversion [IC(S)] | $S_i \to S_1 + \Delta$ | $10^{-11}$–$10^{-14}$ | $k_{IC(S)}[S_i]$ |
| Internal conversion [IC(T)] | $T_i \to T_1 + \Delta$ | $10^{-11}$–$10^{-14}$ | $k_{IC(T)}[T_i]$ |
| Intersystem crossing [ISC(S)] | $S_1 \to T_1 + \Delta$ | $10^{-8}$–$10^{-11}$ | $k_{ISC(S)}[S_1]$ |
| Internal conversion (IC) | $S_1 \to S_0 + \Delta$ | $10^{-7}$–$10^{-9}$ | $k_{IC}[S_1]$ |
| Fluorescent ($F$) emission | $S_1 \to S_0 + h\nu_F$ | $10^{-6}$–$10^{-11}$ | $k_F[S_1]$ |
| Intersystem crossing [ISC(T)] | $T_1 \to S_0 + \Delta$ | $10^{2}$–$10^{-3}$ | $k_{ISC(T)}[T_1]$ |
| Phosphorescent ($P$) emission | $T_1 \to S_0 + h\nu_P$ | $10^{2}$–$10^{-3}$ | $k_P[T_1]$ |

[a] In this column, $S_0, S_1$, and $T_1$ represent the ground, singlet, and triplet states, respectively; $\Delta$ indicates the heat energy, that is, the energy, that is redistributed as vibrational energy during the IC process; $\nu$, $\nu_F$, and $\nu_P$ represent the frequencies of the incident, fluorescent, and phosphorescent photons, respectively.

[b] In this column, the square brackets are used to represent the population of the molecules in the state that is denoted by the enclosed symbol. The rate constants are denoted by $k$'s with various subscripts; for example, in the expression $k_F[S_1]$, $[S_1]$ represents the population of the molecules in the state $S_1$ and $k_F$ is the rate constant for the fluorescence process. $I'_A$ is the number of photons absorbed per unit time per unit volume by the molecules that are in the state $S_0$.

quenching [Rabek (1982)]. This table also shows the lifetimes of the photophysical processes, and the rates of disappearance of various states in each of these processes. It should be noted that the photophysical reactions are considered to be first-order reactions, and, for this reason the reciprocal of the lifetime represents the reaction rate constant or, conversely, the lifetime represents the reciprocal of the reaction rate constant [Rabek (1982)].

Let us consider that some molecules have been excited to the singlet $S_1$ and triplet $T_1$ states by irradiation with a laser beam of photons, and that then the irradiation is stopped. We want to analyze the decay of these singlet and triplet states. To begin with, that is, at time $t = 0$, the populations of the molecules in the singlet and triplet states are given by $[S_1]_0$ and $[T_1]_0$, respectively. From Table 3.6, we find that the $S_1$ and $T_1$ states will decay due to three and two types of transitions, respectively. Thus, the rates of disappearance of the $S_1$ and $T_1$ states can be written [Rabek (1982, 1987)] as

$$-\frac{d[S_1]}{dt} = k_{IC}[S_1] + k_F[S_1] + k_{ISC(S)}[S_1] \tag{3.37}$$

and

$$-\frac{d[T_1]}{dt} = k_{ISC(T)}[T_1] + k_P[T_1] \tag{3.38}$$

The solutions of these equations are:

$$\ln \frac{[S_1]}{[S_1]_0} = -(k_{IC} + k_F + k_{ISC(S)})t \tag{3.39}$$

and

$$\ln \frac{[T_1]}{[T_1]_0} = -(k_{ISC(T)} + k_P)t \tag{3.40}$$

### 3.9.1.1. Unimolecular Lifetime

In unimolecular reactions, the lifetime of the molecules represents the reciprocal of the reaction rate constant. This lifetime refers to the time needed for the population of molecules in an excited state to decrease to $1/e$ of its initial value. From Eqs. (3.39 and 3.40), we obtain for the excited singlet state $S_1$,

$$\ln(1/e) = -(k_F + k_{IC} + k_{ISC(S)})\tau_F$$

or

$$\tau_F = \frac{1}{k_F + k_{IC} + k_{ISC(S)}} \qquad (3.41)$$

and for the excited triplet state $T_1$,

$$\ln(1/e) = -(k_P + k_{ISC(T)})\tau_P$$

or

$$\tau_P = \frac{1}{k_P + k_{ISC(T)}} \qquad (3.42)$$

If there is energy quenching occurring with rate constants $k_{Q(S)}$ and $k_{Q(T)}$ for the singlet and triplet states, respectively, the lifetimes are given by

$$\tau_F = \frac{1}{k_F + k_{IC} + k_{ISC(S)} + k_{Q(S)}} \qquad (3.43)$$

and

$$\tau_P = \frac{1}{k_P + k_{ISC(T)} + k_{Q(T)}} \qquad (3.44)$$

### 3.9.1.2. The True Radiative Lifetime

Expressions (3.41 and 3.42) represent the lifetimes of the molecules in the singlet and triplet states, respectively, which involve the decay time of the respective state due to both radiative and nonradiative transitions. The true radiative lifetime of an excited state, sometimes referred to as the inherent radiative lifetime, corresponds to the time required by a molecule of that excited state to decay to a lower level solely by the emission of radiation. The quantities $\tau_F$ and $\tau_P$ can be related to the true radiative lifetime in the following way:

At steady state, that is, when the rates of the production of $S_1$ and $T_1$ states are equal to the rates of disappearance of the $S_1$ and $T_1$ states, respectively, we can write [Wilkinson (1980), Parker (1968), Arnold et al. (1974)]

$$\frac{d[S_1]}{dt} = I'_A - \{k_F[S_1] + k_{IC}[S_1] + k_{ISC(S)}[S_1]\} = 0$$

or

$$\frac{[S_1]}{I'_A} = \frac{1}{k_F + k_{IC} + k_{ISC(S)}} = \tau_F \tag{3.45}$$

and

$$\frac{d[T_1]}{dt} = k_{ISC(S)}[S_1] - \{k_P[T_1] + k_{ISC(T)}[T_1]\} = 0$$

or

$$\frac{[T_1]}{I'_A} = \frac{1}{k_P + k_{ISC(T)}} \frac{k_{ISC(S)}[S_1]}{I'_A} = \tau_p \frac{k_{ISC(S)}[S_1]}{I'_A} \tag{3.46}$$

In the absence of nonradiative transitions, that is, when the excited states decay only by the emission of radiation, Eqs. (3.41 and 3.42) give the following expressions for the true radiative lifetimes:

$$\tau_F^0 = 1/k_F \tag{3.47}$$

and

$$\tau_P^0 = 1/k_P \tag{3.48}$$

Also, from the definition of quntum yield, we can write

$$\phi_F = \frac{\text{Rate of fluorescent emission per unit volume}}{\text{Rate of excitation (photon absorption) per unit volume}}$$

$$= \frac{k_F[S_1]}{I'_A} \tag{3.49}$$

$$\phi_{IC} = \frac{\text{Rate of internal conversion from } S_1 \text{ to } S_0 \text{ per unit volume}}{\text{Rate of excitation (photon absorption) per unit volume}}$$

$$= \frac{k_{IC}[S_1]}{I'_A} \qquad (3.50)$$

$$\phi_{ISC(S)} = \frac{\text{Rate of intersystem crossing from } S_1 \text{ to } T_1 \text{ per unit volume}}{\text{Rate of excitation (photon absorption) per unit volume}}$$

$$= \frac{k_{ISC(S)}[S_1]}{I'_A} \qquad (3.51)$$

$$\phi_P = \frac{\text{Rate of phosphorescent emission per unit volume}}{\text{Rate of excitation (photon absorption) per unit volume}}$$

$$= \frac{k_P[T_1]}{I'_A} \qquad (3.52)$$

Equations (3.45) and (3.49) through (3.51) yield

$$\phi_F + \phi_{IC} + \phi_{ISC(S)} = 1 \qquad (3.53)$$

From Eqs. (3.45), (3.47), and (3.49), we have

$$\tau_F = \phi_F \tau_F^0 \qquad (3.54)$$

and from Eqs. (3.46), (3.48), (3.51), and (3.53), we obtain

$$\phi_P = \frac{1}{\tau_P^0} \tau_P \phi_{ISC(S)}$$

or

$$\tau_P = \frac{\phi_P}{1 - \phi_F - \phi_{IC}} \tau_P^0 \qquad (3.55)$$

For some molecules, phosphorescence is the dominant mode of decay from the $T_1$ state, and intersystem crossing and fluorescence are the significant modes of decay from the $S_1$ state; that is, $\phi_{IC}$ is essentially zero [Wilkinson

(1980), Parker (1968), Arnold et al. (1974), Wayne (1988)]. Therefore, Eq. (3.55) can be approximately written as [Rabek (1982, 1987)]

$$\tau_P = \frac{\phi_P}{1 - \phi_F} \tau_P^0 \qquad (3.56)$$

Thus far, we have shown how the true radiative lifetime can be related to $\tau_F$ and $\tau_P$ through Eqs. (3.54) and (3.56). It can be obtained from the experimentally determined integrated molar absorption cross section (molar microscopic absorption cross section) of an absorption band by carrying out the integration over that portion of the band which is related to the transition of interest, as given by the following expression [Calvert and Pitts (1966), Rabek (1982, 1987), Wilkinson (1980), Arnold et al. (1974):

$$\tau^0 = \frac{3.5 \times 10^8}{\bar{v}_m^2 \int \bar{\sigma} dv} \text{ s/mol cm}$$

$$\approx \frac{3.5 \times 10^8}{\bar{v}_m^2 \bar{\sigma}_{max} \Delta v_{1/2}} \text{ s/mol cm} \qquad (3.57)$$

where $\bar{v}_m$ is the mean frequency of the absorption in wavenumbers (cm$^{-1}$), $\int \bar{\sigma} dv$ is the integration of the molar microscopic absorption band over the wavenumber $v$ that is related to the transition of interest (the units of $\bar{\sigma}$ and $v$ are cm$^2$/mol and cm$^{-1}$, respectively), $\bar{\sigma}_{max}$ is the maximum peak value of the $\bar{\sigma}$ band related to the transition of interest (cm$^2$/mol), and $\Delta v_{1/2}$ is the half-width of the $\bar{\sigma}$ band at the transition of interest (cm$^{-1}$).

For many polyatomic molecules, the $\Delta v_{1/2}$ is about 5000 cm$^{-1}$, and the absorption band $\bar{\sigma}$ is centered around $\bar{v}_m = 3 \times 10^4$ cm$^{-1}$ in the near-ultraviolet region. The true radiative lifetime $\tau^0$ of such molecules is approximately given by [Calvert and Pitts (1966), Rabek (1982, 1987), Wilkinson (1980), Arnold et al. (1974)]

$$\tau^0 = \frac{10^{-4}}{\bar{\sigma}_{max}} \text{ cm}^2 \text{ s/mol} \qquad (3.58)$$

## 3.9.2. Photophysical Processes in the Presence of a Nonabsorbing Quenching Gas

As noted earlier, if the energy of the absorbed photon is less than the bond dissociation energy, the molecule attains an electronically excited state which decays to lower energy levels either by the emission of light (luminescence), or by nonradiative transitions. For diatomic molecules, the quantum yield of fluorescence is always unity [Okabe (1978), Bixon and Jortner (1969)]. However, the excited molecules can lose their excitation energy through self-quenching, that is, by colliding with their ground state molecules, or through concentration-quenching, that is, by colliding with foreign gas molecules. The quenching phenomenon affects the quantum yield of fluorescence. The relevant photophysical processes and the rates of disappearance of various excited states are given in Table 3.7 [Rabek (1987), Okabe (1978)].

In Table 3.7, $I_A$, $v$, and $\Delta$ have the same meaning as in Table 3.6, and $v'$ is the frequency of the emitted light. $[A^*]$, $[A]$, and $[Q]$ represent the concentrations of the excited molecules, ground state molecules, and the foreign quenching gas molecules, respectively; $k_l$, $k_d$, $k_{sq}$, and $k_{cq}$ denote the rate constants for luminescence, deactivation, self-quenching, and concentration-quenching, respectively; $k_l$ and $k_d$ are in units of $s^{-1}$, and $k_{sq}$ and $k_{cq}$ are in units of (concentration)$^{-1}$.

At steady state, that is, when the rates of production and disappearance of $A^*$ are equal, we can write

$$\frac{d[A^*]}{dt} = I'_A - k_l[A^*] - k_d[A^*] - k_{sq}[A^*][A] - k_{cq}[A^*][Q] = 0$$

or

$$\frac{[A^*]}{I'_A} = \frac{1}{k_l + k_d + k_{sq}[A] + k_{cq}[Q]} \tag{3.59}$$

Table 3.7. Photophysical Processes with Quenching Phenomena

| Photophysical step or process | Photophysical reaction | Rate of disappearance of various states |
|---|---|---|
| Excitation (photon absorption) | $A + hv \rightarrow A^*$ | $I'_A$ |
| Light emission (luminescence) | $A^* \rightarrow A + hv'$ | $k_l[A^*]$ |
| Deactivation | $A^* \rightarrow A + \Delta$ | $k_d[A^*]$ |
| Self-quenching | $A^* + A \rightarrow 2A$ | $k_{sq}[A^*][A]$ |
| Concentration-quenching (quenching by a foreign gas molecule, $Q$) | $A^* + Q \rightarrow A + Q$ | $k_{cq}[A^*][Q]$ |

From the definition of quantum yield, we have the following expression: The quantum yield for luminescence from the excited molecule $A^*$ in the absence of any quenching is given by

$$\phi_0 = \frac{k_l[A^*]}{I'_A} = \frac{k_l}{k_l + k_d} \tag{3.60}$$

and the quantum yield for luminescence from $A^*$ in the presence of quenching processes is given by

$$\phi_q = \frac{k_l[A^*]}{I'_A} = \frac{k_l}{k_l + k_d + k_{sq}[A] + k_{cq}[Q]} \tag{3.61}$$

From Eqs. (3.60 and 3.61), we obtain

$$\frac{\phi_0}{\phi_q} = \frac{l_l + k_d + k_{sq}[A] + k_{cq}[Q]}{k_l + k_d} = 1 + \tau\{k_{sq}[A] + k_{cq}[Q]\} \tag{3.62}$$

where $\tau = 1/(k_l + k_d)$ is the lifetime of $A^*$ in the absence of quenching.

If the quenching due to only one quencher is important, say, $k_{cq}[Q]$ is dominant compared to $k_{sq}[A]$, Eq. (3.62) can be written as

$$\phi_0/\phi_q = 1 + \tau k_{cq}[Q] \tag{3.63}$$

which is known as the Stern–Volmer relation [Rabek (1987), Wilkinson (1980), Okabe (1978), Arnold et al. (1974), Callear and Smith (1963), Myers et al. (1966)]. Kinetic analyses of photochemical reactions involving the quenching of more than one excited state can be found in Dalton and Turro (1970) and Wagner (1971).

The Stern–Volmer plot, that is, the plot of $\phi_0/\phi_q$ vs. $[Q]$, is a straight line whose slope is equal to $\tau k_{cq}$. It should be noted from Eq. (3.63) that $[Q]_{1/2} = 1/\tau k_{cq}$, where $[Q]_{1/2}$ is the concentration of the quencher that reduces $\phi_0$ to one-half of its original value, that is, $\phi_0/\phi_q = 1/0.5$ when $[Q] = [Q]_{1/2}$. The Stern–Volmer plot can be used to analyze the quenching processes [Rabek (1982, 1987)]. The slope of the plot is zero when there is no quenching because the reaction rate is much faster than the quenching rate or the quenching is endothermic. The value of the slope changes with the concentration $[Q]$ when two different excited states, such as the singlet state and the triplet state, participate in the quenching reaction.

### 3.9.3. Photodissociation and Predissociation of Diatomic Molecules

When the energy of the absorbed photon is more than the bond dissociation energy of the molecule, there is dissociation of the molecule. The photochemical processes can be divided into two types of reactions [Rabek (1987)]:

1. Primary photochemical reaction: This type of reaction does not depend on the reactant temperature, and it occurs directly due to the absorbed photon involving electronically excited states.
2. Secondary photochemical reaction: This type of reaction is sometimes called the dark reaction, and it involves the reactions of ions, radicals, radical ions, and electrons that are generated during the primary photochemical reaction.

It should be noted that in thermal reactions, the ground state molecules jump to higher vibrational energy levels of the electronic ground state due to the collisions with other molecules or with the walls of the reaction chamber, and dissociation occurs when the vibrational energy exceeds the bond dissociation energy. The photochemical reactions, on the other hand, involve molecules that are in electronically excited states. The photodissociation of diatomic molecules can be classified into the following three groups:

1. When the photon energy is equal to the bond dissociation energy, the molecule dissociates according to the following reaction:

$$XY \rightarrow X + Y$$

and in this case, the photochemical reaction occurs in the electronic ground state.

2. When the photon energy is slightly higher than the bond dissociation energy, the following processes take place: (a) The molecule enters into a repulsive state and dissociates immediately, or (b) the molecule jumps into an electronically excited stable state from the ground state, and then either decays to the ground state by fluorescence, or enters into a repulsive state by nonradiative transitions and undergoes predissociation as given by

$$XY \rightarrow X + Y$$

It should be noted that predissociation refers to the dissociation that occurs sometime after the absorption of light, and it leads to the broadening of the vibrational or rotational lines. However, in order to detect the occurrence of predissociation by observing the line broadening in the absorption spectrum, the line broadening by predissociation should exceed the line broadening by the Doppler effect [Okabe (1978)]. The Doppler linewidth is usually about $0.05 \text{ cm}^{-1}$.

3. When the photon energy is much higher than the bond dissociation energy, the following processes take place: (a) the molecule enters into a repulsive state, and dissociates according to the following reaction:

$$XY \rightarrow X + Y$$

or (b) the molecule enters into a higher repulsive state, and dissociates and fluoresces according to the following reactions:

$$XY \rightarrow X + Y^*$$
$$Y^* \rightarrow Y + hv_1$$

or

$$XY \rightarrow X^* + Y^*$$
$$X^* \rightarrow X + hv_2$$
$$Y^* \rightarrow Y + hv_3$$

or (c) the molecule jumps into an electronically excited stable state, and dissociates and fluoresces according to reactions that are same as in (b).

### 3.9.4. Photodissociation Processes for Polyatomic Molecules

Polyatomic molecules have more chemical bonds than diatomic molecules and, consequently, there are several different mechanisms for the photodissociation of polyatomic molecules, depending on which bond breaks.

#### 3.9.4.1. Photodissociation of Triatomic Molecules

The photodissociation of triatomic molecules can be achieved in the following three ways:

RUPTURE OF THE WEAKER BOND: In the mechanism, only one bond, which is the weaker of the two bonds of the triatomic molecule, breaks according to the following reaction:

$$XYZ + h\nu \rightarrow X + YZ$$

where the $X-Y$ bond is weaker than the $Y-Z$ bond.

SIMULTANEOUS RUPTURE OF BOTH BONDS AND THE FORMATION OF A NEW BOND: The following reaction is encountered in this mechanism:

$$XYZ + h\nu \rightarrow Y + XZ$$

McNesby *et al.* (1962) observed this type of reaction in the photolysis of $H_2O$,

$$H_2O + h\nu \rightarrow H_2 + O$$

at 123.6 nm. This type of dissociation is common in the vacuum-ultraviolet region, and less common the near-ultraviolet region of the electromagnetic spectrum.

RUPTURE OF BOTH BONDS AND THE FORMATION OF THREE ATOMS: This mechanism involves the following reaction:

$$XYZ + h\nu \rightarrow X + Y + Z$$

This type of dissociation is usually not observed, although its occurrence is energetically possible in the vacuum-ultraviolet region.

### 3.9.4.2. Photodissociation of Polyatomic Molecules

The photodissociation mechanisms of polyatomic molecules can be divided into the following two categories:

RUPTURE OF THE WEAKEST BOND: In this mechanism, only one bond, which is the weakest bond of the polyatomic molecule, breaks according to the following reaction:

$$RX + h\nu \rightarrow R + X$$

where the $R$–$X$ bond is the weakest one. If the photon energy is higher than the bond dissociation energy, the excess energy, that is, the energy beyond that required to break the $R$–$X$ bond, appears mainly as the translational energies of the reaction products.

RUPTURE OF TWO BONDS AND THE FORMATION OF A NEW BOND: Mahan and Mandal (1962), and Okabe and McNesby (1961) observed this type of photodissociation for $CH_4$ and $C_2H_6$ molecules, respectively, in the vacuum-ultraviolet region as shown below:

$$CH_4 + h\nu \rightarrow CH_2 + H_2$$
$$C_2H_6 + h\nu \rightarrow C_2H_4 + H_2$$

DeGraff and Calvert (1967) observed the following dissociation reaction in the near-ultraviolet region:

$$H_2CO + h\nu \rightarrow CO + H_2$$

#### 3.9.4.3. Predissociation

Triatomic as well as other polyatomic molecules predissociate when they are raised to an electronically excited stable state which overlaps the dissociation continuum or the repulsive state. The predissociation mechanism of such molecules is similar to the photodissociation of diatomic molecules.

### 3.9.5. Correlation Rules for Photodissociation

As discussed above, there are several possibilities for the formation of different types of product molecules in the photodissociation of polyatomic molecules. The correlation rules limit the number of possibilities of the formation of products from electronically excited states. The absorption spectra of many polyatomic molecules are diffuse or continuous in the region of photochemical reaction; consequently, it is difficult to use such spectra to determine the symmetry species of the excited state undergoing dissociation. However, if the multiplicity of the excited state is known, the spin correlation or spin conversion rules can be applied between the excited reactant and the product molecules.

Wigner's spin rule [Wilkinson (1980), Okabe (1978)] states that if two molecules $A$ and $B$, which have spin angular momentum quantum numbers

PHOTOLYTIC LCVD                                                              179

$S_A$ and $S_B$, respectively, are combined, the total spin angular momentum quantum number $S_T$ is given by vector addition:

$$S_T = S_A + S_B, \quad S_A + S_B - 1, \quad S_A + S_B - 2, \ldots, |S_A - S_B|$$

where the right-hand side of the above expression includes all the values from $S_A + S_B$ to $|S_A - S_B|$ differing by unity.

Similarly, the total spin angular momentum quantum number $S'_T$ of the products $P$ and $Q$ which are formed with the spin angular momentum quantum numbers $S_P$ and $S_Q$, respectively, in the reaction

$$A + B \rightarrow P + Q$$

is given by

$$S'_T = S_P + S_Q, \quad S_P + S_Q - 1, \quad S_P + S_Q - 2, \ldots, |S_P - S_Q|$$

According to Wigner's spin rule, the above reaction is spin-allowed if $S_T$ and $S'_T$ have at least one value in common. This spin correlation rule can be applied to molecules containing light atoms, where the spin–orbit interaction is small, and may not hold good for molecules containing heavy atoms [Okabe (1978)].

EXAMPLE 1: Consider the following reaction:

$$R \rightarrow P + Q$$

(i) If $S_P = S_Q = \frac{1}{2}$, $S'_T = 1$ or 0 (triplet or singlet state). Thus, $S_T$ will be either 1 or 0, that is, the reactant $R$ will be either in the triplet or singlet state.
(ii) If $S_P = \frac{1}{2}$ and $S_Q = 1$, $S'_T = \frac{3}{2}$ or $\frac{1}{2}$ (quartet or doublet state). Thus, $S_T$ will be either $\frac{3}{2}$ or $\frac{1}{2}$, that is, the reactant $R$ will be either in the quartet or doublet state. These types of transitions are shown in Table 3.8.

EXAMPLE 2: Consider the following reaction:

$$A(T_1) + B(T) \rightarrow P(S_0) + Q(S)$$

for the reactants $S_A = S_B = 1$ and

$$S_T = S_A + S_B, S_A + S_B - 1, \ldots, |S_A - S_B| = 2, 1, 0$$

Table 3.8. Wigner's Spin Rule for the Reaction of the Type $R \leftrightarrow P + Q$

| R | P | Q |
|---|---|---|
| singlet | singlet + singlet | |
| doublet | singlet + doublet | |
| singlet, triplet | doublet + doublet | |
| doublet, quartet | doublet + triplet | |
| triplet, quintet | doublet + quartet | |
| singlet, triplet, quintet | triplet + triplet | |

Table 3.9. Application of Wigner's Spin Rule to Processes of the $A + B \leftrightarrow P + Q^a$

| A | B | P | Q |
|---|---|---|---|
| singlet + singlet | $\leftrightarrow$ | singlet + singlet | |
| singlet + singlet | $\xleftarrow{\text{forbidden}}\rightarrow$ | singlet + triplet | |
| singlet + singlet | $\leftrightarrow$ | triplet + triplet | |
| singlet + singlet | $\xleftarrow{\text{forbidden}}\rightarrow$ | singlet + multiplet* | |
| singlet + multiplet | $\leftrightarrow$ | triplet + multiplet | |
| singlet + triplet | $\leftrightarrow$ | triplet + singlet | |
| singlet + triplet | $\leftrightarrow$ | triplet + triplet | |

$^a \leftrightarrow$ Indicates spin-allowed process in either direction; $\xleftarrow{\text{forbidden}}\rightarrow$ indicates spin-forbidden process in either direction; multiplet* represents any spin multiplicity other than unity.

for the products, $S_P = S_Q = 0$, and $S'_T = 0$. Thus, this reaction is spin-allowed. Table 3.9 shows reactions between various states that satisfy Wigner's spin rule.

## 3.10. GAS AND SURFACE PHASE PROCESSES IN PHOTOLYTIC LCVD

Photolytic LCVD is a useful technique for high-resolution "direct writing" by depositing metal films. Usually, a substrate is enclosed in a few Torr of an organometallic gas, such as dimethyl gallium, and irradiated with a focused ultraviolet continuous wave laser beam. Metal films with submicron features have been deposited [Ehrilich et al. (1980a,b, 1981), Solanki et al. (1981), Deutsch et al. (1979)] in this way. The "direct writing" method is useful for large-scale integrated (LSI) fabrication, mask and wafer repair

[Deutsch et al. (1979)], etc. However, the writing rate is slow, approximately 20 Å/s in a 50-μm spot [Deutsch et al. (1979)]. In order to increase this rate, an understanding of the mechanism of the deposition process is necessary. The following two mechanisms [Wood et al. (1983), Deutsch et al. (1979)] are considered to be important in the deposition of films by the photolytic LCVD technique.

1. Adsorption and photodissociation at the substrate surface: The organometallic precursor molecules are adsorbed at the surface of the substrate, and these adsorbed molecules absorb photons and dissociate to form the metal film.
2. Gas phase photodissociation: In this mechanism, photodissociation occurs in the gas phase, and then the metal particles migrate to the substrate, where they stick to the surface of the substrate by absorption.

It should be noted that in the first mechanism, absorption occurs first and then photodissociation takes place. On the other hand, in the second mechanism photodissociation occurs first and then adsorption takes place. Ehrlich et al. (1981) have developed a mathematical model by incorporating these two mechanisms, that is, by assuming that the absorbed precursor molecules are photodissociated first to form critical nuclei, and then the photodissociation occurs in the gas phase to generate most of the depositing material. Wood et al. (1983) have shown how these two mechanisms can be differentiated by analyzing the dependence of the film deposition rate on laser spot size, and they have concluded that most of the film material is produced due to the gas phase photodissociation, and that this photodissociation reaction is the rate-limiting step. In any case, the adsorption phenomena, such as physisorption and chemisorption, are important in photolytic LCVD because it is such phenomena that cause the atoms of the film material to stick to the substrate surface.

### 3.10.1. Adsorption

When a gas (the term gas is used to denote gas or vapor) comes into contact with a solid or liquid surface, the concentration of gas molecules is found to be higher in the immediate vicinity of the surface than in the bulk of the gas phase, which is away from the interface of the gas and the solid or liquid. The process that leads to this increase in the concentration of gas molecules at the interface is known as adsorption. In the bulk of the solid or liquid, that is, at any point which is inside the solid or liquid and away from the surface, the atoms or molecules are subject to essentially balanced forces because they are surrounded by almost similar species in all directions. However, at the surface, the atoms experience unbalanced forces because they are not surrounded by similar species in all directions, and the balance of forces is partially restored by the adsorption of gas molecules or atoms. Since the surface atoms or molecules are energetically favorable to combine with other species to achieve a balance of forces, adsorption is a spontaneous process that occurs with the liberation of heat; that is, adsorption is an exothermic process. It should be noted that adsorption is different from absorption because absorption involves penetration of the gas into the bulk of the solid or liquid by diffusion. The term sorption is used when adsorption and absorption occur simultaneously.

Table 3.10. Differences between Physical and Chemical Adsorptions

| Physical adsorption | Chemical adsorption |
| --- | --- |
| 1. The heat of physical adsorption is of the same order of magnitude as the heat of liquefaction of the adsorbate and is rarely more than twice or three times as large. | 1. The heat of chemisorption is of the same order as that of the corresponding bulk chemical reaction. In some cases, however, exceptionally low heats of chemisorption are found. |
| 2. Physical adsorption, like condensation, is a general phenomenon and will occur with any gas–solid system provided only that the conditions of temperature and pressure are suitable. | 2. Chemisorption will take place only if the gas is capable of forming a chemical bond with the surface atoms. |
| 3. A physically adsorbed layer may be removed by reducing the pressure at the temperature at which adsorption took place, although the process may be slow on account of diffusion effects. | 3. The removal of a chemisorbed layer often requires much more rugged conditions, especially on metal surfaces where very high temperatures or positive ion bombardment are needed. |
| 4. Under suitable conditions of temperature and pressure, physically adsorbed layers several molecular diameters in thickness are frequently found. | 4. Chemisorption is complete once a monomolecular layer is built up, although physical adsorption may occur on the top of the chemisorbed monolayer. |
| 5. Since physical adsorption is related to the process of liquefaction, it only occurs to an appreciable extent at the pressures and temperatures close to those required for liquefaction. | 5. Chemisorption often proceeds at much lower pressures and much higher temperatures. |
| 6. Physical adsorption is instantaneous. | 6. Chemisorption may be instantaneous but there are many systems where chemisorption involves an activation energy. |

The adsorption of reactant molecules at the surface of a solid was first pointed out by Faraday (1834), who studied the adsorption of gases at surfaces and its effects on surface-catalyzed reactions. Adsorption phenomena can be classified as follows:

PHYSISORPTION: In this type of adsorption, the gas molecules are attached to the substrate surface by the weak molecular attraction forces, which are physical in nature and similar to the van der Waals forces. For this reason, this type of adsorption is also known as physical adsorption or van der Waals adsorption. Physisorption can be considered to be similar to the condensation of a vapor to form a liquid. The heat evolved during physisorption, which is from 0 to 20 kJ/mol [Laidler (1987)], is of the same order of magnitude as that of liquefaction. When the physically adsorbed layer is many molecular diameters thick, it behaves like a two-dimensional liquid in many respects [Young and Crowell (1962)].

CHEMISORPTION: This type of adsorption was first studied by Langmuir (1916), who suggested that the adsorbed molecules are bound to the surface by covalent forces which are similar to the force between atoms in molecules. The heat evolved in this type of adsorption is about 300–500 kJ/mol [Laidler (1987)], which is similar to the heat evolved during chemical bonding and, for this reason, this process is called chemical adsorption or chemisorption. Chemisorption essentially involves the transfer of electrons between the solid (or adsorbent) and the gas (or adsorbate), which leads to the formation of a chemical compound between the adsorbate and the outermost layer of the adsorbent atoms [Young and Crowell (1962)].

Langmuir (1916) suggested that a single layer of adsorbed molecules is formed on the substrate surface by chemisorption, and that this layer saturates the surface by covering all of the available sites. Any additional adsorption that occurs on this layer is due to physisorption. Taylor (1931) pointed out that chemisorption is slower than physisorption because the former process involves an activation energy as do chemical reactions, whereas the latter does not. Adsorption depends on the smoothness of the surface. Since solid surfaces are not fully smooth at the atomic level, adsorption occurs more strongly at some sites than at other sites [Taylor (1925), Constable (1925)]. Similarly, chemical reactions occur preferentially at some sites on the surface, and these sites are called active centers [Taylor (1925)].

In view of the above discussion, the differences between physical and chemical adsorptions can be listed as in Table 3.10. Various aspects of the adsorption processes can be found in Young and Crowell (1962), Ross and Olivier (1964), Dash (1975), Haywood and Trapnell (1964), Miller (1949), Adamson (1982), and Somerjai (1981).

## 3.10.2. Langmuir (Ideal) Adsorption Isotherm

An adsorption isotherm relates the quantity of gas molecules adsorbed and the pressure or concentration of those molecules in the bulk of the gas phase at a constant temperature. Langmuir (1916, 1918) obtained a simple expression for such a relationship under the following assumptions:

1. The adsorbed atoms or molecules occupy single sites on the substrate surface.
2. The adsorbed molecules do not dissociate.
3. The adsorbed molecules do not interact with one another.
4. The substrate surface is perfectly smooth.

These assumptions are difficult to fulfill in practical situations, so the Langmuir adsorption isotherm can be considered as an ideal. To derive an expression for this isotherm, let us consider a surface $S$ containing active adsorption sites of concentration $S_0$, which adsorbs the species $X$ according to the following reaction:

$$X + \overset{|}{\underset{|}{S}}  \rightleftharpoons \overset{X}{\underset{|}{\overset{|}{S}}}$$

or

$$X + S \rightleftharpoons SX \qquad (3.64)$$

Under equilibrium conditions, the equilibrium constant $K_c$ is given by

$$K_c = \frac{[SX]}{[X][S]} \qquad (3.65)$$

where $[SX]$, $[X]$, and $[S]$ denote the concentrations of the occupied adsorption sites, adsorbate species $X$, and unoccupied adsorption sites, respectively. Thus, Eq. (3.65) can be written as

$$K_c = \frac{[SX]}{[X]\{S_0 - [SX]\}} = \frac{[SX]/S_0}{[X]\{1 - [SX]/S_0\}} = \frac{\theta}{[X](1 - \theta)} \qquad (3.66)$$

Since $[X] \propto p_X$, Eq. (3.66) takes the following form in terms of the partial pressure:

$$K_p = \frac{\theta}{p_X(1 - \theta)} \qquad (3.67)$$

Equation (3.66) can be written in the following two forms:

$$\theta = \frac{K_c[X]}{1 + K_c[X]} \qquad (3.68)$$

and

$$1 - \theta = \frac{1}{1 + K_c[X]} \qquad (3.69)$$

Equation (3.68) is the usual form of the Langmuir adsorption isotherm. For low concentrations, $K_c[X] \ll 1$, and Eq. (3.68) yields $\theta \approx K_c[X]$, that is, $\theta \propto [X]$. At high concentrations, $K_c[X] \gg 1$, and Eq. (3.69) yields $1 - \theta \approx 1/K_c[X]$.

### 3.10.3. Adsorption with Dissociation

Consider that the above surface $S$ now adsorbs a species $X_2$ which dissociates into two species after adsorption according to the following reaction:

$$X_2 + \underset{|}{\overset{|}{S}}-\underset{|}{\overset{|}{S}} \rightleftharpoons \underset{|}{\overset{X}{S}}-\underset{|}{\overset{X}{S}}$$

or

$$X_2 + 2S \rightleftharpoons 2SX \qquad (3.70)$$

Under equilibrium conditions, we have

$$K_c = \frac{[SX]^2}{[X_2][S]^2} = \frac{[SX]^2}{[X_2]\{S_0 - [SX]\}^2} = \frac{\{[SX]/S_0\}^2}{[X_2]\{1 - [SX]/S_0\}^2} = \frac{\theta^2}{[X_2](1 - \theta)^2}$$

which leads to

$$\frac{\theta}{1 - \theta} = \sqrt{K_c[X_2]} \qquad (3.71)$$

Equation (3.71) can be written as

$$\theta = \frac{\sqrt{K_c[X_2]}}{1 + \sqrt{K_c[X_2]}} \qquad (3.72)$$

and

$$1 - \theta = \frac{1}{1 + \sqrt{K_c[X_2]}} \qquad (3.73)$$

For low concentrations, $\sqrt{K_c[X_2]} \ll 1$, and Eq. (3.72) yields $\theta \approx \sqrt{K_c[X_2]}$, that is, $\theta \propto [X_2]^{1/2}$. At high concentrations, $\sqrt{K_c[X_2]} \gg 1$, and Eq. (3.73) yields $1 - \theta \approx 1/\sqrt{K_c[X_2]}$.

### 3.10.4. Adsorption of Two Types of Species

Consider that the above surface $S$ now adsorbs two types of species $X$ and $Y$ according to the following reactions:

$$X + {-}\overset{|}{S}{-} \rightleftharpoons {-}\overset{\overset{X}{|}}{S}{-}$$

or

$$X + S \rightleftharpoons SX \qquad (3.74)$$

and

$$Y + {-}\overset{|}{S}{-} \rightleftharpoons {-}\overset{\overset{Y}{|}}{S}{-}$$

or

$$Y + S \rightleftharpoons SY \qquad (3.75)$$

Under equilibrium conditions, we have

$$K_X = \frac{[SX]}{[X][S]} = \frac{[SX]}{[X]\{S_0 - [SX] - [SY]\}} = \frac{[SX]/S_0}{[X]\{1 - [SX]/S_0 - [SY]/S_0\}}$$

$$= \frac{\theta_X}{[X](1 - \theta_X - \theta_Y)} \qquad (3.76)$$

and

$$K_Y = \frac{[SY]}{[Y][S]} = \frac{[SY]}{[Y]\{S_0 - [SX] - [SY]\}} = \frac{\theta_Y}{[Y](1 - \theta_X - \theta_Y)} \quad (3.77)$$

Equations (3.76) and (3.77) can be solved to obtain the following isotherms for the species $X$ and $Y$:

$$\theta_X = \frac{K_X[X]}{1 + K_X[X] + K_Y[Y]} \quad (3.78)$$

$$\theta_Y = \frac{K_Y[Y]}{1 + K_X[X] + K_Y[Y]} \quad (3.79)$$

### 3.10.5. Nonideal Adsorption

In many practical situations, the adsorption isotherms are found to deviate from the Langmuir (ideal) adsorption isotherm because the substrate surface is not perfectly smooth, and the adsorbed species interact with one another. Also, deviations can occur because there is adsorption in more than one layer. When a molecule or an atom is attached to a surface, it may enhance or inhibit the adsorption at a neighboring site. To account for such nonideal behaviors, Freundlich, and later Slugin and Frumkin, proposed the following relationships [Laidler (1987)]:

$$x = kc^n \quad (3.80)$$

and

$$\theta f = \ln ac \quad (3.81)$$

respectively, where $k$, $n$, $f$, and $a$ are empirical constants, $x$ is the amount of the species adsorbed, and $c$ is the concentration of the adsorbate species in the bulk of the gas phase.

Expression (3.80) indicates that the amount of material adsorbed will increase indefinitely as the concentration of the adsorbate increases; that is, the surface will never be saturated with the adsorbed species. Equations (3.80) and (3.81) have both been derived theoretically by considering the distribution of adsorption sites of different energies [Laidler (1987)].

## 3.11. CONCENTRATION AND TEMPERATURE MEASUREMENTS

To understand the mechanisms of adsorption and the deposition and growth of a thin film on a substrate, the temperature and the concentration of various species involved in the LCVD process should be known. Mathematical modeling of the LCVD process allows us to determine the concentration, temperature, film shape and thickness, etc. Experimental techniques, such as laser spectroscopy, can be used to determine the concentration and temperature in both photolytic and pyrolytic LCVD processes, in order to establish the mechanism of the dissociation of the precursor molecules and to obtain various chemical kinetics data, which are very important for developing a reliable mathematical model. In this section, we will briefly discuss a laser spectroscopic technique for concentration and temperature measurements.

### 3.11.1. Laser-Induced Fluorescence

*L*aser-*i*nduced *f*luorescence (LIF) is one of the laser spectroscopic techniques which can be used for concentration and temperature measurements. In this technique, the atoms or molecules of interest are irradiated with a laser beam whose frequency matches that of an absorption line of the atoms or molecules that are being studied. Due to the absorption of a proton, the atom or molecule rises to an excited state, and then decays to a lower state by the fluorescence process. The fluorescent radiation can be detected by standard spectroscopic techniques [Demtroder (1982), Crosley (1982), Lakowicz (1983), Wolfbeis (1993)], and important information about the system can be obtained from the relative intensities of various fluorescence lines. LIF experiments can be carried out in two ways [Crosley (1982)]:

EXCITATION SCAN: In this method, a tunable (dye) laser is used to scan the absorption spectrum of an atom or a molecule of interest. Fluorescence occurs when the laser frequency coincides with that of an absorption line of the atom or molecule that is being investigated. In this way, the excitation scan produces the absorption spectrum of the atom or molecule of interest.

FLUORESCENCE SCAN: In this method, the frequency of the laser is kept fixed at a particular value, which excites the atoms or molecules of interest by inducing a single transition, and the resulting fluorescence is scanned using a monochromator. Only the pumped level radiates if the excited species is isolated. However, other energy levels can become populated as a result of the collisional energy transfer in the presence of the colliding species.

Donnelly and Karlicek (1982) have studied the fluorescence from the fragments of InCl, GaClPH$_3$, and AsH$_3$. Pasternack *et al.* (1982) have used LIF to monitor the time evolution of photofragments during laser photolysis of various hydrocarbons, and determined the reaction rates from the observed data.

### 3.11.2. Advantages of LIF

LIF has several advantages compared to other diagnostic probes as listed below:

1. It provides a positive signal on a null background, whereas a small reduction in the transmitted signal is usually observed in the absorption spectroscopy.
2. It is nonintrusive and highly sensitive.
3. LIF spectra are relatively simple, which makes species identification less difficult. Also, considerable selectivity is possible because the lower and upper states are well-defined.
4. High temporal and spatial resolution can be achieved in the LIF technique.
5. Due to the availability of high-intensity lasers, high population densities can be obtained in excited states to produce strong fluorescent signals.

### 3.11.3. Basic Theory of LIF

The fundamentals of fluorescence spectroscopy and its various applications can be found in Lakowicz (1983) and Wolfbeis (1993), and the basic phenomena involved in LIF processes have been discussed by Eckbreth (1988), Daily (1977), and Lucht and Laurendeau (1979). Various transition phenomena and the associated Einstein coefficients (rate constants) and the rate expressions of the absorption, the spontaneous and stimulated emissions of the radiation, and quenching are given in Table 3.11. In the absorption and the stimulated emission rate expressions, the spectral radiative energy density $W_\nu$ is used instead of the total radiative energy density, which represents the total amount of the radiative energy per unit volume

Table 3.11. Various Transition Phenomena and the Associated Einstein Coefficients and Rate Expressions (see Appendix A).[a]

| Transition process | Energy level diagram | Rate constant[b] | | Rates of disappearnce from various energy levels |
| --- | --- | --- | --- | --- |
| | | Einstein coefficient | Quenching rate constant | |
| Absorption | $\rightarrow \stackrel{u}{\underset{l}{\uparrow}}$<br>$N_l + h\nu \rightarrow N_u$ | $B_{lu}$ | | $B_{lu} W_\nu N_l$ |
| Spontaneous emission (fluorescence) | $u$<br>$l$<br>$N_u \rightarrow N_l + h\nu$ | $A_{ul}$ | | $A_{ul} N_u$ |
| Stimulated emission[c] | $u$<br>$l$<br>$N_u + h\nu \rightarrow N_l + 2h\nu$ | $B_{ul}$ | | $B_{ul} W_\nu N_u$ |
| Collisonal quenching[d] | $u$<br>$l$<br>$N_u + N_{q_i} \rightarrow N_l + N_{q_j}$ | | $Q_{ul} = k_q N_{qi}$ | $k_q N_u N_{qi}$<br>$= Q_{ul} N_u$ |

[a]In this table, $l$ and $u$ refer to the lower and upper energy levels, respectively.
[b]It should be noted that the quenching rate constant $Q_{ul}$ depends on the composition and temperature [Eckbreth, (1988), p. 309], $A_{ul}$ and $Q_{ul}$ are in units of s$^{-1}$, and $B_{lu}$ and $B_{ul}$ are in units of m$^3$ Hz J$^{-1}$ s$^{-1}$.
[c]In this process, the emitted photons have the same characteristics, such as frequency, polarization, etc., as the incident photons that cause this transition.
[d]Loss of energy due to collisions between the excited state and other molecules.

and is given by $\int W_\nu d\nu$, because the laser spectrum is sometimes broader than the atomic or molecular absorption linewidth. The permissible values of the lower and upper energy states $l$ and $u$, respectively, that is, the allowed transitions for fluorescence, can be determined by analyzing the Einstein coefficient $A_{ul}$, as discussed below.

### 3.11.3.1. Selection Rules for Fluorescence

From wave mechanics, the expression [Shimoda (1983)] for the Einstein probability of spontaneous emission (fluorescence), that is, the Einstein coefficient $A_{ul}$ can be written as [Calvert and Pitts (1966), pp. 51, 170–173]

$$A_{ul} = \left(\frac{16\pi^3 v^3}{3\varepsilon_0 hc^3}\right) g_l D_{ul} \qquad (3.82)$$

where $v = (E_u - E_l)/h$, and $E_u$ and $E_l$ are the energies of an atom or molecule in the upper and lower energy levels, respectively. The quantity $g_l$ is the degeneracy of the $l$th level, that is, the number of possible states in the $l$th level that have the same energy even though their wavefunctions are not the same. The dipole strength $D_{ul}$ is given by

$$D_{ul} = |\mathbf{m}|^2 = \left|\int \psi_u^* \mathbf{M} \psi_l d\tau\right|^2 \qquad (3.83)$$

where the integration element $d\tau$ includes all nuclear and electronic coordinations; $\mathbf{m}$, the value of the integral, is called the dipole (or transition) moment integral; $\psi_l$ and $\psi_u^*$ are the total wavefunction and the complex conjugate of the total wavefunction of the lower and upper energy levels, respectively; and $\mathbf{M}$ is known as the dipole moment vector or dipole moment operator, and given by

$$\mathbf{M} = \sum_{i=1}^{n} e r_i \qquad (3.84)$$

Here $r_i$ represents the distance of the $i$th electron from the nucleus, and the summation is carried out by considering the electrons ($n$ is the total number of electrons) involved in the transition process. It should be noted that the transition moment integral $\mathbf{m}$ is a measure of the charge displacement as the transition occurs [Calvert and Pitts (1966), p. 53].

A transition between the two levels $u$ and $l$ is allowed only if the matrix element $D_{ul}$ given by Eq. (3.83) is not zero [Herzberg (1966), p. 128]. Although Mulliken and Rieke (1941) have evaluated the transition moment integrals for a series of simple molecules, the accurate determination of $D_{ul}$ for many molecules, even for moderately complex molecules, is a very complicated and difficult task. However, by using the Born–Oppenheimer approximation, the wavefunction $\psi$ may be separated into its electronic, vibrational, and rotational components [Okabe (1978), pp. 51–56] such that

$$\psi = \psi_{\text{el}} \psi_{\text{vib}} \psi_{\text{rot}}$$

With this simplification, Eq. (3.82) can be written as

$$A_{ul} = \left(\frac{16\pi^3 v^3}{3\varepsilon_0 hc^3}\right) g_l M_{el}^2 M_{vib}^2 M_{rot}^2 \tag{3.85}$$

where $M_{el}$, $M_{vib}$, and $M_{rot}$ are the electronic matrix element, the Franck–Condon factor, and the Honl–London factor, respectively. Fluorescence occurs when all of these three probability factors are not zero.

The Honl–London factor is nonzero when the change in the total angular momentum $J$ due to the transition of the species from one energy level to another is 0 or $\pm 1$, that is

$$\Delta J = -1, \; 0, \; 1 \tag{3.86}$$

Selection rule (3.86) indicates that a fluorescence spectrum can have at most three lines for a single excited state.

For atoms and diatomic molecules [Herzberg (1950), p. 240] a transition is allowed if selection rule (3.86) is satisfied under the condition that a state with $J = 0$ cannot decay to another $J = 0$ state. Due to the approximations used to determine the wavefunction for evaluating the transition moment integrals, the theoretically predicted forbidden transitions sometimes do occur, especially with large polyatomic molecules [Herzberg (1966)]. For this reason, it is difficult to distinguish between allowed and forbidden transitions in large polyatomic molecules. An allowed transition is usually characterized by an intense absorption band.

### 3.11.3.2. Determination of the Population Densities

We will now express the population densities $N_l$ and $N_u$ in term of the Einstein coefficients in the absence of quenching. The expressions for $N_l$ and $N_u$ in the presence of quenching are obtained in Section (3.11.5).

The rate of absorption of radiation (see Table 3.11), that is, the rate of disappearance of atoms from state $l$ to $u$ is

$$-\frac{dN_l}{dt} = B_{lu} W_v N_l \tag{3.86}$$

PHOTOLYTIC LCVD

Also, the total rate of spontaneous and stimulated emissions of radiation (see Table 3.11), that is, the rate of disappearance of atoms from state $u$ to state $l$ is

$$-\frac{dN_u}{dt} = A_{ul}N_u + B_{ul}W_v N_u \tag{3.87}$$

At thermal equilibrium, the two rates are equal, and Eqs. (3.86) and (3.87) lead to the following relationship:

$$\frac{N_l}{N_u} = \frac{A_{ul} + B_{ul}W_v}{B_{ul}W_v} \tag{3.88}$$

Also, in thermal equilibrium at temperature $T$, we can use the Boltzmann distribution to write [Eastham (1989)]

$$\frac{N_u}{N_l} = \frac{g_u}{g_l}\exp\left(-\frac{E_u - E_l}{k_B T}\right) = \frac{g_u}{g_l}\exp\left(-\frac{h\nu}{k_B T}\right) \tag{3.89}$$

Equations (3.88) and (3.89) can be solved to express $N_u$ and $N_l$ in terms of the Einstein coefficients and $W_v$. Multiplying Eqs. (3.88) and (3.89), we obtain the following expression for $W_v$:

$$W_v = \frac{g_u A_{ul}}{g_l B_{lu}\exp(h\nu/k_B T) - g_u B_{ul}} \tag{3.90}$$

The expression for $A_{ul}$ is given by Eq. (3.82), and $B_{ul}$ is given [Shimoda (1983)] by

$$B_{ul} = \frac{2\pi^2}{3\varepsilon_0 h^2} g_l D_{ul} \tag{3.91}$$

where $D_{ul}$ is defined by Eq. (3.83). From Eqs. (3.82) and (3.91), we can write

$$A_{ul} = \frac{8\pi h\nu^3}{c^3} B_{ul} \tag{3.92}$$

To satisfy the condition that $W_v \to \infty$ as $T \to \infty$, we find [Einstein (1916)] from Eq. (3.90) that

$$g_l B_{lu} = g_u B_{ul} \tag{3.93}$$

From Eqs. (3.90), (3.92), and (3.93), we can write

$$W_v = \frac{8\pi v^2}{c^3} \frac{hv}{\exp(hv/k_B T) - 1} \tag{3.94}$$

which is the same as Planck's radiation law. It should be noted that we have to consider $B_{ul} \neq 0$, that is, assume that the stimulated emission exists, in order to show that Eq. (3.90) is identical to Planck's law.

Regarding the population density, it can be seen from Eq. (3.89) that under conditions of thermal equilibrium $g_l N_u$ is always less than $g_u N_l$ irrespective of the temperature of the system. Due to the difference in state populations, the loss of radiation as a result of absorption exceeds the production of radiation by stimulated emission, that is, $N_l B_{lu} \gg N_u B_{ul}$. As the production of laser light requires that $g_l N_u$ be larger than $g_u N_l$, that is, there must be population inversion. Eq. (3.89) suggests that laser cannot be generated under conditions of thermal equilibrium. For this reason, external sources, such as electrical discharges, chemical reactions, or visible radiation pumping, are used to create nonequilibrium systems to generate laser.

### 3.11.4. Linewidth Broadening Phenomena

Although the absorption and emission of radiation involve transitions of an atom or a molecule between quantum-mechanically well-defined states, the transitions are not infinitely instantaneous, that is, not fully monochromatic. Due to the collisions, thermal motion, and finite lifetime of excited species [Lengyel (1971)], the transitions occur over a finite frequency range that is referred to as the linewidth. The magnitude of the linewidth or line broadening is usually represented by the half-intensity width, or half-breadth of absorption or emission, which refers to the width of the line at the intensity that is one-half of the maximum intensity of the absorption or emission line of interest, and is usually called the *full width at half-maximum* or *half-height* (FWHM or FWHH). While observing fluorescence, the fluorescent radiation linewidth may be broadened as a result of several different phenomena, described below [Calvert and Pitts (1966), Eckbreth (1988), Okabe (1978)] and, consequently, the observed spectrum has to be corrected to obtain the true fluorescence spectrum.

## 3.11.4.1. Natural Linewidth

Owing to the finite lifetime of the excited state, a transition occurs from an upper to a lower state over a finite frequency range that is known as the natural linewidth. According to Heisenberg's uncertainty principle, there are uncertainties $\Delta E$ and $\Delta t$ in the energy and lifetime of an electronic state, respectively, which are related by

$$\Delta E \Delta t = h/2\pi \quad (3.95)$$

setting $\Delta E = h\Delta v_n$ and $\Delta t = \tau_m$ in Eq. (3.95), where $\tau_m$ is the mean radiative lifetime of the excited state, the expression for the natural width of a spectral line can be written as

$$\Delta v_N = 1/2\pi\tau_m \quad (3.96)$$

Since $v = c/\lambda$, and $v = c\bar{v}$, where $\bar{v}$ is the wavenumber, we have $\Delta v = -c\Delta\lambda/\lambda^2$ and $\Delta v = c\Delta\bar{v}$. Therefore, in terms of wavenumber and wavelength, the natural linewidth is given by

$$\Delta\bar{v}_N = 1/2\pi c\tau_m \quad \text{and} \quad -\Delta\lambda_N = \lambda_c^2/2\pi c\tau_m$$

where $\lambda_c$ is the wavelength at the line center. The natural broadening, $\Delta v_N$, $\Delta\bar{v}_N$, or $-\Delta\lambda_N$, of a spectral line is an intrinsic property of the atom, does not depend on the external effects, and increases as the radiative lifetime decreases. The natural linewidth is usually of the order of 0.001 cm$^{-1}$.

## 3.11.4.2. Doppler Broadening

Doppler broadening is due to the thermal motion of the radiation-emitting atom. When an electronically excited atom, which emits radiation, moves with a velocity $v$ at an angle $\theta$ between the direction of motion of the atom of interest and the line of sight of the observer, the shift in frequency of the emitted radiation is given by [White (1934)]

$$\Delta v' = v_0 \frac{v \cos\theta}{c} \quad (3.97)$$

where $v_0$ is the central frequency (s$^{-1}$), which refers to the frequency at the line center of the emitted radiation when the atom is at rest, that is, $v = 0$.

Assuming that the Maxwell–Boltzmann velocity distribution holds good for the atom of interest, the Doppler broadening of an absorption or emission line, $\Delta v_D$, is given by

$$\Delta v_D = 2\sqrt{\ln 2}\,\frac{v_0}{c}\sqrt{\frac{2RT}{M}} = 1.665\,\frac{v_0}{c}\sqrt{\frac{2RT}{M}} \qquad (3.98)$$

Taking the universal gas constant $R = 8.314 \times 10^7$ ergs/K mol, the speed of light $c = 2.9979 \times 10^{10}$ cm/s, and expressing the temperature $T$ and the atomic weight $M$ in units of K and g/mol, respectively, Eq. (3.98) can be written as

$$\Delta v_D = 7.162 \times <10^{-7}\,v_0\sqrt{T/M} \qquad (3.99)$$

Doppler broadening can be written in terms of the wavenumber and the wavelength, as

$$\Delta \bar{v}_D = 7.162 \times 10^{-7}\,\bar{v}_0\sqrt{T/M} \quad \text{and} \quad -\Delta \lambda_D = 7.162 \times 10^{-7}\,\lambda_0\sqrt{T/M}$$

where $\bar{v}_0$ and $\lambda_0$ are the wavenumber and wavelength, respectively, at the line center when $v = 0$.

Since Doppler broadening depends on temperature, the temperature of a system can be determined from Eq. (3.99) by measuring the width of a line that is broadened mainly due to the Doppler effect. At low pressures, Doppler broadening is found to be dominant over all other line broadening phenomena. On the other hand, Doppler broadening can be eliminated in the observed spectrum by observing the atom of interest at a right angle to its motion, so that there is no motion of the atom along the line of sight, that is $\theta = \pi/2$, which leads to $\cos\theta = 0$ and, consequently, $\Delta v_D = \Delta \bar{v}_D = \Delta \lambda_D = 0$. The Doppler linewidth is usually of the order of $0.05\,\text{cm}^{-1}$ at 300 K. Neglecting the natural broadening, and considering only the Doppler broadening, the absorption coefficient $\alpha_{\bar{v}}$ per unit wavenumber interval can be written as [Mitchell and Zemansky (1961), p. 99]

$$\alpha_{\bar{v}} = \alpha_0 \exp\left[-\frac{2(\bar{v}-\bar{v}_0)}{\Delta \bar{v}_D}\sqrt{\ln 2}\right] \qquad (3.100)$$

and the total absorption coefficient $\alpha$ is given by

$$\alpha = \int_0^\infty \alpha_{\bar{v}}\,d\bar{v} = \frac{1}{2}\sqrt{\frac{\pi}{\ln 2}}\,\alpha_0 \Delta \bar{v}_D \qquad (3.101)$$

where $\alpha_0$ is the maximum absorption coefficient at the central wavenumber $\bar{v}_0$.

### 3.11.4.3. Instrumental Broadening

This effect arises due to the finite resolution of the instrument, such as the spectrograph, spectrometer, etc., that is used to observe the spectrum [Okabe (1978), pp. 29 and 60]. The resolution $R_I$ of such an instrument is given by

$$R_I = \lambda/|\Delta\lambda_I|$$

or

$$\Delta\bar{v}_I = 1/\lambda R_I = \bar{v}/R_I \qquad (3.102)$$

where $\Delta\lambda_I$ and $\Delta\bar{v}_I$ are the magnitudes of the smallest changes in the wavelength and wavenumber, respectively, that the instrument can detect. With a high-resolution instrument, say, $R_I = 5 \times 10^5$, the instrumental broadening for $\lambda = 0.2\,\mu\text{m}$, which is in the near-ultraviolet region, can be shown to be $\Delta\bar{v}_I = 0.1\,\text{cm}^{-1}$, which is larger than the typical Doppler linewidth at 300 K. Since Eq. (3.102) indicates that $\Delta\bar{v}_I$ decreases as $R_I$ increases, the Doppler broadening becomes comparable with the instrumental broadening when the resolution of the instrument is very high.

### 3.11.4.4. Pressure Broadening

At low pressures, usually below 10 mTorr in [Okabe, (1978), p. 29] the linewidth is broadened mainly due to the natural and Doppler broadening phenomena. At high pressures, the line is further broadened due to the collisions of the electronically excited species with other species. Also, the line shape becomes asymmetric due to the formation of quasi-molecules at high pressures. Collisional broadening phenomena are classified as Stark, Lorentz, or Holtsmark broadening, depending on the nature of the species involved in the collision process as discussed below.

*3.11.4.4a. Stark Broadening.* Stark broadening of the absorption or emission line is due to the collisions of atoms or molecules with electrons and ions. The electric field of the electrons and ions splits the emitted line, which is similar to the Stark effect which causes the line to be shifted and broadened. Although, the line splitting is very small, the Stark broadening is comparable to the Doppler broadening. Stark broadening can also occur due to the interatomic fields of the atoms or molecules having quadruple moments. However, the linewidth is broadened considerably due to the Stark effect in most discharges, sparks, arcs, and plasmas at high temperatures.

Various applications of the Stark effect for plasma diagnostics, such as temperature and concentration measurements, can be found in Griem (1964). Experimental data for the Stark linewidth $\Delta\lambda_S$ and shift $(\Delta\lambda)_{\text{shift}}$ are given in Konjevic and Roberts (1976), and Konjevic and Wiese (1976) for several atoms. The Stark linewidth and shift can be approximately determined by the following expressions [Wiese (1965)]:

$$\Delta\lambda_S \approx 2 \times 10^{-16}[1 + 1.75 \times 10^{-4} N_e^{1/4} \alpha(1 - 0.068 N_e^{1/6} T^{-1/2})]wN_e \quad (3.103)$$

$$(\Delta\lambda)_{\text{shift}} \approx 10^{-16}[(d/w) \pm 2 \times 10^{-4} N_e^{1/4} \alpha(1 - 0.068 N_e^{1/6} T^{-1/2})]wN_e \quad (3.104)$$

where $\Delta\lambda_S$ is the Stark linewidth (Å), $(\Delta\lambda)_{\text{shift}}$ is the Stark line shift (Å), $N_e$ is the electron density (number of electrons/cm$^3$), $T$ is the electron temperature (K), $\alpha N_e^{1/4} = \alpha_i$ is the ion broadening parameter (the values of $\alpha$ are tabulated for several atoms in Griem (1964, 1974), where $\alpha = \alpha_i/N_e^{1/4}$), and $w$ and $d$ are, respectively, the linewidth and lineshift due to the impact of the electrons [Å, given in Griem (1964, 1974) for several atoms].

Expressions (3.103) and (3.104) are sufficiently accurate when the following conditions are satisfied, that is,

$$10^{-4} \times \alpha N_e^{1/4} < 0.5, \quad R_1 > 1, \quad \text{and } R_2 < 0.8$$

where

$$R_1 = 8 \times 10^{-2} w \lambda^{-2} (T/M)^{-1/2} N_e^{2/3}$$

$$R_2 = 9 \times 10^{-2} N_e^{1/6} T^{-1/2}$$

$\lambda$ is the wavelength in Å and $M$ is the atomic weight of the atom of interest in g mol$^{-1}$.

It should be noted that $\alpha$, $w$, and $d$ are given in Griem (1964, 1974) for $N_e = 10^{16}$ electrons cm$^{-3}$. Expressions (3.103) and (3.104) can be applied to the lines of singly ionized atoms by changing 0.068 to 0.11 in both expressions. In Eq. (3.104), the minus sign should be used after the $d/w$ term at high temperatures only for those lines that have $d/w < 0$ at low temperatures, and the plus sign should be used for all other cases.

*3.11.4.4b. Lorentz Broadening.* This broadening occurs due to the collisions of atoms with foreign gas atoms. In terms of wavenumber, Lorentz broadening is given by Mitchell and Zemansky (1961, p. 160)

$$\Delta \bar{v}_L = Z_L/\pi c \qquad (3.105)$$

where $Z_L$ is the number of collisions with foreign gas atoms per second per absorbing atom, and is directly proportional to the foreign gas pressure. Lorentz broadening causes an asymmetry and a shift of the spectral line, which is explained by considering that the atom of interest, $A$, combines with the foreign gas atom, $F$, to form a quasi-molecule, and that the transition occurs between the ground and excited states of the quasi-molecule instead of the atom $A$ [Mitchell and Zemansky (1961), p. 175], that is,

$$A-F \xleftrightarrow{hv} A^*-F \quad \text{occurs}$$

$$A \xleftrightarrow{hv} A^* \quad \text{does not occur}$$

in Lorentz broadening, where $A^*-F$ and $A^*$ are the excited states of the quasi-molecule, $A-F$, and the atom of interest, $A$, respectively.

*3.11.4.4c. Holtsmark Broadening.* This broadening occurs due to the collisions between atoms of the same kind. In terms of wavenumber, Holtsmark broadening is given by [Hutcherson and Griffin (1973)]

$$\Delta \bar{v}_H = K \frac{e^2 f N}{c^2 m \bar{v}_c} \qquad (3.106)$$

where $K$ is a constant, $e$ and $m$ are the charge and mass of an electron, $\bar{v}_c$ is the wavenumber at the line center, $N$ is the number density of the atoms of interest, that is, the number of atoms per unit volume, and $f$ is the oscillator strength. The value of $K$ depends on the theory used to derive Eq. (3.106). Weisskopf (1932) obtained $K=1$ based on simple collision theory; Margenau and Watson (1936) and Furssow and Wlassow (1936), respectively, found it to be $\pi/3$ and $4/3$; and other theoretical studies [Reck *et al.* (1965), Lindholm (1945), Foley (1946), Byron and Foley (1964)] have produced similar values. The oscillator strength $f$ is given by [Griffin and Hutcherson (1969), Barrow (1962)]

$$f = \frac{mc^2}{N\pi e^2} \int \alpha_{\bar{v}} d\bar{v} = \frac{mc}{8\pi e^2 \bar{v}_c^2} \frac{1}{\tau_m} \frac{g_u}{g_l} \qquad (3.107)$$

where $\alpha_{\bar{v}}$ has the same meaning as in Eq. (3.100), $\tau_m$ is the mean lifetime of the resonance level, and $g_u$ and $g_l$ are the degeneracies of the upper and lower states, respectively.

### 3.11.4.5. Predissociation Broadening

Predissociation, which is defined as the dissociation that occurs sometime after the absorption of radiation, can take place during photolytic LCVD, and it leads to the broadening of vibrational or rotational lines [Okabe (1978), pp. 60, 61]. Predissociation broadening is difficult to detect in the absorption spectrum because it is usually smaller than the Doppler broadening. However, the emission spectrum can be used to detect predissociation because the spectral region, which is usually a continuum and weak-radiation-emitting region in the absence of predissociation, becomes broadened with discrete energy levels, and relatively strong emission lines are generated due to predissociation broadening.

## 3.11.5. Concentration Measurements Using LIF

The concentrations of various species provide important information regarding the mechanisms of pyrolytic and photolytic reactions during LCVD. The intensity of the fluorescent radiation can be related to the concentration and, therefore, concentration can be determined by measuring the fluorescent radiation. To derive such an expression, we will consider a two-level state with a lower energy level $l$ and an upper energy level $u$ in the presence of quenching (see Table 3.11). The rates of accumulation of the species of interest in these two levels are given by the following two equations:

$$\frac{dN_l}{dt} = -B_{lu}W_v N_l + A_{ul}N_u + B_{ul}W_v N_u + Q_{ul}N_u \qquad (3.108)$$

$$\frac{dN_u}{dt} = B_{lu}W_v N_l - A_{ul}N_u - B_{ul}W_v N_u - Q_{ul}N_u \qquad (3.109)$$

Assuming that there were no atoms or molecules in the upper level $u$ prior to laser excitation, we can write

$$N_u + N_l = N_0 \qquad (3.110)$$

Using Eq. (3.110) in Eq. (3.108), we can write the solution of Eq. (3.108) as [Eckbreth (1988)]

$$N_u(t) = \frac{N_0 B_{lu} W_v}{s}[1 - \exp(-st)] \qquad (3.111)$$

where $s = B_{lu}W_v + B_{ul}W_v + A_{ul} + Q_{ul}$. At the steady state, that is, when $t \to \infty$, Eq. (3.111) can be written as

$$N_u^* = N_u(t)|_{t \to \infty}$$

$$= N_0 \frac{B_{lu}}{B_{lu} + B_{ul}} \frac{1}{1 + (1/W_v)[(A_{ul} + Q_{ul})/(B_{lu} + B_{ul})]}$$

$$= N_0 \frac{B_{lu}}{B_{lu} + B_{ul}} \frac{1}{1 + W_v^{sat}/W_v} = N_0 \frac{B_{lu}}{B_{lu} + B_{ul}} \frac{1}{1 + I_v^{sat}/I_v} \quad (3.112)$$

where the saturation spectral laser energy density is

$$W_v^{sat} = \frac{A_{ul} + Q_{ul}}{B_{lu} + B_{ul}} \quad (3.113)$$

and the saturation spectral laser intensity $I_v^{sat} = W_v^{sat}c$.

To relate $N_u^*$, which is given by Eq. (3.112), to the experimental fluorescence signal, we find that the total number of fluorescent photons, $S_f$, that are collected from the laser-excited probe volume $V$ is given by [Allen (1987)]

$$S_f = \left(\frac{\Omega}{4\pi} V\right)(\tau_p\eta)(N_u^* A_{ul}) \quad (3.114)$$

$$= \left(\frac{\Omega}{4\pi} V\tau_p\eta\right) \frac{N_0 B_{lu}}{B_{lu} + B_{ul}} \frac{A_{ul}}{1 + I_v^{sat}/I_v} \quad (3.115)$$

which can be expressed in terms of the total collected fluorescent energy $F$ by the following equation:

$$F = S_f h\nu = h\nu\left(\frac{\Omega}{4\pi} V\tau_p\eta\right) \frac{N_0 B_{lu}}{B_{lu} + B_{ul}} \frac{A_{ul}}{1 + I_v^{sat}/I_v}$$

At low laser excitation intensities, that is, when $I_v \ll I_v^{sat}$, Eq. (3.115) can be written as

$$S_f = \left(\frac{\Omega}{4\pi} V\tau_p\eta\right) \frac{N_0 B_{lu}I_v}{c} \frac{A_{ul}}{A_{ul} + Q_{ul}} \quad (3.116)$$

Equation (3.116) represents the linear excitation regime, where the fluorescence varies linearly with the incident laser intensity and the initial concentration of the probe molecules. The term $A_{ul}/(A_{ul} + Q_{ul})$, which is a ratio of the spontaneous emission or fluorescence rate to the total atomic or molecular energy decay rate (note that such decays do not include the stimulated emission), is known as the fluorescent efficiency or the Stern–Vollmer factor [Kychakoff et al. (1984)]. Since $Q_{ul}$ appears in Eq. (3.116), we have to obtain the quenching rate in order to determine the concentration by using Eq. (3.116) in this linear regime.

At high laser excitation intensities, that is, when $I_v \gg I_v^{sat}$, Eq. (3.115) can be written as

$$S_f = \left(\frac{\Omega}{4\pi} V \tau_p \eta\right) \frac{N_0 B_{lu} A_{ul}}{B_{lu} + B_{ul}} \qquad (3.117)$$

Equation (3.117) respresents the saturation regime, where the absorption and stimulated emission rates are much higher than the spontaneous emission and quenching rates. In the saturation regime, the concentration can be determined by using Eq. (3.117) without knowing the quenching rate. However, complete saturation is difficult to achieve in practice due to the spatial and temporal variations of the incident laser intensity. Since the intensity is less at the edge of the laser beam, saturation does not occur at the outer edges of the probe volume $V$. Also, due to the change in laser intensity with time, the probe volume does not remain saturated during the entire duration of the laser pulse.

### 3.11.6. Temperature Measurements Using LIF

Temperature measurements allow us to gain insight into the reaction temperature, the activation and bond dissociation energies required for pyrolytic and photolytic reaction, and the temperature distribution inside the deposition chamber during LCVD, to relate the experimental fluorescence signal to the temperature of the laser-excited probe volume $V$, we will assume that the probe volume is under a *local thermodynamic equilibrium* (LTE) condition, that is, all the species, such as ions, electrons, neutral atoms, and excited and unexcited species, are at the same temperature. Under this condition, we can write the following Boltzmann equation

[Vincenti and Kruger (1965)]:

$$N_i = N \frac{g_i}{Q} \exp\left(-\frac{E_i}{k_B T}\right) \qquad (3.118)$$

where $N_i$ is the number density of the species, that is, the number of species per unit volume, in the $i$th state of energy $E_i$, $N$ is the total number density, $T$ is the temperature of the system, $g_i$ is the degeneracy of the $i$th state ($g_i$ indicates the number of atoms or molecules which can occupy the $i$th energy state), $E_i$ is the energy of the $i$th state, and $Q$ is the partition function, which describes the way in which the atoms or molecules are partitioned, or separated, into various energy levels, and is given by

$$Q = \sum_i g_i \exp\left(-\frac{E_i}{k_B T}\right)$$

Taking $i = u$ in Eq. (3.118) and noting that $N = N_0$ and $N_i = N_u^*$, we obtain the following expression from Eqs. (3.114) and (3.118):

$$S_f = \frac{\Omega}{4\pi} \tau_p \eta \left[ A_{ul} N_0 \frac{g_u}{Q} \exp\left[\left(-\frac{E_u}{k_B T}\right)\right] \right] \qquad (3.119)$$

or

$$\ln\left(\frac{S_f}{\tau_p A_{ul} g_u}\right) = -\frac{E_u}{k_B T} + \ln\left(\frac{\Omega}{4\pi} V \eta \frac{N_0}{Q}\right) \qquad (3.120)$$

If we measure $S_f$ for a number of spectral lines of the same species, we can determine the temperature $T$ from Eq. (3.120) by plotting $\ln(S_f/\tau_p A_{ul} g_u)$ vs. $E_u$, which is known as the Boltzmann plot [Adrain (1984)]. The magnitude of the slope of this plot is equal to $(k_B T)^{-1}$.

## 3.12. PRECURSOR MOLECULES

Alkyls, carbonyls, halides, and hydrides are the usual precursor molecules for both pyrolytic and photolytic LCVD. The chemical kinetics and the activation and bond dissociation energies of many such compounds can be found in Bamford and Tipper (1972). A review of several precursors for

LCVD of various materials can be found in Ehrlich and Tsao (1983). Various studies concerning the deposition of metal films, semiconductors, and dielectrics have been discussed by Eden (1992), who has presented the precursor molecules, optical sources and their wavelengths, and the experimental conditions and results for the following materials:

METAL FILMS

Group IB: Cu, Au
Group IIB: Zn, Cd
Group IIIA: Al, Ga, In, Tl
Group IVB, VIB, and VIIB transition metals: Ti, Cr, Mo, W, Mn
Group IVA metals: Sn, Pb
Group VIII transition metals Fe, Ir, Ni, Pt
Group VIA: Se

SEMICONDUCTOR FILMS

Elemental films: C, Ge, Si
IIB–VIA binary and ternary compounds: CdTe, $Cd_xHg_{1-x}Te$, HgTe, ZnO, ZnS, ZnSe; and IIIA-VA compound films: GaAs, AlGaAs, GaAlAs, AlAs, GaN, GaP, InP, InSb.

DIELECTRIC FILMS

$Al_2O_3$, $Cr_2O_3$, $CrO_2$, $Cr_2O_3$, $P_3N_5$, $SiO_2$, $SiO_xN_y$, $Si_xO_yN_z$, $SiN_x$, $Si_xN_y$, $Si_3N_4$, $SnO_2$, $TaO_x$, TiN, $VO_x$.

FILMS OF OTHER MATERIALS

$Bi_2O_3$, CaO, $CaCO_3$, CuO, SrO, $SrCO_3$, polymers, $TiB_2$, siloxane and phthalateorganic films.

Thin film technology is now used extensively for electronic applications, but it can be utilized as well to deposit oxidation, corrosion, and wear resistant films on many materials for other applications. This technology is also, important for producing optical devices. InP/InGaPAs film is used to fabricate optoelectronic devices for optical fiber communication. The composition of the film $In_{1-x}Ga_xP_{1-y}As_y$ may vary according to the values of $x$ and $y$ depending on the deposition conditions. However, it remains lattice-matched to InP, and its band gap [Antypas and Moon (1973)] corresponds to photon wavelengths in the range of 0.92 to 1.35 $\mu$m. Since the dispersion minimum [Payne and Grambling (1975)] of fused silica fibers is about 1.2 to 1.3 μm, the InP/InGaPAs system is suitable for long-distance high-bit-rate optical fiber communication. Studies on some of the optical devices can be found in Bogatov et al. (1975), Hitchens et al. (1975), Pearsall et al. (1976), Wieder et al. (1977), and Hurwitz and Hsieh (1978).

PRECURSOR SELECTION CRITERIA

The precursors for photolytic LCVD can be selected based on the following criteria:

FLUIDITY: The precursor should be such that it can be delivered to the deposition chamber easily. For this reason, it must be a gas at room temperature or a liquid with a suitable vapor pressure below its thermal decomposition temperature. The pressure for photolytic LCVD is usually in the range of 10 mTorr to 1 Torr.

THERMAL DISSOCIATION TEMPERATURE: The thermal dissociation temperature of the precursor should be relatively high, in order that (a) the substrate surface can be raised to a suitable temperature to control the film growth and microstructure, and (b) there is sufficient time for the film material to rearrange and form a dense film with a homogeneous microstructure. Usually, the precursors are so chosen that their thermal decomposition temperature lies in the range from 100 to 300°C.

ABSORPTIVITY: The absorption spectrum of the reactant molecules should have a peak at the wavelength of the laser, that is used for carrying out photolytic LCVD in order to ensure photochemical reaction.

CORROSIVENESS: The precursor should not be corrosive to prevent any damage to the equipment used for LCVD.

TOXICITY: The precursors and the reaction products should be environmentally safe and nonhazardous to health. It should be noted that although TMA fulfills the above requirement, it should not be used to deposit Al film because it is a pyrophoric liquid which reacts spontaneously in air to produce $Al_2O_3$ and flammable hydrocarbon gases [Walsh and Warsop (1962)]. Considerable research has been carried out with conventional CVD processes to find replacements for the toxic group-V-bearing dopant gases [Hsu et al. (1983)] such as $AsH_3$ and $PH_3$.

## NOMENCLATURE

| | |
|---|---|
| $a$ | Area on which the photon beam strikes the target perpendicularly |
| $A$ | Avogadro's number |
| $A_{ul}$ | Einstein coefficient for spontaneous emission |
| $B_{ul}$ | Einstein coefficient for stimulated emission |
| $B_{lu}$ | Einstein coefficient for absorption |
| $c$ | Velocity of the electromagnetic radiation |
| $d$ | Thickness of the target |
| $D_0(X - Y)$ | Bond dissociation energy of the molecule XY |
| $D_{ul}$ | Dipole strength |
| $e$ | Electronic charge |
| $E$ | Energy of a photon or a quantum |
| $E^*$ | Electronic energy of $Y^*$ |
| $E_d$ | Dissociation energy for the molecule when it is in its electronic ground state |
| $E_d^*$ | Dissociation energy for the electronically excited stable molecule |
| $E_m$ | Energy of 1 mol of quanta or photons |
| $F$ | Total collected fluorescent energy |
| $g_l$ | Degeneracy of the $l$th energy level, that is, the number of possible states in the $l$th level that have the same energy even though their wave functions are not the same |
| $g_u$ | Degeneracy of the $u$th energy level, that is, the number of possible states in the $u$th level that have the same energy even though their wave functions are not the same |
| $\bar{G}_T^0$ | Gibbs energy function at the standard state (1 atm pressure) at temperature T K |
| $\Delta G_T^0$ | Change in the Gibbs free energy at the standard state (1 atm pressure) at temperature T K |
| $h$ | Planck's constant ($= 6.626 \times 10^{-34}$ Js) |
| $\hbar$ | Planck's constant ($= h/2\pi$) |
| $\Delta H_{f0}^{\circ}(i, g)$ | Standard enthalpy or heat of formation of the gaseous species $i$, where $i = X$, $Y$, or $XY$. The superscript °, the degree sign, refers to the thermodynamic standard state of 1 atm pressure, and the subscript 0, zero, refers to the 0 K temperature |
| $\Delta H_0^0$ | Increase in enthalpy for a gas phase reaction in the thermodynamic standard state of 1 atm pressure at 0 K |
| $I$ | Intensity of photons or photon flux (number of photons per unit area per unit time) |

| | |
|---|---|
| $I_0$ | Intensity of a photon beam that strikes the target (number of photons striking the target per unit area per unit time) |
| $I_v$ | Spectral intensity ($I_v = W_v C$, see Appendix A) |
| $I'_A$ | Number of photons absorbed per unit time per unit volume of the reactant molecules, R, where R = $S_0$ in Table 3.4, and R = A in Table 3.5 |
| $J$ | Total angular momentum quantum number |
| $k_B$ | Boltzmann constant |
| $k_q$ | Quenching rate constant due to collisions with species $q$ |
| $K_c$ | Equilibrium constant in terms of concentration |
| $K_p$ | Equilibrium constant in terms of partial pressure (see Section 3.10.2) |
| $K_P$ | Equilibrium constant in terms of total pressure (see Section 3.7.1) |
| $K_X$ | Equilibrium constant in terms of concentration for the adsorption |
| $K_Y$ | Equilibrium constant in terms of concentration for the adsorption of Y |
| $l$ | Orbital angular momentum quantum number |
| $L$ | Total orbital angular momentum quantum number |
| $m_e$ | Mass of an electron |
| $m_l$ | Orbital magnetic quantum number |
| $\mu_M$ | Attenuation coefficient of a mixture |
| $m_s$ | Spin magnetic quantum number |
| $M$ | Spin multiplicity ($=2S+1$) |
| $M_L$ | Total orbital magnetic quantum number |
| $n$ | Principal quantum number |
| $N$ | Total number of molecules within the interaction volume which is equal to the volume of the laser beam inside the reaction chamber |
| $N_0$ | Concentration of the atoms or molecules of interest in the lower energy level $l$ prior to laser excitation |
| $N_l$ | Concentration of the atoms or molecules of interest in the lower energy level $l$ |
| $N_u$ | Concentration of the atoms or molecules of interest in the upper energy level $u$ |
| $N_u^*$ | Value of $N_u$ at steady state, that is, as the time $t \to \infty$ |
| $\bar{N}_d$ | Average number of dissociated molecules |
| $N_{q_i}$ | Concentration of the quenching species $q$ in the $i$th energy level |
| $N_{q_j}$ | Concentration of the quenching species $q$ in the $j$th energy level |
| $p_X$ | Partial pressure of the adsorbate species $X$ in the bulk of the gas phase (in Section 3.10) |

| | |
|---|---|
| $P_a$ | Atomic density of the target (number of atoms per unit volume of the target) |
| $P_i$ | Pressure of the $i$th species for $i = X$, $Y$, or $XY$. $P_i$ must be in units of atmosphere |
| $Q_{ul}$ | Quenching rate constant due to collisions with the species $q$, $Q_{ul} = k_q N_{q_i}$. |
| $R$ | Universal gas constant |
| $\dot{R}$ | Rate of the absorption of photons by the target as the photon beam passes through the target (number of photons absorbed per unit time) |
| $s$ | Spin angular momentum quantum number |
| $S$ | Total spin angular momentum quantum number (in Sections 3.2–3.9) |
| $S_i$ | Singlet states, $i = 0, 1, 2, 3, \ldots, S_0$ refers to the ground singlet state (in Sections 3.2–3.9) |
| $S_0$ | Concentration of the active adsorption sites on the surface $S$ before the adsorption begins (in Section 3.10) |
| $S_f$ | Total number of fluorescent photons that are collected from a laser-excited probe volume |
| $[S]$ | Concentration of the unoccupied adsorption sites (in Section 3.10) |
| $[SX]$ | Concentration of the adsorption sites occupied by $X$ |
| $[SY]$ | Concentration of the adsorption sites occupied by $Y$ |
| $T$ | Chemical reaction temperature (K) (in Section 3.7) |
| $T_i$ | Triplet states $i = 1, 2, 3, \ldots$ |
| $T_r$ | Reference temperature (K) which can be taken as 298 K |
| $v$ | Velocity of an electron |
| $V$ | Laser-excited probe volume, which is given by the product of the cross-sectional area of the laser beam and the length of the beam that is focused onto the detector |
| $W_0$ | Work function of the metal |
| $W_v$ | Spectral radiative energy density, that is, radiative energy per unit volume per unit oscillation frequency interval (Jm$^{-3}$Hz$^{-1}$) (see Appendix A) |
| $x$ | Amount of adsorbed species |
| $[X_2]$ | Concentration of the adsorbate species $X_2$ in the bulk of the gas phase (in Section 3.10) |
| $[X]$ | Concentration of the adsorbate species $X$ in the bulk of the gas phase (in Section 3.10) |
| $[Y]$ | Concentration of the adsorbate species $Y$ in the bulk of the gas phase (in Section 3.10) |

## GREEK SYMBOLS

| | |
|---|---|
| $\beta$ | Two-photon absorption constant |
| $\eta$ | Transmissivity of the fluorescence photon collection optics |
| $\eta$ | Dissociation yield |
| $\theta$ | Fraction of the active adsorption sites occupied by the species $X$ ($\theta = [SX]/S_0$) |
| $\theta_X$ | Same as $\theta$ |
| $\theta_Y$ | Fraction of the active adsorption sites occupied by $Y$ ($\theta_Y = [SY]/S_0$) |
| $\lambda$ | Wavelength of the electromagnetic radiation |
| $\mu$ | Attenuation coefficient, or macroscopic absorption cross section |
| $\mu_{m_i}$ | Attenuation coefficient of the $m_i$th type of molecule for $i = 1$ and 2. |
| $\mu_M$ | Attenuation coefficient of a mixture |
| $\nu$ | Frequency of a photon |
| $\nu_f$ | Minimum frequency of the probe laser or radiation required to produce the fluorescing species $Y^*$ |
| $\nu_i$ | Minimum frequency of the probe laser or radiation required to produce the ion $X^+$ |
| $\rho_{m_1}$ | Molecular density of the molecule $m_1$ (Number of molecules of type $m_1$ per unit volume of the mixture of the molecules $m_1$ and $m_2$) |
| $\rho_{m_2}$ | Molecular density of the molecule $m_2$ (Number of molecules of type $m_2$ per unit volume of the mixture of the molecules $m_1$ and $m_2$) |
| $\rho_a$ | Atomic density of the target (number of atoms per unit volume of the target) |
| $\sigma$ | Photon absorption cross section of an atom |
| $\sigma_d$ | Dissociation cross section |
| $\bar{\sigma}_d$ | Effective cross section for dissociation |
| $\sigma_m$ | Photon absorption cross section of a molecule |
| $\sigma_X$ | Photon absorption cross section of atom $X$ |
| $\sigma_Y$ | Photon absorption cross section of atom $Y$ |
| $\sigma_{m_1}$ | Photon absorption cross section of molecule $m_1$ |
| $\sigma_{m_2}$ | Photon absorption cross section of molecule $m_2$ |
| $\tau$ | Laser irradiation time |
| $\tau_p$ | Laser pulse duration (see Section 3.11.5) |
| $\tau_F$ | Lifetime for the excited singlet state ($S_1$) |
| $\tau_P$ | Lifetime for the excited triplet state ($T_1$) (see Section 3.9.1) |
| $\tau^0$ | True radiative lifetime ($\tau^0 = \tau_F^0$ or $\tau_P^0$) |

| | |
|---|---|
| $\tau_F^0$ | True radiative lifetime for the excited singlet state $S_1$ |
| $\tau_P^0$ | True radiative lifetime for the excited triplet state $T_1$ |
| $\phi$ | Laser fluence ($= I\tau$) |
| $\phi_i$ | Ionization potential of the species $X$ |
| $\phi_B$ | Quantum yield defined in terms of the product molecules |
| $\phi_A$ | Quantum yield defined in terms of the reactant molecules |
| $\phi_F$ | Quantum yield of fluorescence, or fluorescence efficiency |
| $\phi_{IC}$ | Quantum yield of internal conversion from $S_1$ to the ground state, or internal conversion efficiency |
| $\phi_{ISC(S)}$ | Quantum yield of intersystem crossing from the singlet $S_1$ state to the triplet $T_1$, state, or triplet formation efficiency |
| $\phi_P$ | Quantum yield of phosphorescence, or phosphorescence efficiency |
| $\Omega$ | Solid angle subtended by the fluorescence photon collection system at the laser-excited probe volume |

# REFERENCES

Adamson, A. W. (1982), *Physical Chemistry of Surfaces*, 4th Ed., Wiley, New York.
Adrain, R. S. (1984), Some Industrial Uses of Laser-Induced Plasmas, *Industrial Applications of Lasers*, Koebner, H., ed., Wiley, New York, pp. 135–176.
Allen, M. G. (1987), *Digital Imaging Techniques for Single and Multi-phase Reacting Flow-Fields*, HTGL Report T-259, Stanford University, April 1987.
Allen, S. D., and Bass, M. (1979), *J. Vac. Sci Technol.* **16**, 431.
Andrews, D. L. (1986), *Lasers in Chemistry*, Springer-Verlag, New York, pp. 118–127.
Antypas, G., and Moon, R. (1973), Growth and Characterization of InP–InGaAsP Lattice Matched Heterojunctions, *J. Electrochem. Soc.* **120**, 1574.
Arnold, D. R., Baird, N. C., Bolton, J. R., Brand, J. C. D., Jacobs, P. W. M., de Mayo, P., and Ware, W. R. (1974), *Photochemistry–An Introduction*, Academic, New York, pp. 95–104.
Bamford, C. H., and Tipper, C. F. H., eds. (1972), *Comprehensive Chemical Kinetics*, Vol 4, *Decomposition of Inorganic and Organometallic Compounds*, Elsevier, New York.
Barrow, G. M. (1962), *Introduction to Molecular Spectroscopy*, McGraw-Hill, New York, p. 81.
Bäuerle, D. (1986), *Chemical Processing with Lasers*, Springer-Verlag, New York, pp. 22–35.
Bauman, R. P. (1962). *Absorption Spectroscopy*, Wiley, New York, pp. 270–274.
Bixon, M., and Jortner, J. (1969), *J. Chem. Phys.* **50**, 3284.
Bodenstein, M., and Lutkemeyer, H. (1924), *Z. Phys. Chem.* **114**, 208.
Bogatov, A., Dolginov, L., Druzhinina, L., Eliseev, P., Sverdlov, B., and Shevchenko, E. (1975), Heterojunction Lasers Made of GaInAsP and AlGaSbAs Solid Solutions, *Sov. J. Quantum Electron.* **4**, 1281.
Breckenridge, W. H., and Stewart, G. M. (1986), Pulsed Laser Photolysis of Chromium Hexacarbonyl in the Gas Phase, *J. Am. Chem. Soc.* **108**, 364.
Bunsen, R. W., and Roscoe, H. E. (1859), *Pogg. Ann.* **101**, 193.
Byron, Jr., F. W., and Foley, H. M. (1964), *Phys. Rev.* A **134**, 625.
Callear, A. B., and Smith, I. W. M. (1963), *Trans. Faraday Soc.* **59**, 1720.

Calvert, J. G., and Pitts, Jr., J. N. (1966), *Photochemistry*, Wiley., New York, pp. 1–26.
Compton, A. H. (1923a), *Phys. Rev.* **21**, 715.
Compton, A. H. (1923b), *Phys. Rev.* **22**, 409.
Condon, E. U. (1928), *Phys. Rev.* **32**, 858.
Constable, F. H. (1925), *Proc. Roy. Soc. Lond.* **A108**, 355.
Crosley, D. R. (1982), *J. Chem. Ed.* **59**, 447.
Daily, J. W. (1977), Use of Rate Equations to Describe Laser Excitation in Flames, *Appl. Opt.* **16**, 2322.
Dalton, J. C., and Turro, N. J. (1970), Kinetic Analyses of Photochemical Reactions Which Involve Quenching of More Than One Excited State: Quenching of Both Singlet and Triplet States Molecular Photochemistry, *Mole. Photochem.* **2**, 133.
Dash, J. C. (1975), *Films on Solid Surfaces*, Academic, New York.
Davis, D. D., and Okabe, H. (1968), Determination of Bond Dissociation Energies in Hydrogen Cyanide, Cyanogen and Cyanogen Halides by the Photodissociation Method, *J. Chem. Phys.* **49**, 5526.
DeGraff, B. A., and Calvert, J. G. (1967), *J. Am. Chem. Soc.* **89**, 2247.
Demtroder, W. (1982), *Laser Spectroscopy*, Springer-Verlag, New York.
Deutsch, T. F., Ehrlich, D. J., and Osgood, Jr., R. M. (1979), *Appl. Phys. Lett.* **35**, 175.
Ditchburn, R. W. (1957), *Light*, Interscience, New York, p. 82.
Donnelly, V. M., and Karlicek, R. F. (1982), *J. Appl. Phys.* **53**, 6399.
Draper, J. W. (1841), *Philos. Mag.* **19**, 195.
Eastham, D. A. (1989), *Atomic Physics of Lasers*, Taylor and Francis, Philadelphia, pp. 75–82, and 34–35.
Eckbreth, A. C. (1988), *Laser Diagnostics for Combustion Temperature and Species*, Abacus, Massachusetts.
Eden, J. G. (1992), *Photochemical Vapor Deposition*, Wiley, New York.
Ehrlich, D. J., Osgood, Jr., R. M., and Deutsch, T. F. (1980a), *Appl. Phys. Lett.* **36**, 698.
Ehrlich, D. J., Osgood, Jr., R. M., and Deutsch, T. F. (1980b), *IEEE J. Quantum Elect.* **QE-16**, 1233.
Ehrlich, D. J., Osgood, Jr., R. M., and Deutsch, T. F. (1981), *Appl. Phys. Lett.* **38**, 946.
Ehrlich, D. J., Osgood, Jr., R. M., and Deutsch, T. F. (1982), *J. Vac. Sci. Technol.* **21**, 23.
Ehrlich, D. J., and Tsao. J. Y. (1983), A Review of Laser-Microchemical Processing, *J. Vac. Sci. Technol.* **B1**, 969.
Einstein, A. (1905a), *Ann. Phys.* **17**, 549.
Einstein, A. (1905b), *Ann. Physik* **17**, 132.
Einstein, A. (1906), *Ann. Phys.* **19**, 371.
Einstein, A. (1912a), *Ann. Phys.* **37**, 832.
Einstein, A. (1912b), *Ann. Phys.* **38**, 881.
Einstein, A. (1913), *J. Phys.* **3**, 227.
Einstein, A. (1916), Strahlungs-emission und-absorption nach der quantertheorie, *Verh. Dtsch. Phys. Ges.* **18**, 318.
Faraday, M. (1834), *Philos. Trans.* **124**, 55.
Foley, H. M. (1946), *Phys. Rev.* **69**, 616.
Franck, J. (1925), *Trans. Faraday Soc.* **21**, 536.
Furssow, W., and Wlassow, A. (1936), *Physik. Z. Sowjetunion* **10**, 378.
Griem, H. R. (1964), *Plasma Spectroscopy*, McGraw-Hill, New York.
Griem, H. R. (1974), *Spectral Line Broadening by Plasmas*, Academic, New York.
Griffin, P. M., and Hutcherson, J. W. (1969), Oscillator Strengths of the Resonance Lines of Krypton and Xenon, *J. Opt. Soc. Am.* **59**, 1607.
Hayward, D. O., and Trapnell, B. M. W. (1964), *Chemisorption*, Butterworths, London.
Hertz, H. (1887), *Ann. Physik* **31**, 983.

Herzberg, G. (1945), *Molecular Spectra and Molecular Structure II: Infrared and Raman Spectra of Polyatomic Molecules*, Van Nostrand, New York.
Herzberg, G. (1950), *Molecular Spectra and Molecular Structure I: Spectra of Diatomic Molecules*, 2nd Ed., Van Nostrand, Princeton.
Herzberg, G. (1966), *Molecular Spectra and Molecular Structure III: Electronic Spectra and Electronic Structure of Polyatomic Molecules*, Van Nostrand, Princeton.
Hess, P. (1989), *Photoacoustic, Photothermal and Photochemical Processes at Surfaces and in Thin Films*, Hess, P., ed., Springer-Verlag, New York, pp. 1–9.
Hitchens, W., Holonyak, Jr., N., Wright, P., and Coleman, J. (1975), Low Threshold LPE InGaPAs Yellow Double-Heterojunction Laser Diodes, *Appl. Phys. Lett.* **27**, 245.
Hsu, C., Cohen, R., and Stringfellow, G. (1983), OMVPE Growth of InP Using TMIn, *J. Cryst. Growth* **63**, 8.
Hurwitz, C., and Hsieh, J. (1978), GaInAsP/InP Avalanche Photodiodes, *Appl. Phys. Lett.* **32**, 487.
Hutcherson, J. W., and Griffin, P. M. (1973), Self-Broadened Absorption Linewidths for the Krypton Resonance Transitions, *J. Opt. Soc. Am.* **63**, 338.
Incropera, F. P., and DeWitt, D. P. (1985), *Fundamentals of Heat and Mass Transfer*, Wiley, New York, 2nd Ed., p. 548.
Jonah, C., Chandra, P., and Bersohn, R. (1971), *J. Chem. Phys.* **55**, 1903.
Kompa, K. L., and Smith, S. D., eds. (1979), *Laser-Induced Processes in Molecules*, Springer Series in Chem. Phys., Vol. 10, Springer-Verlag, New York.
Konjevic, N., and Roberts, J. R. (1976), A Critical Review of the Stark Widths and Shifts of Spectral Lines from Non-Hydrogenic Atoms, *J. Phys. Chem. Ref. Data* **5**, 209–218.
Konjevic, N., and Wiese, W. L. (1976), Experimental Stark Widths and Shifts for Non-Hydrogenic Spectral Lines of Ionized Atoms (A Critical Review and Tabulation of Selected Data), *J. Phys. Chem. Ref. Data* **5**, 259–277.
Kychakoff, G., Howe, R. D., and Hanson, R. K. (1984), Quantitative Flow Visualization Technique for Measurements in Combustion Gases, *Appl. Opt.* **23**, 704.
Laidler, K. J. (1987), *Chemical Kinetics*, Harper and Row, New York, 3rd Ed., pp. 348–376, and 229–275.
Lakowicz, J. R. (1983), *Principles of Fluorescence Spectroscopy*, Plenum, New York.
Lamarsh, J. R. (1966), *Introduction to Nuclear Reactor Theory*, Addison-Wesley, Massachusetts, pp. 17–22.
Langmuir, I. (1916), *J. Am. Chem. Soc.* **38**, 2221.
Langmuir, I. (1918), *J. Am. Chem. Soc.* **40**, 1361.
Lengyel, B. A. (1971), *Lasers*, Interscience, New York.
Lewis, G. N. (1926), *Nature* **118**, 874.
Lewis, G. N., Randall, M., Pitzer, K. S., and Brewer, L. (1961), *Thermodynamics*, McGraw-Hill, New York.
Lide, D. R., editor-in-chief (1992), *CRC Handbook of Chemistry and Physics*, CRC Press, Ann Arbor, 73rd ed. pp. **10**-298, **10**-299.
Lindholm, E. (1945), *Ark. Mat. Ast. Phys.* **A32**, 1.
Long, L. H. (1961), *Pure Appl. Chem.* **2**, 61.
Lucht, R. P. and Laurendeau, N. M. (1979), Two-Level Model for Near Saturated Fluorescence in Diatomic Molecules, *Appl. Opt.* **18**, 856.
Mahan, B. H., and Mandal, R. (1962), *J. Chem. Phys.* **37**, 207.
Margenau, H., and Watson, W. W. (1936), *Rev. Mod. Phys.* **8**, 22.
McNesby, J. R., Tanaka, I., and Okabe, H. (1962), *J. Chem. Phys.* **36**, 605.
Miller, A. R. (1949), *The Adsorption of Gases on Solids*, Cambridge University Press, New York.
Millikan, R. A. (1916), *Phys. Rev.* **7**, 355.
Mitchell, A. C. G., and Zemansky, M. W. (1961), *Resonance Radiation and Excited Atoms*,

Cambridge University Press, London.
Mulliken, R. S., and Rieke, C. A. (1941), *Repts. Progr. Phys.* **8**, 231.
Myers, G. H., Silver, D. M., and Kaufman, F. (1966), *J. Chem. Phys.* **44**, 718.
Okabe, H. (1974), *Chemical Spectroscopy and Photochemistry in the Vacuum Ultraviolet*, Sandorfy, C., Ausloos, P. J., and Robin, M. B., eds., Reidel, Boston, p. 513.
Okabe, H. (1978), *Photochemistry of Small Molecules*, Wiley, New York, pp. 46–56, 61–70, 81–106.
Okabe, H., and Dibeler, V. H. (1973), Photon Impact Studies of $C_2HCN$ and $CH_3CH$ in the Vacuum Ultraviolet: Heats of Formation of $C_2H$ and $CH_3CN$, *J. Chem. Phys.* **59**, 2430.
Okabe, H., and McNesby, J. R. (1961), *J. Chem. Phys.* **34**, 668.
Parker, C. A. (1968), *Photoluminescence of Solutions with Applications to Photochemistry and Analytical Chemistry*, Elsevier, New York, pp. 86–89.
Pasternack, L., Nelson, H. H., and McDonald, J. R. (1982), *J. Chem. Ed.* **59**, 456.
Payne, D., and Grambling, W. (1975), Zero Material Dispersion in Optical Fibers, *Electron Lett.* **11**, 176.
Pearsall, T., Miller, B., and Capik, R. (1976), Efficient Lattice-Matched Double-Heterostructure LED's at $1.1\,\mu m$ from GaInAsP, *Appl. Phys. Lett.* **28**, 499.
Pearse, R. W. B., and Gaydon, A. G. (1963), *The Identification of Molecular Spectra*, Wiley, New York.
Perlman, M. L., and Rollefson, G. K. (1941), The Vapor Density of Iodine at High Temperatures, *J. Chem. Phys.* **9**, 362.
Planck, M. (1901), *Ann. Physik* **4**, 553.
Price, S. J. W. (1972), *Decomposition of Inorganic and Organometallic Compounds*, Bamford, C. H., and Tipper, C. F. H., eds., Chemical Kinetics, Vol. 4, Elsevier, Amsterdam, p. 197.
Rabek, J. F. (1982), *Experimental Methods in Photochemistry and Photophysics*, Part 2, Wiley, New York, pp. 727–745.
Rabek, J. F. (1987), *Mechanisms of Photophysical Processes and Photochemical Reactions in Polymers: Theory and Applications*, Wiley, New York.
Reck, G. P., Takebe, H., and Mead, C. A. (1965), *Phys. Rev.* A137, 683.
Rosen, B. (1970), *Spectroscopic Data Relative to Diatomic Molecules*, International Tables of Selected Constants, Vol. 17, Pergamon Press, New York.
Ross, S., and Olivier, J. P. (1964), *On Physical Adsorption*, Interscience, New York.
Rothschild, M. (1989), Spectroscopy and Photochemistry of Gases, Adsorbates, and Liquids, *Laser Microfabrication: Thin Film Processes and Lithography*, Ehrlich, D. J., and Tsao, J. Y., eds., Academic, New York, pp. 163–230.
Shimoda, K. (1983), *Introduction to Laser Physics*, 2nd. Ed., Springer-Verlag, New York, pp. 78–84.
Solanki, R., and Collins, G. J. (1983), *Appl. Phys. Lett.* **42**, 662.
Solanki, R., Boyer, P. K. Mahan, J. E., and Collins, G. J. (1981), *Appl. Phys. Lett.* **38**, 572.
Solanki, R., Ritchie, W. H., and Collins, G. J. (1983), *Appl. Phys. Lett.* **43**, 454.
Somerjai, G. A. (1981), *Chemistry in Two Dimensions: Surfaces*, Cornell University Press, Ithaca.
Spinks, J. W. T., and Woods, R. J. (1990), *An Introduction to Radiation Chemistry*, Wiley, New York, 3rd Ed., pp. 39–49.
Stark, J. (1908), *Phys. Z* **9**, 889 and 894.
Steiner, U. E., and Wolff, H. J. (1991), Magnetic Field Effects in Photochemistry, *Photochemistry and Photophysics*, Vol. IV, Rabek, J. F., ed., CRC Press, Boston, pp. 1–130.
Stull, D. R., and Prophet, H. (Project Directors) (1971, *JANAF Thermochemical Tables*, 2nd Ed., Nat. Stand. Ref. Data Ser., National Bureau of Standards (US), Vol. 37.
Stull, D. R., Westrum, Jr., E. F., and Sinke, G. C. (1969), *The Chemical Thermodynamics of Organic Compounds*, Wiley, New York.

Taylor, H. S. (1925), *Proc. Roy. Soc. Lond.* **A108**, 105.
Taylor, H. S. (1931), *J. Am. Chem. Soc.* **53**, 578.
Technical Staff, February 28, (1966), Electric Field Raises Polymerization Rate, *Chem. Eng. News*, **44** (9), 37.
Thompson, H. W., and Linnett, J. W. (1936), *Proc. Roy. Soc. Lond.* **A156**, 108.
Trotman-Dickenson, A. F., and Milne, G. S. (1967), *Tables of Bimolecular Gas Reactions*, Vol. 9, Nat. Stand. Ref. Data Ser., National Bureau of Standards, USA.
Turro, N. J. (1967), *Molecular Photochemistry*, Benjamin, New York, pp. 30–42.
Vincenti, W. G., and Kruger, C. H. (1965), *Introduction to Physical Gas Dynamics*, Wiley, New York.
von Grotthuss, C. J. D. (1819), *Ann. Phys.* **61**, 50.
Wagman, D. D., Evans, W. H., Parker, V. B., Halow, I., Bailey, S. M., and Schumm, R. H. (1968), *Selected Values of Chemical Thermodynamic Properties*, Tech. Note, National Bureau of Standards (US), New York, pp. 270–273.
Wagner, P. J. (1971), Kinetic Analyses of Photochemical Reactions that Involve Quenching of Two Excited States, *Mol. Photochem.* **3**, 23.
Walsh, A., and Warsop, R. (1962), The Ultraviolet Spectra of Hydrides of Group V Elements, *Advances in Molecular Spectroscopy*, Proc., IVth Int. Meeting Molecular Spectroscopy, Vol. 2, Mangini, A., ed., Pergamon, Oxford, pp. 582–591.
Wayne, R. P. (1988), *Principles and Applications of Photochemistry*, Oxford University Press, New York, p. 80.
Weisskopf, V. F. (1932), *Z. Physik* **75**, 287.
White, H. E. (1934), *Introduction to Atomic Spectra*, McGraw-Hill, New York.
Wieder, H., Clawson, A., and McWilliams, G. (1977), InGaAsP/InP Heterojunction Photodiodes, *Appl. Phys. Lett.* **31**, 468.
Wiese, W. L. (1965), Line Broadening, *Plasma Diagnostic Techniques*, Huddlestone, R. H. and Leonard, S. L., eds., Academic, New York, pp. 265–317.
Wilkinson, F. (1980), *Chemical Kinetics and Reaction Mechanisms*, Van Nostrand-Reinhold, New York, pp. 249–291.
Wolfbeis, O. S., ed. (1993), *Fluorescence Spectroscopy: New Methods and Applications*, Springer-Verlag, New York.
Wood, T. H., White, J. C., and Thacker, B. A. (1983), UV Photodecomposition for Metal Deposition: Gas vs. Surface Phase Processes, *Laser Diagnostics and Photochemical Processing for Semiconductor Devices*, Osgood, R. M., Brueck, S. R. J., and Schlossberg, H. R., eds., North-Holland, New York, pp. 35–41.
Young, D. M., and Crowell, A. D. (1962), *Physical Adsorption of Gases*, Butterworths, London.

# FOUR

# PYROLYTIC LCVD MODELING

## 4.1. INTRODUCTION

A mathematical model of a given LCVD process is a theoretical representation which allows us to analyze various aspects of the process. It can be used to evaluate the relative importance of the different process parameters, such as the laser power, wavelength and beam diameter and the pressure of the gas inside the deposition chamber, etc. The models also provide a basic understanding of the chemical kinetics and the film deposition mechanisms. The results of the mathematical models can be used to minimize the number of experiments required to obtain the maximum amount of information, to select a proper set of the process parameters for experiments, and to design an optimized LCVD system.

In pyrolytic LCVD, the laser beam heats up the substrate to create a localized hot spot which raises the internal energy of the reactant molecules as they collide with the substrate surface at that spot. This process eventually leads to thermal dissociation of the reactant at the gas–substrate interface and to the deposition of a thin film on the substrate surface. In some cases, the reactant is so chosen that it absorbs the laser beam to rise in temperature, and dissociates in the gas phase due to the thermal decomposition reactions, with a thin film eventually being deposited on the substrate. In both cases, the temperature of the reactant molecules is raised to induce the chemical reactions and, therefore, the heat and mass transfer processes are coupled to each other. In the case where the substrate is heated, the absorptivity of the surface changes as the film is deposited, which alters the temperature of the localized hot spot and, consequently, the chemical reaction and the film deposition rate are affected. This phenomenon also couples the heat and mass transfer processes during pyrolytic LCVD. For this reason, the heat and mass conservation equations have to be solved simultaneously to model the LCVD process. However, if the chemical reaction rate is very slow, the heat and mass transfer equations can be solved separately by assuming that these two transfer processes are uncoupled. Even in other cases when the chemical reaction rate is not very slow, the solutions obtained by solving the decoupled heat and mass transfer equations can be utilized sequentially to set up an iterative scheme for modeling the LCVD process.

To model the pyrolytic LCVD process, we have to solve the mass, energy, and momentum conservation equations in a particular coordinate system depending on the configuration of the system, such as the shape and direction of propagation of the laser beam and the orientation of the substrate with respect to the laser beam, as well as the process conditions, such as the reactant flow rate and the pressure inside the reaction chamber. The laser beam and substrate are usually perpendicular or parallel to each other as shown in Figs. 4.1a and b, respectively. The parallel arrangement [Flint et al. (1984), Gattuso et al. (1983), Meunier et al. (1983), Bilenchi et al. (1982), West and Gupta (1984)] is useful to deposit a film over a large area by inducing thermal decomposition of the reactant molecules in the gas phase, and in this configuration the film qualities, such as the film density and microstructure, are controlled by monitoring the temperature of the substrate (see Fig. 4.1b).

LCVD reactors can be operated in the batch or continuous mode. In the batch mode, the deposition chamber is filled with the reactant molecules, the inlet and outlet valves of the chamber are closed, the film is deposited for a certain period of time, the reaction product and other gases are removed from the reactor, and then the chamber is charged with fresh reactant molecules to repeat the steps of the deposition process. Due to the absence of the gas flow into and out of the LCVD reactors during the deposition process, the forced convection does not exist in batch reactors. In the continuous operation mode, however, the reactant gas flows continuously into and out of the LCVD reactors during the deposition process, and the effect of forced convection can be important. At low pressures inside the LCVD reactors, the natural convection is usually not important, and diffusion is the most dominant transport mechanism in the absence of the forced convection.

To model the pyrolytic LCVD process, it is convenient to divide the deposition chamber into at least two regions, such that the concentration and temperature change more rapidly in one region than in the other. This type of multiregion analysis reduces the number of computational

# PYROLYTIC LCVD MODELING

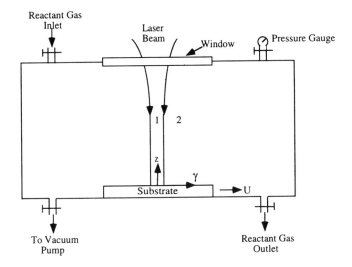

(a) The laser beam is perpendicular to the substrate.

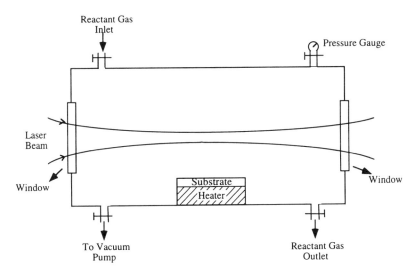

(b) The laser beam is parallel to the substrate.

Figure 4.1. Schematic diagram for typical LCVD chambers.

grids for numerical modeling, and also allows us to obtain analytical solutions by selecting a suitable geometry for the inner region that surrounds the laser beam.

## 4.2. GOVERNING EQUATIONS AND BOUNDARY CONDITIONS

There can be several species, that is, different types of molecules, in the LCVD chamber, whose concentration, velocity, and temperature distributions have to be determined in order to model the deposition process. Usually, the mass conservation equation is solved for each species, and by assuming that all of the species are under local thermodynamic equilibrium, the momentum and energy conservation equations are solved to determine the velocity and temperature fields, respectively, at various points in the gas phase. The derivations of the mass, momentum, and energy conservation equations for a single-component system as well as for a mixture of gases can be found in Bird et al. (1960). In this section, we will briefly discuss the governing equations and boundary conditions for the configurations shown in Figs. 4.1a and b.

### 4.2.1. Mass Conservation Equation

The equation for the conservation of the mass of $i$th species, where $i = 1, 2, 3, \ldots, i^*$, with $i^*$ being the total number of species inside the deposition chamber, in the $r$th region, where $r = 1, 2, 3, \ldots, r^*$, with $r^*$ being the total number of regions created in the deposition chamber for the purpose of modeling, is given by

$$\frac{\partial C_{ir}}{\partial t} + \nabla \cdot (C_{ir}\mathbf{u}) = -\nabla \cdot \mathbf{J}_{ir} + S_{p_{ir}}, \qquad (4.1a)$$

where the source term $S_{p_{ir}}$ is zero for the configuration of Fig. 4.1a, and represents the mass of the $i$th species produced per unit time per unit volume in the $r$th region for the configuration of Fig. 4.1b. It should be noted that the mass of the $i$th species produced per unit time per unit area in the $r$th region is considered in the boundary condition for Fig. 4.1a, as discussed in Section 4.4. For a multicomponent system, the mass flux $\mathbf{J}_{ir}$, is given by [Bird et al. (1960), p. 567]

$$\mathbf{J}_{ir} = \mathbf{J}_{ir_c} + \mathbf{J}_{ir_T} + \mathbf{J}_{ir_p} + \mathbf{J}_{ir_F} \qquad (4.1b)$$

where $\mathbf{J}_{ir_c}$, $\mathbf{J}_{ir_T} + \mathbf{J}_{ir_p}$, and $\mathbf{J}_{ir_F}$ are the mass fluxes of the $i$th species in the $r$th region due to the concentration (ordinary) diffusion that obeys Fick's law of mass diffusion, thermal diffusion (which is known as the Soret effect), pressure diffusion, and forced diffusion, respectively. $\mathbf{J}_{ir_p}$ is negligible when the pressure gradient is small. The pressure diffusion is used in centrifuge separations by establishing a large pressure gradient. $\mathbf{J}_{ir_F}$ is important for ionic systems, where the product of the local electric field strength and the ionic charge represents the external force on an ion. $\mathbf{J}_{ir_F}$ is zero if gravity is the only external force. The thermal diffusion flux $\mathbf{J}_{ir_T}$ is important for a large temperature gradient, which is usually encountered during laser-aided materials processing; otherwise, this flux is very small. Neglecting $\mathbf{J}_{ir_p}$ and $\mathbf{J}_{ir_F}$ in Eq. (4.1b), we can write

$$\mathbf{J}_{ir} = \mathbf{J}_{ir_c} + \mathbf{J}_{ir_T} = -D_{im}\nabla C_{ir} - D_{iT}\nabla \ln T \quad (4.1c)$$

The boundary conditions for Fig. 4.1a are discussed in Section 4.4. For the boundary conditions for Fig. 4.1b, the concentration and mass flux are considered to be continuous at the interface of two regions in the gas phase, and the concentration or mass flux is taken to be zero, or the mass flux is taken to be proportional to the concentration at the solid wall of the deposition chamber, as discussed in Section 5.2.

### 4.2.2. Momentum Conservation Equation

The momentum conservation equation in the $r$th region is given by

$$\rho\left[\frac{\partial \mathbf{u}}{\partial \mathbf{r}} + (\mathbf{u}\cdot\nabla)\mathbf{u}\right] = -\nabla\cdot\tau - \nabla\cdot P\delta + \rho(P, T)\mathbf{g} \quad (4.1d)$$

for a multicomponent system. Here we have omitted the subscript $r$, which denotes the $r$th region, from the variables, and the shear stress tensor $\tau$ is given by [Bird *et al.* (1960), p. 565],

$$\tau = -\mu(\nabla\mathbf{u} + (\nabla\mathbf{u})^t) + (\tfrac{2}{3}\mu - \kappa)(\nabla\cdot\mathbf{u})\delta \quad (4.1e)$$

The bulk viscosity $\kappa$ is zero for low-density monatomic gases and is usually insignificant for dense gases and liquids [Bird *et al.* (1960), p. 79]. Expressions for estimating $\kappa$ can be found in Hirschfelder *et al.* (1954, pp. 503, 647).

The unit tensor $\delta$ is represented by

$$\delta = \begin{bmatrix} 1 & 0 & 0 \\ 0 & 1 & 0 \\ 0 & 0 & 1 \end{bmatrix}$$

In Cartesian coordinates $\tau$, $\nabla \mathbf{u}$, $(\nabla \mathbf{u})^t$, and $\nabla \cdot \mathbf{u}$ are given by

$$\tau = \begin{bmatrix} \tau_{xx} & \tau_{xy} & \tau_{xz} \\ \tau_{yx} & \tau_{yy} & \tau_{yz} \\ \tau_{zx} & \tau_{zy} & \tau_{zz} \end{bmatrix}$$

$$\nabla \mathbf{u} = \begin{bmatrix} \dfrac{\partial u}{\partial x} & \dfrac{\partial v}{\partial x} & \dfrac{\partial w}{\partial x} \\ \dfrac{\partial u}{\partial y} & \dfrac{\partial v}{\partial y} & \dfrac{\partial w}{\partial y} \\ \dfrac{\partial u}{\partial z} & \dfrac{\partial v}{\partial z} & \dfrac{\partial w}{\partial z} \end{bmatrix}$$

$$(\nabla \mathbf{u})^t = \begin{bmatrix} \dfrac{\partial u}{\partial x} & \dfrac{\partial u}{\partial y} & \dfrac{\partial u}{\partial z} \\ \dfrac{\partial v}{\partial x} & \dfrac{\partial v}{\partial y} & \dfrac{\partial v}{\partial z} \\ \dfrac{\partial w}{\partial x} & \dfrac{\partial w}{\partial y} & \dfrac{\partial w}{\partial z} \end{bmatrix}$$

$$\nabla \cdot \mathbf{u} = \frac{\partial u}{\partial x} + \frac{\partial v}{\partial y} + \frac{\partial w}{\partial z}$$

For both configurations shown in Figs. 4.1a and b, the boundary conditions for Eq. (4.1d) can be obtained by considering that the velocity is zero at the gas–solid interface, and that the shear stress is continuous at the interface of two regions in the gas phase.

### 4.2.3. Energy Conservation Equation

The energy conservation equation in the $r$th region is given by

$$\rho \left[ \frac{\partial}{\partial r}(\bar{U} + \tfrac{1}{2}u_0^2) + \mathbf{u} \cdot \nabla(\bar{U} + \tfrac{1}{2}u_0^2) \right]$$
$$= -\nabla \cdot \mathbf{q} - \nabla \cdot [(\tau + P\delta)\mathbf{u}] + \rho(P, T)\mathbf{u} \cdot \mathbf{g} + S_T \quad (4.1\text{f})$$

for a multicomponent system. Equation (4.1f) is written for the gas phase for both configurations of Fig. 4.1 by again omitting the subscript $r$ from the variables. The thermal source term $S_T$ is zero for the configuration of Fig. 4.1a assuming that the laser does not heat up the gas, and it represents the amount of thermal energy produced per unit time per unit volume of the mixture for the configuration of Fig. 4.1b. For a multicomponent system, the heat flux $\mathbf{q}$ is given by [Bird et al. (1960), p. 566]

$$\mathbf{q} = \mathbf{q}_c + \mathbf{q}_i + \mathbf{q}_d \tag{4.1g}$$

where $\mathbf{q}_c$, $\mathbf{q}_i$, and $\mathbf{q}_d$ are the heat fluxes due, respectively, to thermal diffusion that obeys Fourier's law of heat conduction, the interdiffusion of various species, and the Dufour effect, which represents the effect of the concentration gradient. A discussion on the Dufour effect and an expression for $\mathbf{q}_d$ can be found in Hirschfelder et al. (1954, pp. 522, 705). However, Bird (1960), p. 566, has pointed out that the value of $\mathbf{q}_d$ is usually very small. Therefore, neglecting $\mathbf{q}_d$ in Eq. (4.1g), we can write

$$\mathbf{q} = \mathbf{q}_c + \mathbf{q}_i = -k\nabla T + \sum_{i=1}^{i^*} \mathbf{j}_i \bar{H}_i \tag{4.1h}$$

Noting that the enthalpy of the mixture, $H$, depends on $T$, $P$, and $n_i$, that is, $H = H(T, P, n_1, n_2, \ldots, n_{i^*})$, the partial molal enthalpy $\bar{H}_i$ is given by

$$\bar{H}_i = \left(\frac{\partial H}{\partial n_i}\right)\bigg|_{T, P, n_j (j \neq i)} = \bar{H}_i(T, P, n_i, n_2, \ldots, n_{i^*}) \tag{4.1i}$$

where $n_j$ stands for all the mole numbers except $n_i$. $\bar{H}_i$ is also referred to as the partial molar enthalpy, and in this context the terms molal and molar are used interchangeably [de Heer (1986)].

The energy conservation equation for the solid substrate of Fig. 4.1a, and the associated boundary conditions are discussed in Section 4.3. For the solid substrate of Fig. 4.1b, the energy conservation equation is the same as the heat conduction equation used in Section 4.3, and the boundary conditions can be expressed in the same way as in Section 4.3, when the method of heating the substrate is specified. The temperature and heat flux are considered to be continuous at the interface of two regions in the gas phase.

It should be noted that if the mathematical problem is symmetric about a line or a plane, the gradients of various variables, such as the concentration, speed, and temperature, along the normal to the line or plane can be taken to be zero at the line or plane of symmetry. In any case, the mass, momentum, and energy conservation equations have to be solved simultaneously to model the LCVD process. However, because these are coupled nonlinear partial differential equations they are difficult to solve in practice. For this reason, the conservation equations and boundary conditions are simplified by using appropriate approximations in order to model the LCVD process.

## 4.3. MODELING OF THE TEMPERATURE FIELD

Availability of high-power laser and the economical advantages of using laser as a tool for materials processing have led to many interesting applications of laser technology [Mazumder (1987)]. Considerable efforts are being made in various industries to implement laser technology for materials processing, and in various research facilities to understand the physics of such processes. Layer glazing [Breinan and Kear (1983)] and laser cladding [Singh and Mazumder (1987), Kar and Mazumder (1987, 1988)] have already demonstrated the feasibility of synthesizing novel materials in near net shape using laser. These processes can generate relatively thick coatings of metastable materials, whereas LCVD can produce thin (Å to several hundreds of $\mu$m) layers of metals and ceramics [Allen (1981), Chou et al. (1989), Mazumder and Allen (1980)]. Metals and ceramics (such as Ti and TiN) can be codeposited using this technique to produce composites. Heat transfer in the substrate plays an important role in laser processing of materials, and mathematical models of the processes allow us to examine the effect of various parameters on the efficacy of the process and control the manufacturing.

### 4.3.1. Introduction

In the literature, one can find that a lot of work has been done to determine the distribution of laser energy in the substrate materials. These studies consider either constant thermophysical properties for the substrate material or infinite or semi-infinite geometry. One of the earliest works on the energy distribution in the substrate during laser heating is due to Cline and Anthony (1977). They considered infinite geometry and constant thermophysical properties for the substrate material and solved the heat conduction equation by using Green's function for a Gaussian beam moving at a constant velocity. Lax (1977, 1978, 1979) studied the temperature rise under steady-state conditions that resulted from a stationary Gaussian beam in a semi-infinite cylindrical medium. He considered the linear (constant thermophysical properties) case in Lax (1977) and the nonlinear (temperature-dependent thermal conductivity) case by the method of the Kirchhoff transformation in Lax (1978) and Lax (1979). Hess et al. (1980) presented a quasi-steady-state solution for the temperature distribution in a radially infinite cylindrical medium with temperature-dependent thermal conductivity by neglecting the effect of scan velocity of the laser beam. Bell (1979) developed a one-dimensional thermal model for laser annealing over a wide range of laser pulse durations and absorption coefficients. Nissim et al. (1980) studied the effects of scanning elliptical or circular CW laser beams on the temperature distributions in semiconductors. For an infinite medium, they modeled the linear case following Cline and Anthony (1977), and for the nonlinear case they used the Kirchhoff transformation.

Kokorowski et al. (1955) studied the temperature distributions during CW laser annealing with a stationary beam beyond the melt temperature. They solved the steady-state heat conduction equation in a radially infinite cylindrical medium with temperature-dependent thermal conductivity. Moody and Hendel (1982) presented a numerical algorithm for temperature distribution in an infinite medium during laser heating to generalize the models of Lax (1977, 1978, 1979), Cline and Anthony (1977), Hess et al. (1980), Nissim et al. (1980), and Kokorowski et al. (1955). Recently, Kant (1988) modeled the heating of a multilayered cylindrical medium resting on a half-space with a stationary laser beam. He considered the medium to be radially infinite and the materials of the medium to have constant thermophysical properties.

Apart from these works, various studies concerning the laser heating of different types of materials, such as metals, semiconductors, thermally insulating materials, and thin films, can be found in Gibbons and Sigmon (1982), Abraham and Halley (1987), Sanders (1984), Ferrieu and Auvert (1983), Kwong and Kim (1982), Kim et al. (1981), Calder and Sue (1982), Burgener and Reedy (1982), Meyer et al. (1980a, b), El-Adawi and Elshehawey (1986), and Liarokapis and Raptis (1985). It can be seen that a lot of work has been done to model the energy distribution in the substrate due to laser heating by considering the substrate infinite or semi-infinite. Although the results of such models can be applied to substrates of dimensions much larger than the laser spot size, the infinite or semi-infinite medium approximation will not hold good for small substrates. A transient and three-dimensional thermal analysis for laser heating of uniformly moving finite slabs can be found in Kar and Mazumder (1989), and is discussed below.

## 4.3.2. Governing Equations and Boundary Conditions

In LCVD processes, the temperature field has to be obtained to determine the chemical reaction zone on the surface of the substrate. The chemical reaction zone is defined as that region on the surface of the substrate in which the temperature is higher than or equal to the chemical reaction temperature. Also, the temperature field can be used to determine the thermal stress of the deposited film (see Appendix B). The temperature field is obtained by solving the heat conduction equation with temperature-dependent thermophysical properties of the substrate. Figure 4.2a shows the finite-slab geometry and the Cartesian coordinate system under consideration. In this study, the substrate is considered to move at a constant velocity in the negative $x$-direction and the laser beam is stationary. For the purpose of mathematical analysis, the frame of reference is so chosen that the substrate stays stationary and the laser beam moves with respect to the substrate at a constant velocity in the positive $x$-direction. The energy transfer equation in the slab is given by

$$\rho(T)C_p(T)\frac{\partial T}{\partial t}(x, y, z, t) = \nabla[k(T)\nabla T(x, y, z, t)] \quad (4.2a)$$

Equation (4.2a) is solved to satisfy the following initial and boundary conditions at the surfaces of the slab. Both convective and radiative heat

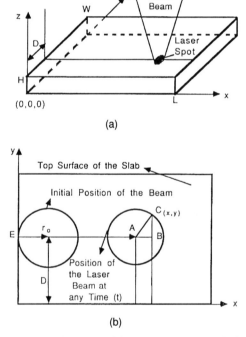

Figure 4.2. (a) Geometric configuration of the substrate and the relative position of the laser spot. (b) Top view of the laser–substrate interaction zone at the initial time and at a later time.

losses through the boundary surfaces of the substrate are considered in this analysis:

$$T(x, y, z, 0) = T_i \tag{4.2b}$$

$$k(T)\frac{\partial T}{\partial x} = h_{x0}(T - T_a) + \sigma\varepsilon(T^4 - T_a^4) \quad \text{at} \quad x = 0 \tag{4.2c}$$

$$-k(T)\frac{\partial T}{\partial x} = h_{xL}(T - T_a) + \sigma\varepsilon(T^4 - T_a^4) \quad \text{at} \quad x = L \tag{4.2d}$$

$$k(T)\frac{\partial T}{\partial y} = h_{y0}(T - T_a) + \sigma\varepsilon(T^4 - T_a^4) \quad \text{at} \quad y = 0 \tag{4.2e}$$

$$-k(T)\frac{\partial T}{\partial y} = h_{yW}(T - T_a) + \sigma\varepsilon(T^4 - T_a^4) \quad \text{at} \quad y = W \tag{4.2f}$$

$$k(T)\frac{\partial T}{\partial z} = h_{z0}(T - T_a) + \sigma\varepsilon(T^4 - T_a^4) \quad \text{at} \quad z = 0 \tag{4.2g}$$

$$k(T)\frac{\partial T}{\partial z} = \frac{2P(1 - R)}{\pi r_0^2} e^{-2r^2/r_0^2} - h_{zH}(T - T_a) - \sigma\varepsilon(T^4 - T_a^4)$$

$$\text{at} \quad z = H \tag{4.2h}$$

where $R$ represents the reflectivity of the composite medium, which is made up of SS 304, Ti film, and $TiBr_4$ vapor. Such a medium is encountered in LCVD of pure Ti from $TiBr_4$ on SS 304, as discussed in the next section. The high-temperature data for the absorptivity $A$, which is $1 - R$, of the composite medium is obtained from the low-temperature data as explained in Chapter 2. $A$ is considered constant and equal to the absorptivity of pure Ti in this analysis. While setting up the boundary condition (4.2h), the loss of the absorbed laser energy in the thin metallic film is considered negligibly small. Also, the effect of the optical absorption depth is ignored in this model because it is usually very small for metallic materials. It should be noted that the location of the laser beam $r$ varies with time. Figure 4.2b shows the top view of the location of the laser beam on the top surface of the substrate at time zero and at a later time $t$. The substrate moves to the left with a uniform velocity $U$, that is, the laser beam moves in the positive x-direction with a constant velocity $U$ with respect to the substrate. Thus, $r$ is given by

$$r^2 = (x - r_0 - Ut)^2 + (y - D)^2 \tag{4.3}$$

where $D$ is the $y$-coordinate of the center of the cross section of the laser beam on the top surface of the substrate.

### 4.3.3. Method of Solution

The above conduction problem is nonlinear because the thermophysical properties of the substrate are considered to vary with temperature and radiation is taken to be one of the modes of heat loss from the substrate. To obtain an analytic solution of this problem, the governing Eq. (4.2a) is linearized by using Kirchhoff's transformation [Carslaw and Jaeger (1986)] and the boundary conditions are linearized by expanding the nonlinear terms in a Taylor series. The linearization process is carried out after reformulating the problem in terms of the following dimensionless variables:

$$x^* = x/L, \qquad y^* = y/W, \qquad z^* = z/H, \qquad t^* = t/\tau$$

where $\tau = L/U$, and

$$T^* = T/T_d, \qquad \bar{T}_a^* = \bar{T}_a/T_d$$

In this model $\bar{T}_a$ is taken to be equal to $T_d$. Also, we consider that the thermal conductivity of the substrate depends on temperature in the following way:

$$k(T) = pT^q$$

where $q$ and $\log p$ are, respectively, the slope and the intercept of $\log k(T)$ vs. $T$ graph. The dimensionless thermal conductivity is defined as

$$k^*(T^*) = \frac{k(T)}{k(T_d)} = \frac{T^q}{T_d^q} = T^{*q} \qquad (4.4)$$

To simplify the governing Eq. (4.2a) using Kirchhoff's transformation, we define a transformed temperature $T'^*$ such that

$$\frac{\partial T'^*}{\partial z^*} = k^*(T^*) \frac{\partial T^*}{\partial z^*} \qquad (4.5)$$

Substituting Eq. (4.4) into Eq. (4.5) and integrating the latter with respect to temperature from 1 to $T^*$, we obtain

$$T^{*q+1} = 1 + (q+1)T'^* \qquad (4.6)$$

The nonlinear term on the left-hand side of Eq. (4.6) is expanded around $\bar{T}_a^*$ by a Taylor series to obtain the following linear relation between the original temperature $T^*$ and the transformed temperature $T'^*$:

$$T^* = \frac{1}{(q+1)\bar{T}_a^{*q}}[1 - \bar{T}_a^{*q+1} + (q+1)T'^*] + \bar{T}_a^* \qquad (4.7)$$

Equation (4.7) is used to simplify the boundary conditions. For this purpose, we first make the boundary conditions (4.2c) through (4.2h) dimensionless and then linearize the radiative heat loss terms by expanding the nonlinear terms around $\bar{T}_a^*$ by a Taylor series. Thus, we obtain the following effective heat transfer coefficients $H_i$ at the boundary surfaces of the substrate:

$$H_i = h_i + \sigma\varepsilon T_d^3(\bar{T}_a^{*2} + T_a^{*2})(\bar{T}_a^* + T_a^*)$$

where $i = 1, 2, 3, 4, 5,$ and 6, which denote the $x = 0$, $x = L$, $y = 0$, $y = W$, $z = 0$ and $z = H$ planes, respectively. The resulting boundary conditions can now be expressed in terms of the transformed temperature $T'^*$. However, we define

$$T'' = T'^* + T_c \qquad (4.8)$$

where

$$T_c = \bar{T}_a^{*q}\left[\bar{T}_a^* - T_a^* + \frac{1 - \bar{T}_a^{*q+1}}{(q+1)\bar{T}_a^{*q}}\right]$$

and

$$X_0 = \frac{H_1 L}{k(T_d)\bar{T}_a^{*q}} \qquad X_L = \frac{H_2 L}{k(T_d)\bar{T}_a^{*q}}$$

$$Y_0 = \frac{H_3 W}{k(T_d)\bar{T}_a^{*q}} \qquad Y_W = \frac{H_4 W}{k(T_d)\bar{T}_a^{*q}}$$

$$Z_0 = \frac{H_5 H}{k(T_d)\bar{T}_a^{*q}} \qquad Z_H = \frac{H_6 H}{k(T_d)\bar{T}_a^{*q}}$$

In terms of these variables, the boundary conditions can be written as

$$\frac{\partial T_s''}{\partial x^*} = X_0 T_s'' \quad \text{at} \quad x^* = 0 \tag{4.9a}$$

$$\frac{\partial T_s''}{\partial x^*} = X_L T_s'' \quad \text{at} \quad x^* = 1 \tag{4.9b}$$

$$\frac{\partial T_s''}{\partial y^*} = Y_0 T_s'' \quad \text{at} \quad y^* = 0 \tag{4.9c}$$

$$\frac{\partial T_s''}{\partial y^*} = Y_W T_s'' \quad \text{at} \quad y^* = 1 \tag{4.9d}$$

$$\frac{\partial T_s''}{\partial z^*} = Z_0 T_s'' \quad \text{at} \quad z^* = 0 \tag{4.9e}$$

$$\frac{\partial T_s''}{\partial z^*} = P^* \exp(-r^{*2}) + Z_H T_s'' \quad \text{at} \quad z^* = 1 \tag{4.9f}$$

where

$$P^* = \frac{2PAH}{\pi L^2 r_0^* k(T_d) T_d}$$

$$r^{*2} = \frac{2}{r_0^{*2}} \left[ (x^* - r_0^* - t^*)^2 + \frac{1}{a^2}(y^* - D^*)^2 \right]$$

$$r_0^* = r_0/L \quad \text{and} \quad D^* = D/W$$

The initial condition (4.2b) simplifies to

$$T_s'' = T_i' \quad \text{at} \quad t^* = 0 \tag{4.9g}$$

where

$$T_i' = T_c + \frac{T_i^{*q+1} - 1}{q+1}$$

and

$$T_i^* = T_i/T_d$$

The governing Eq. (4.2a) can be simplified by using the above dimensionless variables and the transformations defined by Eqs. (4.5) and (4.8) to obtain

$$\frac{\partial^2 T'''^*}{\partial x^{*2}} + a^2 \frac{\partial^2 T'''^*}{\partial y^{*2}} + s^2 \frac{\partial^2 T'''^*}{\partial z^{*2}} = \frac{1}{\text{Fo}} \frac{\partial^2 T''}{\partial t^*} \tag{4.9h}$$

where the aspect ratio $a = L/W$; the slenderness ratio $s = L/H$ and the Fourier number $\text{Fo} = \alpha(T)\tau/L^2$, and $\alpha(T)$ is almost constant for many materials because it varies slowly with temperature. In this study, $\alpha(T)$ is considered constant and is taken to be the temperature-averaged value of the thermal diffusivity over the temperature ranging from the ambient temperature to the melting point of the substrate.

The above problem defined by Eqs. (4.9a) through (4.9h) is solved by successively applying a Fourier transform in the $x$, $y$ and $z$ directions. The Fourier double integral transform [Sneddon (1972)] in the $x$ and $y$ directions is

$$\bar{T}''(\lambda_{lx}, \lambda_{my}, z^*, t^*) = \int_0^1 \int_0^1 dx^* dy^* T''(x^*, y^*, z^*, t^*) K_{lx}(x^*) K_{my}(y^*) \quad (4.10a)$$

and the inversion formula is

$$T''(x^*, y^*, z^*, t^*) = \sum_{l=0}^{\infty} \sum_{m=0}^{\infty} \frac{\bar{T}''(\lambda_{lx}, \lambda_{my}, z^*, t^*)}{N_{lx} N_{my}} K_{lx}(x^*) K_{my}(y^*) \quad (4.10b)$$

where the kernels $[K_{lx}(x^*), K_{my}(y^*)]$, the characteristic equations for the eigenvalues $(\lambda_{lx}, \lambda_{my})$, and the normalization constants $(N_{lx}, N_{my})$ for the above integral transform in each of the $x$ and $y$ directions are, respectively, defined as

$$K_{lx}(x^*) = X_0 \sin(\lambda_{lx} x^*) + \lambda_{lx} \cos(\lambda_{lx} x^*)$$

$$\tan \lambda_{lx} = \frac{\lambda_{lx}(X_0 - X_L)}{\lambda_{lx}^2 + X_0 X_L}$$

$$N_{lx} = \frac{1}{2}\left[X_0 + (\lambda_{lx}^2 + X_0^2) + \left(1 - \frac{X_L}{\lambda_{lx}^2 + X_L^2}\right)\right]$$

and

$$K_{my}(y^*) = Y_0 \sin(\lambda_{my} y^*) + \lambda_{my} \cos(\lambda_{my} y^*)$$

$$\tan \lambda_{my} = \frac{\lambda_{my}(Y_0 - Y_W)}{\lambda_{my}^2 + Y_0 Y_W}$$

$$N_{my} = \frac{1}{2}\left[Y_0 + (\lambda_{my}^2 + Y_0^2) \left(1 - \frac{Y_W}{\lambda_{my}^2 + Y_W^2}\right)\right]$$

Here the kernels and the eigenvalues are so chosen that the boundary conditions (4.9a) through (4.9d) are satisfied. Application of the integral transform (4.10a) to Eq. (4.9h) and the boundary and initial conditions (4.9e) through (4.9g) yields the following: Equation (4.9h) becomes

$$-(\lambda_{lx}^2 + a^2\lambda_{my}^2)\bar{T}'' + s^2\frac{\partial^2 \bar{T}''}{\partial z^{*2}} = \frac{1}{\text{Fo}}\frac{\partial \bar{T}''}{\partial t^*} \qquad (4.11a)$$

The boundary conditions (4.9e) and (4.9f) become

$$\frac{\partial \bar{T}''}{\partial z^*} = Z_0 \bar{T}'' \quad \text{at} \quad z^* = 0 \qquad (4.11b)$$

and

$$\frac{\partial \bar{T}''}{\partial z^*} = Z_H \bar{T}'' + f_{lm}(t^*) \quad \text{at} \quad z^* = 1 \qquad (4.11c)$$

where

$$f_{lm}(t^*) = \int_0^1 \int_0^1 dx^* dy^* K_{lm}(x^*) K_{my}(y^*) P^* e^{-r^{*2}(x^*, y^*, t^*)} \qquad (4.11d)$$

This integral is evaluated at the end of this subsection. The initial condition (4.9g) becomes

$$\bar{T}'' = \bar{T}_i' \qquad (10e)$$

where

$$\bar{T}_i' = T_i' \left[ \sin\lambda_{lx} + \frac{X_0}{\lambda_{lx}}(1 - \cos\lambda_{lx}) \right]\left[ \sin\lambda_{my} + \frac{Y_0}{\lambda_{my}}(1 - \cos\lambda_{my}) \right]$$

Equation (4.11a) will be solved by applying the Fourier integral transform in the $z$-direction after homogenizing the boundary condition (4.11c) by the following substitution:

$$\bar{T}''(\lambda_{lx}, \lambda_{my}, z^*, t^*) = \bar{T}'''(\lambda_{lx}, \lambda_{my}, z^*, t^*) + \bar{T}^{iv}(\lambda_{lx}, \lambda_{my}, z^*, t^*) \qquad (4.12)$$

where

$$\bar{T}^{iv}(\lambda_{lx}, \lambda_{my}, z^*, t^*) = f_{lm}(t^*)\frac{Z_0 z^* + 1}{\Delta_z}$$

# PYROLYTIC LCVD MODELING

and

$$\Delta_z = Z_0(1 - Z_H) - Z_H$$

Substituting Eq. (4.12) into Eq. (4.11a), using the boundary and the initial conditions (4.11b, c, and e), and then applying the following integral transform:

$$\bar{\bar{T}}'''(\lambda_{lx}, \lambda_{my}, \lambda_{nz}, t^*) = \int_0^1 dz^* \bar{T}'''(\lambda_{lx}, \lambda_{my}, z^*, t^*) K_{nz}(z^*)$$

where the inversion formula is

$$\bar{T}'''(\lambda_{lx}, \lambda_{my}, z^*, t^*) = \sum_{m=0}^{\infty} \frac{\bar{\bar{T}}'''(\lambda_{lx}, \lambda_{my}, \lambda_{nz}, t^*)}{N_{nz}} \qquad (4.13)$$

and the kernel ($K_{nz}$), the characteristic equation for the eigenvalues $\lambda_{nz}$ and the normalization constant $N_{nz}$ are

$$K_{nz}(z^*) = Z_0 \sin(\lambda_{nz} z^*) + \lambda_{nz} \cos(\lambda_{nz} z^*)$$

$$\tan \lambda_{nz} = \frac{\lambda_{nz}(Z_0 - Z_H)}{\lambda_{nz}^2 + Z_0 Z_H}$$

$$N_{nz} = \frac{1}{2}\left[Z_0 + (\lambda_{nz}^2 + Z_0^2)\left(1 - \frac{Z_H}{\lambda_{nz}^2 + Z_H^2}\right)\right]$$

we obtain the following results: Equation (4.11a) becomes

$$-(\lambda_{lx}^2 + a^2\lambda_{my}^2)[\bar{\bar{T}}''' + \psi(t^*)] - \lambda_{nz}^2 s^2 \bar{\bar{T}}''' + \frac{1}{Fo}\frac{d}{dt^*}[\bar{\bar{T}}''' + \psi(t^*)] \qquad (4.14)$$

where

$$\psi(t^*) = \int_0^1 dz^* K_{nz}(z^*) \bar{T}^{iv}(\lambda_{lx}, \lambda_{my}, z^*, t^*)$$

$$= \frac{f_{lm}(t^*)}{\Delta_z}\left[\frac{Z_0}{\lambda_{nz}}(1 - \cos \lambda_{nz}) + \sin \lambda_{nz}\right]$$

$$+ \frac{Z_0 f_{lm}(t^*)}{\Delta_z}\left[Z_0\left(\frac{\sin \lambda_{nz}}{\lambda_{nz}^2} - \frac{\cos \lambda_{nz}}{\lambda_{nz}}\right) + \lambda_{nz}\left(\frac{\cos \lambda_{nz} - 1}{\lambda_{nz}^2}\right)\right]$$

To obtain Eq. (4.14), the kernel $K_{nz}(z^*)$ and the eigenvalues $\lambda_{nz}$ are chosen in such a way that the boundary conditions (4.11b) and (4.11c) are satisfied. By adding and subtracting $s^2\lambda_{nz}^2\psi(t^*)$ on the left-hand side of Eq. (4.14), and letting

$$\lambda_{lmn}^2 = \text{Fo}(\lambda_{lx}^2 + a^2\lambda_{my}^2 + s^2\lambda_{nz}^2)$$

we obtain the following ordinary differential equation

$$\frac{d\psi_1(t^*)}{dt^*} + \lambda_{lmn}^2\psi_1(t^*) = \text{Fo}\, s^2\lambda_{nz}^2\psi(t^*) \qquad (4.15)$$

where

$$\psi_1 = \bar{T}'''(\lambda_{lx}, \lambda_{my}, \lambda_{nz}, t^*) + \psi(t^*) \qquad (4.16)$$

The solution of Eq. (4.15) is given by

$$\psi_1(t^*) = \psi_1(0)\exp(-\lambda_{lmn}^2 t^*) + \exp(-\lambda_{lmn}^2 t^*)\int_0^{t^*} \text{Fo}\, s^2\lambda_{nz}^2\exp(\lambda_{lmn}^2 \bar{t}^*)\psi(\bar{t}^*)d\bar{t}^* \qquad (4.17)$$

$\psi_1(t^*)$ is determined from Eq. (4.16) by utilizing the results obtained after taking the Fourier transform in the z-direction of the expressions (4.11e) and (4.12). This yields

$$\psi_1(0) = \bar{T}_i''\left[\sin\lambda_{nz} + \frac{Z_0}{\lambda_{nz}}(1-\cos\lambda_{nz})\right]$$

By using Eq. (4.17) in the expression (4.16), we obtain

$$\bar{T}'''(\lambda_{lx}, \lambda_{my}, \lambda_{nz}, t^*) = \psi_1(0)\exp(-\lambda_{lmn}^2 t^*) + \psi_2(t^*) \qquad (4.18)$$

where

$$\psi_2(t^*) = \bar{\psi}_2(t^*) - \psi(t^*)$$

and

$$\bar{\psi}_1(t^*) = \exp(-\lambda_{lmn}^2 t^*)\int_0^{t^*} \text{Fo}\, s^2\lambda_{nz}^2\psi(\bar{t}^*)\exp(\lambda_{lmn}^2 t^*)d\bar{t}^* - \psi(t^*) \qquad (4.19)$$

PYROLYTIC LCVD MODELING    233

An explicit expression for $\bar{\psi}_1(t^*)$ is obtained at the end of this subsection. By applying the inversion formula (4.13) to Eq. (4.18), and then using the expression (4.12), the inversion formula (4.10b), and the expression (4.8), we obtain

$$T'^*(x^*, y^*, z^*, t^*) = \sum_{l=0}^{\infty} \frac{K_{lx}(x^*)}{N_{lx}} \left\{ \sum_{m=0}^{\infty} \frac{K_{my}(y^*)}{N_{my}} \left\{ \bar{T}^{iv}(\lambda_{lx}, \lambda_{my}, x^*, t^*) \right. \right.$$

$$\left. \left. \times \sum_{n=0}^{\infty} [\psi_1(0)\exp(-\lambda_{lmn}^2 t^*) + \psi_2(t^*)] \frac{K_{nz}(z^*)}{N_{nz}} \right\} \right\} - T_c$$

(4.20)

From the transformed temperature $T'^*$ given by Eq. (4.20), the dimensionless temperature $T^*$ is determined by using Eq. (4.7).

### 4.3.3.1. Evaluation of $f_{lm}(t^*)$

The expression for $f_{lm}(t^*)$ is given by Eq. (4.11d) as follows:

$$f_{lm}(t^*) = P^* \int_0^1 dy^* K_{my}(y^*) \exp\left[\frac{-2(y^* - D^*)^2}{a^2 r_0^{*2}}\right]$$

$$\times \int_0^1 dx^* K_{lx}(x^*) \exp\left[\frac{-2[x^* - (r_0^* + t^*)]^2}{r_0^{*2}}\right]$$

It should be noted that due to the exponential terms the integrands of the above two integrals will be very small at a radial distance larger than $r_0^*$ from the center of the laser beam on the top surface of the substrate. Therefore, if the center of the laser beam is one beam radius away from any of the four edges of the top surface of the substrate the limits 0 and 1 of both integrations can be replaced by $-\infty$ and $\infty$, respectively. Because of this, the first integration of the above equation can be approximated by

$$Y = \int_{-\infty}^{\infty} dy^* \exp\left[\frac{-2y^* - D^*)^2}{(a^2 r_0^{*2})}\right] [Y_0 \sin(\lambda_{my} y^*) + \lambda_{my} \cos(\lambda_{my} y^*)]$$

$$= \left(\frac{\pi r_0^{*2} a^2}{2}\right) \exp\left(\frac{-\lambda_{my}^2 r_0^{*2} a^2}{8}\right) [Y_0 \sin(D_{my}^*) + \lambda_{my} \cos(D_{my}^*)]$$

Similarly, the second integration in the expression for $f_{lm}(t^*)$ can be approximated by

$$X = \left(\frac{\pi r_0^{*2} a^2}{2}\right) \exp\left(\frac{-\lambda_{my}^2 r_0^{*2} a^2}{8}\right) [X_0 \sin(r_0^* + t^*) + \lambda_{lx} \cos \lambda_{lx}(r_0^* + t^*)]$$

Finally, $f_{lm}(t^*)$ can be approximately written as

$$f_{lm}(t^*) = P^* Y X$$

### 4.3.3.2. Evaluation of $\bar{\psi}_1(t^*)$

The expression for $\bar{\psi}_1(t^*)$ is given by Eq. (4.19) as follows:

$$\psi_1(t^*) = \exp(-\lambda_{lmn}^2 t^*) \int_0^{t^*} \text{Fo} \, s^2 \lambda_{nz}^2 \exp(\lambda_{lmn}^2 \bar{t}^*) \psi(\bar{t}^*) d\bar{t}^*$$

$$= \text{Fo} \, \frac{\lambda_{nz}^2 \exp(-\lambda_{lmn}^2 t^*)}{H^{*2}} \int_0^{t^*} \exp(\lambda_{lmn}^2 \bar{t}^*) K_1 f_{lm}(\bar{t}^*) d\bar{t}^*$$

where

$$K_1 = \frac{1}{\Delta z}\left(Z_0 \frac{1 - \cos \lambda_{nz}}{\lambda_{nz}} + \sin \lambda_{nz}\right)$$

$$+ \frac{1}{\Delta z}\left[Z_0\left(\frac{\sin \lambda_{nz}}{\lambda_{nz}^2} - \frac{\cos \lambda_{nz}}{\lambda_{nz}^2}\right) + \frac{\cos \lambda_{nz} - 1}{\lambda_{nz}} + \sin \lambda_{nz}\right]$$

Using the expression for $f_{lm}(t^*)$, we can rewrite the above equation as

$$\bar{\psi}_1(t^*) = K_2 \exp\left(-\frac{\lambda_{lx}^2 r_0^{*2}}{8}\right) \exp(-\lambda_{lmn}^2 t^*) \int_0^{t^*} d\bar{t}^* \exp(\lambda_{lmn}^2 \bar{t}^*)$$

$$\times [X_0 \sin \lambda_{lz}(r_0^* + \bar{t}^*) + \lambda_{lx} \cos \lambda_{lx}(r_0^* + \bar{t}^*)]$$

where

$$K_2 = \sqrt{\frac{\pi}{2}} \, r_0^* K_1 \text{Fo} \, s^2 \lambda_{nz}^2 P^* Y$$

Carrying out the above integration, we obtain

$$\bar{\psi}_1(t^*) = K_2 \left\{ \frac{X_0}{(\lambda_{lmn}^4 + \lambda_{lx}^2)} \exp\left(-\frac{\lambda_{lx}^2 r_0^{*2}}{8}\right) \right.$$

$$\times \left[ \lambda_{lmn}^2 \sin \lambda_{lx}(t^* + r_0^*) - \lambda_{lx} \cos \lambda_{lx}(t^* + r_0^*) \right]$$

$$+ \frac{\lambda_{lx}}{\lambda_{lmn}^4 + \lambda_{lx}^2} [\lambda_{lmn}^2 \cos \lambda_{lx}(t^* + r_0^*) - \lambda_{lx} \sin \lambda_{lx}(t^* + r_0^*)] \exp\left(-\frac{\lambda_{lx}^2 r_0^{*2}}{8}\right)$$

$$- \frac{X_0}{\lambda_{lmn}^4 + \lambda_{lx}^2} \exp\left(-\lambda_{lmn}^2 t^* - \frac{\lambda_{lx}^2 r_0^{*2}}{8}\right) [\lambda_{lmn}^2 \sin(\lambda_{lx} r_0^*) - \lambda_{lx} \cos(\lambda_{lx} r_0^*)]$$

$$\left. - \frac{\lambda_{lx}}{\lambda_{lmn}^4 + \lambda_{lx}^2} \exp\left(-\lambda_{lmn}^2 t^* - \frac{\lambda_{lx}^2 r_0^{*2}}{8}\right) [\lambda_{lmn}^2 \cos(\lambda_{lx} r_0^*) - \lambda_{lx} \sin(\lambda_{lx} r_0^*)] \right\}$$

### 4.3.4. Results and Discussion

Equation (4.20) can be used to carry out parametric studies of laser heating of finite slabs. Various parameters such as the wavelength of the laser beam; laser power; the shape, diameter, and, speed of the laser beam relative to the substrate; the dimensions, and, the thermophysical and optical properties of the substrate; and the conditions of the medium surrounding the substrate can affect the temperature distribution in the substrate. Some typical results are presented in Figs. 4.3 through 4.14. Also, two simple expressions for the variation of peak temperature with laser power and with substrate velocity are obtained.

For all these results, the value of $r_0$ is taken to be 2 mm. The laser beam is located at $y^* = 0.5$ and moves in the positive $x$-direction with respect to the substrate. The values of the heat transfer coefficients in the boundary conditions (4.2c) through (4.2h) are determined from the following considerations: The slab considered in this study is very thin in the $z$-direction. Its length, width, and height are 1 cm, 1 cm, and 0.1587 cm, respectively. Since the slab is exposed to the laser beam at the $z = H$ plane, the surfaces at $z = 0$ and $z = H$ are expected to heat up more than the other four surfaces for the same ambient conditions at all six sides of the slab. Thus, the convective activity will be more at the $z = 0$ and $z = H$ planes than at the other surfaces. In this study, the convective heat loss from the surfaces at $x = 0$, $x = L$, $y = 0$, and $y = W$ is considered to be due to free convection, and the heat transfer coefficients are taken to be 5 W/m²-K [Incropera and Dewitt (1985)] at these four surfaces. The heat transfer coefficients at the

$z = 0$ and $z = H$ planes are determined by assuming that the Biot (Bi) numbers are equal at all the surfaces, where the characteristic dimensions in Bi numbers are taken to be the length, width, and height of the slab for the surfaces perpendicular to the $x$, $y$ and $z$ axes, respectively.

Figures 4.3 through 4.5 represent the temperature distribution on the top surface of the substrate. Although the computation for the surface temperature distribution is carried out according to the coordinate system of Fig. 4.2a, the three-dimensional Figs. 4.3 through 4.5 are plotted by letting $x^*$, $y^*$, and the temperature $T^*$ increase along $ox^*$, $oy^*$, and $oT^*$ directions, respectively, as indicated in these three figures. Here the point $o$ represents one of the corners of the square containing the lines $o'v$ and $o''v$ as two of its sides on the plane of $o'vo''$. For improved clarity in representing the surface temperature field in three-dimensional plots, views from the two planes, one from the plane $T'o'v$ located at $x^* = 1$ and the other from $T''o''v$ located at $y^* = 1$ are shown. Figures 4.3 through 4.5 shows the temperature fields for laser power $P = 600$ W, laser scanning speed relative to the substrate $U = 0.25$ cm/s, and for the dimensionless time $t^* = 0.2$, 0.5, and 0.8, respectively.

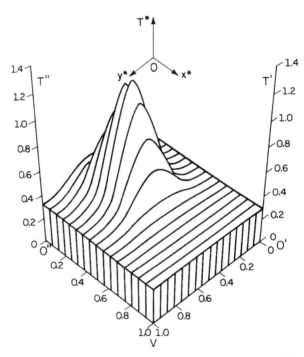

Figure 4.3. Surface temperature distribution along $y^* = 0.5$ at $t^* = 0.2$ for a 600-W laser beam moving in the x-direction at a constant velocity of 0.25 cm/s relative to the substrate.

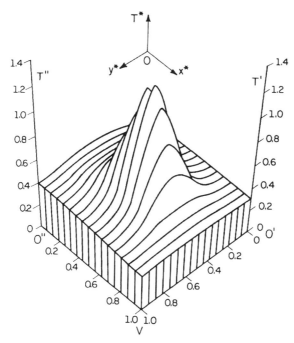

Figure 4.4. Surface temperature distribution along $y^* = 0.5$ at $t^* = 0.5$ for a 600-W laser beam moving in the $x$-direction at a constant velocity of 0.25 cm/s relative to the substrate.

It can be seen from these figures that the shape of the surface temperature field has a Gaussian like structure, because of the consideration of Gaussian laser beam as the source of heat in this study. The temperature and the length of the heated zone ahead of the laser beam in the $x^*$-direction are found to increase as $t^*$ increases. This is so because at a low scanning speeds Fo is large; that is, the conduction rate is higher than the heat storage rate. Consequently, the substrate material which is in front of the laser beam is heated by the heat conducted away from the laser heated spot. Hence, the laser energy is progressively imparted to points on the substrate which are at higher temperatures than the preceding points. For the very same reason, the laser heated zone in the $y^*$-direction increases as $t^*$ increases for low scanning speeds.

Knowing the width of the laser heated zone is very important when depositing film by such processes as LCVD. The chemical reaction that generates the film-forming material will take place wherever the temperature is greater than or equal to the chemical reaction temperature. As explained above, the width of the heated zone will increase as the laser beam scans the substrate at a low scanning speed. This means that the width of the zone,

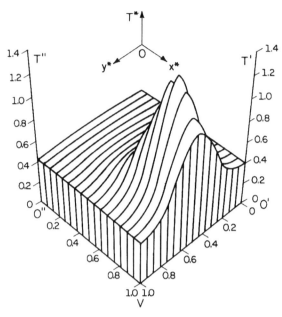

Figure 4.5. Surface temperature distribution along $y^* = 0.5$ at $t^* = 0.8$ for a 600-W laser beam moving in the $x$-direction at a constant velocity of 0.25 cm/s relative to the substrate.

to be referred to as the chemically reactive zone, over which the temperature can be greater than or equal to the film-forming chemical reaction temperature will increase progressively along the scanning direction for low scanning speeds of the laser beam. Thus, nonuniform and wide films will be deposited on the substrate for low scanning speeds of the laser beam. On the other hand, if the scanning speed is high, the Fourier number will be small; that is, the conduction rate will be lower than the heat storage rate. This reduces the area of the chemically reactive zone because there is less conduction of heat from the laser heated spot. As a result, a narrow and uniform film can be deposited on the substrate by increasing the scanning speed of the laser beam. This concept is reflected in the results presented in Figs. 4.6 through 4.8. These figures show the width of the chemically reactive zone and its variation in the scanning direction for various scanning speeds and laser powers. The chemically reactive zones are symmetrical around the line $y^* = 0.5$ because in this study the laser beam is located at $y^* = 0.5$. In each of the Figs. 4.6 through 4.8, the chemically reactive zones are bounded by the curves A, B, C, and D for scanning speeds 0.35, 0.25, 0.167, and 0.125 cm/s, respectively. For a given laser power, it can be seen from each of these figures that the chemically reactive zone becomes wider and less

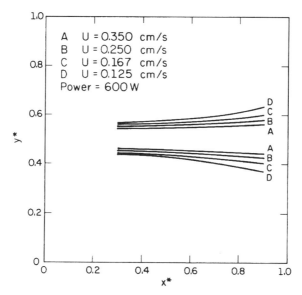

Figure 4.6. Variation of the width of the heated zone at the surface of the substrate along the direction of motion of the laser beam for various scanning speeds of the beam relative to the substrate and for laser power of 600 W.

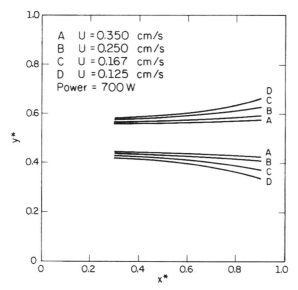

Figure 4.7. Variation of the width of the heated zone at the surface of the substrate along the direction of motion of the laser beam for various scanning speeds of the beam relative to the substrate and for laser power of 700 W.

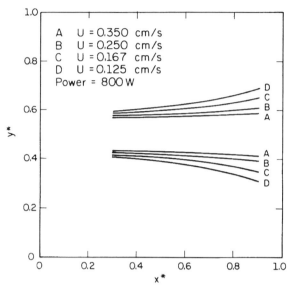

Figure 4.8. Variation of the width of the heated zone at the surface of the substrate along the direction of motion of the laser beam for various scanning speeds of the beam relative to the substrate and for laser power of 800 W.

uniform in width as the scanning speed decreases. The same can be found for a given scanning speed and different powers of the laser beam by comparing the results of Figs. 4.6 through 4.8.

It is established from Figs. 4.6 through 4.8 that the chemically reactive zone becomes narrower as the laser scanning speed increases. However, there will be a critical scanning speed at which the two boundary curves of the chemically reactive zone will collapse into one giving rise to the narrowest possible film deposition region. With any scanning speed higher than the critical speed, the width of the chemically reactive zone cannot be reduced any further. Thus, the film deposition process has to be operated at a scanning speed lower than the critical speed. The critical scanning speed is defined as the one at which the dimensionless peak temperature $T_p^*$ is unity, where peak temperature refers to the temperature at the center of the laser beam on the top surface of the substrate. The line $T_p^* = 1$ is referred to as the line of the narrowest chemical reaction zone in Figs. 4.9 through 4.11. These figures are plotted on logarithmic scales which show the linear variation of the $T_p^*$ with the laser scanning speed $U$ for different powers of the laser beam at various locations on the top surface of the substrate. The points of intersection of the line of the narrowest chemical reaction zone with the curves of Figs. 4.9 through 4.11 give the critical scanning speeds of

# PYROLYTIC LCVD MODELING

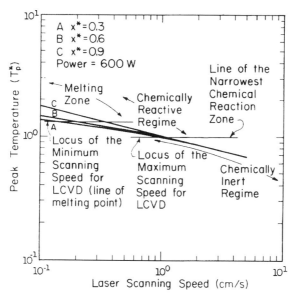

Figure 4.9. Variation of the peak temperature with laser scanning speed relative to the substrate at various locations on the top surface of the substrate for laser power of 600 W.

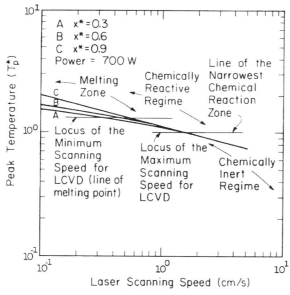

Figure 4.10. Variation of the peak temperature with laser scanning speed relative to the substrate at various locations at the top surface of the substrate for laser power of 700 W.

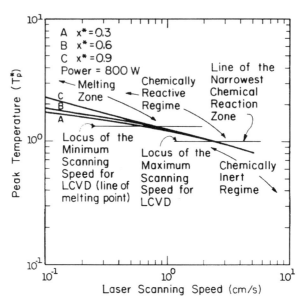

Figure 4.11. Variation of the peak temperature with laser scanning speed relative to the substrate at various locations on the top surface of the substrate for laser power of 800 W.

the laser beam. The region to the right of the critical speed is referred to as the chemically inert regime because the operating conditions of this region do not raise the surface temperature of the substrate to the film-forming chemical reaction temperature. The region to the left of the critical speed is referred to as the chemically reactive regime, where the operating conditions are such that films of finite width can be deposited. However, the surface temperature of the substrate can reach its melting temperature at a low scanning speed for a given operating condition, as can be seen from Figs. 4.12 through 4.14. Since melting the substrate is not desirable in LCVD processes, the scanning speed has to be higher than the upper limit of the speed at which melting occurs. The line $T_p^* = T_m/T_d$, where $T_m$ is the melting temperature of the substrate, is referred to as the melting point line in Figs. 4.9 through 4.11. The points of intersection of the melting point line with the curves of Figs. 4.9 through 4.11 give the scanning speeds above which the process has to be operated to avoid melting the substrate. Thus, from thermal considerations the operating regime for an LCVD process is bounded by the melting point line and the line of narrowest chemical reaction zone and their points of intersection with the curve of Figs. 4.9 through 4.11.

Figures 4.12 through 4.14 represent the variation of the peak temperature with laser power at various locations on the surface of the substrate for

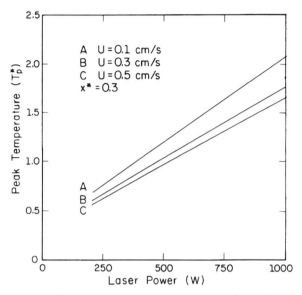

Figure 4.12. Variation of the peak temperature with laser power for various scanning speeds relative to the substrate at $x^* = 0.3$ on the top surface of the substrate.

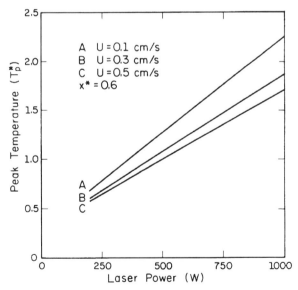

Figure 4.13. Variation of the peak temperature with laser power for various scanning speeds relative to the substrate at $x^* = 0.6$ on the top surface of the substrate.

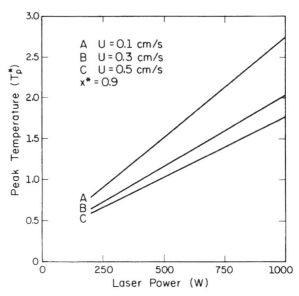

Figure 4.14. Variation of the peak temperature with laser power for various scanning speeds relative to the substrate at $x^* = 0.9$ on the top surface of the substrate.

different scanning speeds of the laser beam. These figures show that the peak temperature increases as the scanning speed decreases for a given operating condition. It should be observed that the peak temperature $T_p^*$ varies linearly with the laser power $P$ on linear scales and with $U$ on logarithmic scales in the chemically reactive regime. These relations can be expressed as

$$T_p^* = \gamma U^\delta$$

at different locations on the top surface of the substrate for various powers of the laser beam, and

$$T_p^* = \eta P + \beta$$

at different locations on the top surface of the substrate for various scanning speeds of the laser beam.

The values of $\gamma$, $\delta$, $\eta$, and $\beta$ are given in Tables 4.1 and 4.2 for various operating conditions. It should be noted that the expression $T_p^* = \gamma U^\delta$ is applicable in the chemically reactive regime. Since the substrate temperature decreases as the scanning speed increases, there will be a critical scanning speed, say $U^*$, at and above which the substrate temperature will remain at its initial temperature. Thus the slopes of the curves of Figs. 4.9 through 4.11

will be zero for scanning speeds higher than $U^*$. This physical aspect is not reflected by the equation $T_p^* = \gamma U^\delta$ because this expression is obtained from the results of the chemically reactive regime.

It should be noted that when $P = 0$, the peak temperature must be equal to the initial temperature of the substrate. This is indeed the case as is evident by the values of $\beta$ in Table 4.2 because the initial temperature $T_i$ of the substrate and the thermal decomposition temperature $T_d$ of the film-forming chemical reaction are taken to be 343 and 1173 K, respectively, in this study, for which the dimensionless initial temperature of the substrate is 0.2925.

Table 4.1. Values of $\gamma$ and $\delta$ for the Expression $T_p^* = \gamma U^\delta$

| Power (W) | $x^*$ | $\gamma$ | $\delta$ |
|---|---|---|---|
| 600 | 0.3 | 0.985 | −0.141 |
|  | 0.6 | 1.006 | −0.163 |
|  | 0.9 | 1.006 | −0.244 |
| 700 | 0.3 | 1.101 | −0.146 |
|  | 0.6 | 1.125 | −0.169 |
|  | 0.9 | 1.127 | −0.253 |
| 800 | 0.3 | 1.217 | −0.150 |
|  | 0.6 | 1.244 | −0.174 |

Table 4.2. Values of $\eta$ and $\beta$ for the Expression $T_p^* = \eta P + \beta$

| $x^*$ | $U$(cm/s) | $\eta$ | $\beta$ |
|---|---|---|---|
| 0.9 | 0.1 | 0.1783(−2)[a] | 0.2925 |
|  | 0.2 | 0.1480(−2) | 0.2924 |
|  | 0.5 | 0.1342(−2) | 0.2925 |
| 0.6 | 0.1 | 0.1955(−2) | 0.2925 |
|  | 0.3 | 0.1574(−2) | 0.2924 |
|  | 0.5 | 0.1413(−2) | 0.2925 |
| 0.9 | 0.1 | 0.2461(−2) | 0.2924 |
|  | 0.3 | 0.1747(−2) | 0.2925 |
|  | 0.4 | 0.1486(−2) | 0.2924 |

[a]The numbers inside parentheses are order-of-magnitude exponents

### 4.3.5. Summary

The three-dimensional and transient heat conduction equation is solved in Cartesian coordinates for slabs having finite dimensions and moving at a constant velocity. The temperature-dependent thermophysical properties of the material of the slab are considered and both convective and radiative losses of energy from the slab to the surrounding medium are taken into account. The laser beam is considered to be Gaussian in shape. Based on these considerations, an analytic expression for the three-dimensional and transient temperature field is obtained.

The surface temperature field has Gaussian-like structure because of the Gaussian laser beam considered in this study. The width of the chemically reactive zone is found to depend on the scanning speed of the laser beam. The width of this zone becomes more uniform and decreases as the scanning speed increases. The critical scanning speed for the narrowest film deposition is determined. Also, the lower and the upper limits of the scanning speeds for LCVD processes are obtained. The peak temperature is found to vary linearly with the laser power for given scanning speeds and the logarithm of the peak temperature is shown to vary linearly with the logarithm of the scanning speed in the chemically active regime for a given laser power.

## 4.4. MODELING OF METAL (Ti) FILM DEPOSITION

Laser technology has attracted considerable attention because of its economical and technical advantages in manufacturing and materials processing [Mazumder (1987)]. While efforts are being made in various industries to implement laser technology, basic research is also pursued for understanding the mechanisms of such processes. Layer glazing [Breinan and Kear (1983)] and laser cladding [Singh and Mazumder (1987), Kar and Mazumder (1987, 1988)] have already demonstrated the feasibility of synthesizing novel materials. These processes can generate relatively thick coatings whereas LCVD can produce thin layers of metals and ceramics [Allen, 1981, Chou et al. (1989), Mazumder and Allen (1980)]. Metals and ceramics (such as Ti and TiN) can be codeposited using this technique to produce composites. Since heat and mass transfer play an important role in all of these processes, mathematical models of the transport phenomena allow us to understand the effects of various parameters on the performance of these processes in order to optimize and efficiently control various laser-aided manufacturing techniques.

## 4.4.1. Introduction

In LCVD processes, the laser heating aspect of laser–materials interactions will have to be considered in conjunction with the chemical reactions among various species and their distributions inside the LCVD chamber. In conventional CVD processes, the substrate material is heated to a high temperature in the environment of reactant molecules which are usually in the vapor or gaseous state. The reactant molecules undergo thermal decomposition to produce the depositing materials, which subsequently adhere to the surface of the substrate to form a film. Although a similar principle is used in LCVD, it differs from the CVD processes because only a very small area of the substrate is heated during LCVD. LCVD can be achieved by relying on the thermal decomposition reaction (pyrolysis) or on the photochemical dissociation reaction (photolysis) of the reactant molecules. Combination of these two processes can also be used in LCVD. Several studies have been carried out to model the CVD process. Sitrl et al. (1974) performed thermodynamic equilibrium calculations for CVD processes. Bloom and Giling (1978) studied CVD by using diffusion and surface kinetic models. Coltrin et al. (1984, 1986) developed a mathematical model for a CVD process by considering gas phase chemical kinetics coupled with fluid mechanics. Jensen (1987) presented a summary of various studies on different types of CVD processes. Choi et al. (1987) analyzed a heat transfer problem related to MCVD processes. Yarmoff and McFeely (1988) have studied experimentally the process of CVD of W on Si from $WF_6$ hexafluoride. Holstein (1992) has reviewed the design and modeling aspects of CVD reactors.

The LCVD process has also been studied theoretically by several investigators. Piglmayer et al. (1984) studied temperature distributions in laser-induced pyrolytic deposition. Zeiger and Ehrlich (1989) have used a three-dimensional Green function to model a thermally driven localized surface reaction. Herman et al. (1983) developed a model for laser-induced pyrolytic deposition processes. Allen et al. (1986) used a finite difference technique to study laser heating and film formation by considering temperature-dependent thermophysical and optical properties of the substrate material. Skouby and Jensen (1988) used the Galerkin finite element method to examine the effects of mass transfer and chemical kinetics on the shape of the deposit on a semi-infinite substrate in laser-assisted CVD with a Gaussian and stationary laser beam. Copley (1988) has presented a mass transport model to analyze the effects of diffusion and convection during LCVD. McWilliams et al. (1984) have used the LCVD technique to deposit films for the interconnection of very large-scale integrated (VLSI) gate arrays. The topology of the deposited spot in LCVD is of interest. Allen

(1981) found experimentally that the shape of the Ni film formed on quartz substrates in LCVD from nickel carbonyl [Ni(Co)$_4$] is volcanic. Moylan *et al.* (1986) observed that the Cu film deposited on Si substrate in LCVD has a volcanic shape. Similarly, Chou *et al.* (1989) observed volcano-like Ti deposits in LCVD of Ti on SS 304 from TiBr$_4$. The mathematical model of Skouby and Jensen (1988) also yields Ni deposits with volcanic morphology for certain operating parameters.

Following the work of Kar *et al.* (1991), we will discuss a three-dimensional transient model for mass diffusion with chemical reactions during the LCVD of pure Ti on SS 304 substrate from TiBr$_4$ in order to carry out parametric studies for predicting the thickness of the deposited film, and to understand the effects of various physicochemical parameters on the thickness and the morphology of the film.

### 4.4.2. Governing Equations and Boundary Conditions

Heat and mass transfer are two important phenomena that have to be considered for modeling the LCVD processes. Besides this, the mechanisms of the chemical reactions occurring during LCVD and the associated chemical kinetics should also be known along with other parameters, such as the sticking probability, diffusion coefficients, and absorptivity. It has been established [Funaki *et al.* (1961)] that the following reactions occur during thermal decomposition of TiBr$_4$:

(A) $\text{TiBr}_4(g) \longrightarrow \text{TiBr}_2(s) + \text{Br}_2(g)$

(B) $\text{TiBr}_2(s) + \text{TiBr}_4(g) \underset{<623K}{\overset{>623K}{\rightleftarrows}} 2\text{TiBr}_3(s)$

(C) $3\text{TiBr}_2(s) \underset{>773K}{\overset{<773K}{\rightleftarrows}} 2\text{TiBr}(s) + \text{TiBr}_4(g)$

(D) $4\text{TiBr}(s) \underset{>1173K}{\overset{<1173K}{\rightleftarrows}} 3\text{Ti}(s) + \text{TiBr}_4(g)$

These reactions are considered to model the LCVD of Ti films on SS 304 under the following assumptions.

1. Reactions take place only on the substrate surface within a region to be called chemical reaction zone where the temperature is greater than or equal to the Ti film-forming chemical reaction temperature (1173 K).

# PYROLYTIC LCVD MODELING

2. Since the temperature in the chemical reaction zone is usually higher than 1173 K, the forward reaction in (B) and the reverse reactions in (C) and (D) are ignored. In the absence of the forward reaction in (B), there will be no production of $TiBr_3$, which implies that there will be no $TiBr_3$ to induce the reverse reaction. For this reason, the reverse reaction in (B) is also ignored in this model.
3. Reactions that may occur on the substrate surface outside the chemical reaction zone are not considered in this model because the work by Chou et al. (1989) shows that the chemical reactions occurring outside the chemical reaction zone do not have a significant effect on the deposition of Ti film on SS 304 substrate.

Using these assumptions, we can study the following irreversible reactions:

(E) $TiBr_4(g) \xrightarrow{k_1} TiBr_2(s) + Br_2(g)$

(F) $3TiBr_2(s) \xrightarrow{k_2} 2TiBr(s) + TiBr_4(g)$

(G) $4TiBr(s) \xrightarrow{k_3} 3Ti(s) + TiBr_4(g)$

The rate of production of various species is given by

$$\dot{r}_1 = m_1 \left[ \frac{1}{4} k_3 \left( \frac{C_3}{m_3} \right)^{n_3} + \frac{1}{3} k_2 \left( \frac{C_2}{m_2} \right)^{n_2} - k_1 \left( \frac{C_1}{m_1} \right)^{n_1} \right] \quad (4.21)$$

$$\dot{r}_2 = m_2 \left[ k_1 \left( \frac{C_1}{m_1} \right)^{n_1} - k_2 \left( \frac{C_2}{m_2} \right)^{n_2} \right] \quad (4.22)$$

$$\dot{r}_3 = m_3 \left[ \frac{2}{3} k_2 \left( \frac{C_2}{m_2} \right)^{n_2} - k_3 \left( \frac{C_3}{m_3} \right)^{n_3} \right] \quad (4.23)$$

$$\dot{r}_4 = m_4 \left[ \frac{3}{4} k_3 \left( \frac{C_3}{m_3} \right)^{n_3} \right] \quad (4.24)$$

$$\dot{r}_5 = m_5 \left[ k_1 \left( \frac{C_1}{m_1} \right)^{n_1} \right] \quad (4.25)$$

In the above expressions (4.21) through (4.25), as well as in the following expression, the subscripts $i = 1, 2, 3, 4,$ and 5 refer to the species $TiBr_4$, $TiBr_2$, $TiBr$, $Ti$, and $Br_2$, respectively. Thus $\dot{r}_i$ is the mass of the $i$th species produced per unit area of the substrate surface per unit time.

Also, the following Arrhenius equation for the reaction rate constant is used:

$$k_j = k_{0j} \exp[-E_j/(RT(x, y, 0, t)] \quad \text{for } j = 1, 2, \text{ and } 3 \quad (4.26)$$

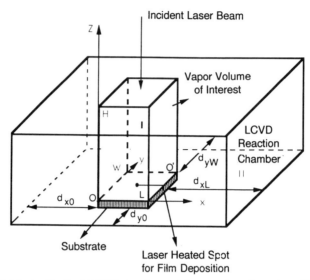

Figure 4.15. An idealized LCVD chamber and the mass transfer region of interest.

The concentration of various species such as $TiBr_4$, $TiBr_2$, $TiBr$, $Ti$, and $Br_2$ inside the LCVD chamber and the thickness of the deposited Ti film are determined by solving five three-dimensional transient, and nonlinear mass diffusion equations. Figure 4.15 shows the LCVD chamber, the substrate, and the Cartesian coordinate system under consideration. $OLO'W$ represents the top surface of the substrate, and the point $O$ the origin of the chosen coordinate system. In Fig. 4.15, the LCVD chamber is divided into two regions. The top surface of the substrate forms the base of region I, which extends up to the top inner surface of the LCVD chamber. Region II constitutes the rest of the chamber; thus, the length, width, and height of region I are $L$, $W$, and $H$, respectively.

The governing equation for mass transfer of the $i$th species in region I is given by

$$\frac{\partial C_i}{\partial t} = D_i \left( \frac{\partial^2 C_i}{\partial x^2} + \frac{\partial^2 C_i}{\partial y^2} + \frac{\partial^2 C_i}{\partial z^2} \right) \qquad \text{for } i = 1, 2, 3, 4, \text{ and } 5 \quad (4.27a)$$

where $C_i$ represents $C_i(x, y, z, t)$, that is, the concentration of the $i$th species at any point $(x, y, z)$ in Cartesian coordinates at any time $t$.

Equation (7a) is solved to satisfy the following boundary and initial conditions:

$$D_i \frac{\partial C_i}{\partial x} = \beta_{ix0}(C_i - C_{i\infty}) \quad \text{at} \quad x = 0 \tag{4.27b}$$

$$-D_i \frac{\partial C_i}{\partial x} = \beta_{ixL}(C_i - C_{i\infty}) \quad \text{at} \quad x = L \tag{4.27c}$$

$$D_i \frac{\partial C_i}{\partial y} = \beta_{iy0}(C_i - C_{i\infty}) \quad \text{at} \quad y = 0 \tag{4.27d}$$

$$-D_i \frac{\partial C_i}{\partial y} = \beta_{iyW}(C_i - C_{i\infty}) \quad \text{at} \quad y = W \tag{4.27e}$$

$$-D_i \frac{\partial C_i}{\partial z} = \dot{r}_i(1 - \gamma_i) \quad \text{at} \quad z = 0 \tag{4.27f}$$

$$\frac{\partial C_i}{\partial z} = 0 \quad \text{at} \quad z = H \tag{4.27g}$$

$$C_i(x, y, z, t_0) = f_i(x, y, z) \tag{4.27h}$$

The mass transfer coefficients $\beta_{ix0}$, $\beta_{ixL}$, $\beta_{iy0}$, and $\beta_{iyW}$ are obtained from the following considerations. We refer to Fig. 4.15, where the diffusion of various species is considered in region I in this model, and region II represents the rest of the vapor volume in the LCVD chamber. At the interfaces of these two regions, the mass flux and the concentration of each species are continuous. Considering these two conditions at $x = 0$, we can write

$$D_i \frac{\partial C_i}{\partial x} = D_{i\text{II}} \frac{\partial C_{i\text{II}}}{\partial x}$$

and

$$C_i = C_{i\text{II}}$$

Assuming that the concentration varies linearly with $x$ in region II, it can be shown from the above two interface conditions that

$$\beta_{ix0} = D_{i\text{II}}/d_{x0}$$

and similarly, the interface conditions at $x = L$, $y = 0$, and $y = W$ yield, respectively,

$$\beta_{ixL} = D_{iII}/d_{xL}, \quad \beta_{iy0} = D_{iII}/d_{y0}, \quad \text{and} \quad \beta_{iyW} = D_{iII}/d_{yW}$$

where the distances $d_{x0}$, $d_{xL}$, $d_{y0}$, and $d_{yW}$ are as shown in Fig. 4.15.

The diffusion coefficients of each species in regions I and II are determined by using the expressions given in Bird et al. (1960), pp. 511, 571). From this reference, the binary diffusion coefficients of the $i$th species with respect to each of the other species are first calculated, and then the diffusion coefficient of the $i$th species in the mixture is determined by taking the concentration of each species to be one-fifth of the initial concentration of $TiBr_4$ (see Section 4.5).

### 4.4.3. Method of Solution

Equation (4.27a) and the boundary and initial conditions (4.27b–h) represent a nonlinear and coupled mass transfer problem because of the mass generation term in the boundary condition (4.27f). The mass transfer problem is also coupled with the heat transfer problem because of the Arrhenius equation (4.26). Since the pyrolytic decomposition of $TiBr_4$ is considered in this model, the heat transfer problem mainly involves the transfer of laser energy into the substrate and its redistribution in the substrate. The deposits in LCVD can affect the transfer of laser energy into the substrate by altering the thermophysical and optical properties of the substrate surface. In particular, the thermal conductivity, density, specific heat, and absorptivity of the substrate often change when films are deposited. For thin and metallic deposits, the loss of the absorbed laser energy in the film can be considered negligibly small. Although the absorptivity depends on the film thickness at the substrate surface, it is taken to be constant in this model because the exact variation of the absorptivity as the deposition continues is not known. This approximation results in one-way coupling between the heat transfer and the mass transfer problems, and it is possible to solve the former before solving the latter. The heat transfer problem is solved by adopting the mathematical technique of Section 4.3 for a stationary laser beam irradiating the center of the top surface of the substrate. From the solution of the heat transfer problem, the temperature distribution at the substrate surface is obtained and is used in Eq. (4.26) in order to solve the mass transfer problem.

The mass diffusion equation (4.27a), and the associated boundary and initial conditions (4.27b–h) are made dimensionless in the following way:

$$C_i^* = C_i/C_{01}, \quad x^* = xL, \quad y^* = y/W, \quad z^* = z/H, \quad D_i^* = D_i/D_1, \quad t^* = t/\tau$$

where $C_{01}$ is the initial concentration of $TiBr_4$ inside the LCVD chamber. Defining

$$C_{i\infty}^* = C_{i\infty}/C_{01}$$

and

$$C_i'^* = C_i^* - C_{i\infty}^*$$

we can reformulate the mass transfer problem as

$$\frac{1}{Fo_m}\frac{\partial C_i'^*}{\partial t^*} = D_i^*\left(\frac{\partial^2 C_i'^*}{\partial x^{*2}} + W^{*2}\frac{\partial^2 C_i'^*}{\partial y^{*2}} + H^{*2}\frac{\partial^2 C_i'^*}{\partial z^{*2}}\right) \quad (4.28a)$$

and the boundary and initial conditions become

$$\frac{\partial C_i'^*}{\partial x^*} = X_{i0} C_i'^* \quad \text{at} \quad x^* = 0 \quad (4.28b)$$

$$\frac{\partial C_i'^*}{\partial x^*} = X_{iL} C_i'^* \quad \text{at} \quad x^* = 1 \quad (4.28c)$$

$$\frac{\partial C_i'^*}{\partial y^*} = Y_{i0} C_i'^* \quad \text{at} \quad y^* = 0 \quad (4.28d)$$

$$\frac{\partial C_i'^*}{\partial y^*} = Y_{iW} C_i'^* \quad \text{at} \quad y^* = 1 \quad (4.28e)$$

$$\frac{\partial C_i'^*}{\partial z^*} = \psi_i(x^*, y^*, 0, t^*) \quad \text{at} \quad z^* = 0 \quad (4.28f)$$

$$\frac{\partial C_i'^*}{\partial z^*} = 0 \quad \text{at} \quad z^* = 1 \quad (4.28g)$$

$$C_i'^*(x^*, y^*, z^*, t_0^*) = f_i'^*(x^*, y^*, z^*) \quad (4.28h)$$

for $i = 1, 2, 3, 4$, and 5, where the following definitions are employed: the mass transfer Fourier number $Fo_m = D_1\tau/L^2$; the aspect ratio $W^* = L/W$; the slenderness ratio $H^* = L/H$; and the dimensionless mass transfer coefficients $X_{i0}$, $X_{iL}$, $Y_{i0}$, and $Y_{iW}$ are $\beta_{ix0}L/D_i$, $-\beta_{ixL}L/D_i$, $\beta_{iy0}W/D_i$, and $-\beta_{iyW}W/D_i$, respectively. The expressions for $\psi_i$'s are given at the end of this section. The initial condition $f_i'^*(x^*, y^*, z^*)$ is given by

$$f_i'^*(x^*, y^*, z^*) = \frac{f_i(x^*, y^*, z^*)}{C_{01} - C_{i\infty}^*}$$

The above mass transfer problem defined by Eqs. (4.28a–h) is solved by successively applying finite medium Fourier integral transforms in the $x$, $y$, and $z$ directions. The Fourier double-integral transform in the $x$ and $y$ directions is

$$\bar{C}_i'^*(\lambda_{ilx}, \lambda_{imy}, z^*, t^*) = \int_0^1 \int_0^1 dx^* \, dy^* \, C_i'^*(x^*, y^*, z^*, t^*) K_{ilx}(x^*) K_{imy}(y^*)$$

(4.29a)

and the inversion formula is

$$C_i'^*(x^*, y^*, z^*, t^*) = \sum_{l=0}^{\infty} \sum_{m=0}^{\infty} \frac{\bar{C}_i'^*(\lambda_{ilx}, \lambda_{imy}, z^*, t^*)}{N_{ilx} N_{imy}} K_{ilx}(x^*) K_{imy}(y^*) \quad (4.29b)$$

where the kernels $K_{ilx}(x^*)$, $K_{imy}(y^*)$, the characteristic equations for the eigenvalues $\lambda_{ilx}$, $\lambda_{imy}$, and the normalization constants $N_{ilx}$, $N_{imy}$ for the above integral transform in each of $x$ and $y$ directions are, respectively, defined as

$$K_{ilx}(x^*) = X_{i0} \sin(\lambda_{ilx} x^*) + \lambda_{ilx} \cos(\lambda_{ilx} x^*)$$

$$\tan \lambda_{ilx} = \frac{\lambda_{ilx}(X_{i0} - X_{iL})}{\lambda_{ilx}^2 + X_{i0} X_{iL}}$$

$$N_{ilx} = \frac{1}{2}\left[X_{i0} + (\lambda_{ilx}^2 + X_{i0}^2)\left(1 - \frac{X_{iL}}{\lambda_{ilx}^2 + X_{iL}^2}\right)\right]$$

and

$$K_{imy}(y^*) = Y_{i0}\sin(\lambda_{imy}y^*) + \lambda_{imy}\cos(\lambda_{imy}y^*)$$

$$\tan\lambda_{imy} = \frac{\lambda_{imy}(Y_{i0} - Y_{iw})}{\lambda_{imy}^2 + Y_{i0}Y_{iw}}$$

$$N_{imy} = \frac{1}{2}\left[Y_{i0} + (\lambda_{imy}^2 + Y_{i0}^2)\left(1 - \frac{Y_{iw}}{\lambda_{imy}^2 + Y_{iw}^2}\right)\right]$$

The application of the integral transform (4.29a) to Eq. (4.28a) and the boundary and initial conditions (4.28b–h) yields the following: Equation (4.28a) becomes

$$\frac{1}{\text{Fo}_m}\frac{\partial \bar{C}_i'^*}{\partial t^*} = D_i^*\left(-\lambda_{ilx}^2 \bar{C}_i'^* - \lambda_{imy}^2 W^{*2}\bar{C}_i'^* + H^{*2}\frac{\partial^2 \bar{C}_i'^*}{\partial z^{*2}}\right) \quad (4.30\text{a})$$

and the boundary conditions (4.28f and g) and the initial condition (4.28h) become

$$\frac{\partial \bar{C}_i'^*}{\partial z^*} = \bar{\psi}_i(\lambda_{ilx}, \lambda_{imy}, 0, t^*) \quad \text{at} \quad z^* = 0 \quad (4.30\text{b})$$

$$\frac{\partial \bar{C}_i'^*}{\partial z^*} = 0 \quad \text{at} \quad z^* = 1 \quad (4.30\text{c})$$

$$\bar{C}_i'^*(\lambda_{ilx}, \lambda_{imy}, 0, t_0^*) = \bar{f}_i'^*(\lambda_{ilx}, \lambda_{imy}, z^*) \quad (4.30\text{d})$$

where the source term $\bar{\psi}_i(\lambda_{ilx}, \lambda_{imy}, 0, t^*)$ and the initial condition $\bar{f}_i'^*(\lambda_{ilx}, \lambda_{imy}, z^*)$ are given by

$$\bar{\psi}_i(\lambda_{ilx}, \lambda_{imy}, 0, t^*) = \int_0^1\int_0^1 dx^*\,dy^*\,K_{ilx}(x^*)K_{imy}(y^*)\psi_i(x^*, y^*, 0, t^*) \quad (4.31)$$

and

$$\bar{f}_i'(\lambda_{ilx}, \lambda_{imy}, z^*) = \int_0^1\int_0^1 dx^*\,dy^*\,K_{ilx}(x^*)K_{imy}(y^*)f_i'^*(x^*, y^*, z^*) \quad (4.32)$$

It should be noted that the source term $\Psi_i(\lambda_{ilx}, \lambda_{imy}, 0, t^*)$ is nonlinear, which is taken care of by solving Eq. (4.30a) in small time interval steps $\Delta t^*$. At each time step, the source term $\bar{\psi}_i(\lambda_{ilx}, \lambda_{imy}, 0, t^*)$ is evaluated numerically from Eq. (4.31) by using the concentration and the temperature values at $t_0^*$, that is, at the beginning of the time step, and is considered to remain unchanged over a small time interval $\Delta t^*$. Now the integral transform

$$\bar{\bar{C}}_i'^*(\lambda_{ilx}, \lambda_{imy}, \lambda_{inz}, t^*) = \int_0^1 dz^* K_{inz}(z^*) \bar{C}_i'^*(\lambda_{ilx}, \lambda_{imy}, z^*, t^*) \quad (4.33a)$$

with the inversion formula

$$\bar{C}_i'^*(\lambda_{ilx}, \lambda_{imy}, z^*, t^*) = \sum_{n=0}^{\infty} \frac{\bar{\bar{C}}_i'^*(\lambda_{ilx}, \lambda_{imy}, \lambda_{inz}, t^*) K_{inz}(z^*)}{N_{inz}} \quad (4.33b)$$

can be applied to the problem defined by Eqs. (4.30a–d). Here the kernel $K_{inz}(z^*)$, the eigenvalues $\lambda_{inz}$, and the normalization constant $N_{inz}$ are given by

$$K_{inz}(z^*) = \cos(\lambda_{inz} z^*)$$

$$\lambda_{inz} = n\pi, \quad n = 0, 1, 2, \ldots, \infty$$

$$N_{inz} = \tfrac{1}{2}(1 + \delta_{n0})$$

where $\delta_{n0}$ is the Kronecker delta.

By applying the integral transform (4.33a) to the Eqs. (4.30a–d), solving the resulting first-order ordinary differential equation, and using the inversion formulas (4.33b) and (4.29b), the expression for $C_i^*$ can be written after some rearrangement as

$$C_i^*(x^*, y^*, z^*, t^*) = C_{i\infty}^* + \sum_{l=0}^{\infty} \sum_{m=0}^{\infty} \frac{K_{ilx}(x^*) K_{imy}(y^*)}{N_{ilx} N_{imy}} \left\{ \bar{C}_i'''(\lambda_{ilx}, \lambda_{imy}, z^*) \right.$$

$$+ \sum_{n=0}^{\infty} [\bar{f}_i'(\lambda_{ilx}, \lambda_{imy}, \lambda_{inz}) - \bar{\bar{C}}_i'''^*(\lambda_{ilx}, \lambda_{imy}, \lambda_{inz})]$$

$$\left. \times \frac{K_{inz}(z^*)}{N_{inz}} \exp[-\lambda_{lmn}^2 (t^* - t_0^*)] \right\} \quad (4.34)$$

for $t_0^* \leq t^* \leq t_0^* + \Delta t^*$, where

$$\bar{C}_i''(\lambda_{ilx}, \lambda_{imy}, z^*) = \frac{\bar{\psi}_i(\lambda_{ilx}, \lambda_{imy}, 0, t^*)}{\lambda_{ilm}\{\exp(-2\lambda_{ilm}) - 1\}} \{\exp(-\lambda_{ilm}z^*) + \exp[-\lambda_{ilm}(2-z^*)]\}$$

$$\lambda_{ilm}^2 = \lambda_{ilx}^2 + \frac{W^{*2}\lambda_{ilm}^2}{H^{*2}}$$

$$\bar{C}_i'''^*(\lambda_{ilx}, \lambda_{imy}, \lambda_{inz}) = -\bar{\psi}_i \frac{(\lambda_{ilx}, \lambda_{imy}, 0, t_0^*)}{\lambda_{ilm}^2 + \lambda_{inz}^2}$$

$$\lambda_{lmn}^2 = D_i^* \mathrm{Fo}_m(\lambda_{ilx}^2 + W^{*2}\lambda_{imy}^2 + H^{*2}\lambda_{inz}^2)$$

$$\bar{\bar{f}}_i'(\lambda_{ilx}, \lambda_{imy}, \lambda_{inz}) = \int_0^1 dz^* K_{inz}(z^*) \bar{f}_i'(\lambda_{ilx}, \lambda_{imy}, z^*)$$

Equation (4.34) is used for determining the dimensionless concentration of various species in region I. Also, the dimensionless initial condition for the time step $t_0^* + \Delta t^* \leq t^* \leq t_0^* + 2\Delta t^*$ is obtained from Eq. (4.34) by evaluating $C_i^*$ at $t^* = t_0^* + \Delta t^*$.

EXPRESSIONS FOR $\psi_i$

The surface source terms $\psi_i$, $i = 1, 2, 3, 4$, and 5 for TiBr$_4$, TiBr$_2$, TiBr, Ti, and Br$_2$, respectively, which appear in the boundary condition (4.28f) are given by the following expression:

$$\psi_1 = -m_1(1 - \gamma_1) \left[ \frac{D_3^*}{4m_3^{n_3}} \mathrm{Dg}\, C_3^{*n_3} \exp\left(\frac{-E_3^*}{T_s^*}\right) + \frac{D_2^*}{3m_2^{n_2}} \mathrm{Df}\, C_2^{*n_2} \exp\left(\frac{-E_2^*}{T_s^*}\right) \right.$$
$$\left. - \frac{1}{m_1^{n_1}} \mathrm{De}\, C_1^{*n_1} \exp\left(\frac{-E_1^*}{T_s^*}\right) \right]$$

where the Damkohler numbers De, Df, and Dg for the chemical reactions E, F, and G, respectively, are

$$\mathrm{De} = C_{01}^{n_1-1} k_{01} H/D_1$$

$$\mathrm{Df} = C_{01}^{n_2-1} k_{02} H/D_2$$

$$\mathrm{Dg} = C_{01}^{n_3-1} k_{03} H/D_3$$

The dimensionless activation energies $E_j^*, j = 1, 2, 3$ are given by

$$E_j^* = E_j/RT_d$$

Here $j = 1, 2,$ and 3 refer to the reactions E, F, and G, respectively; $T_d$ is the thermal decomposition temperature of reaction G; and $T_s^*$ is the dimensionless temperature of the substrate surface:

$$\psi_2 = -m_2(1-\gamma_2)\left[\frac{1}{m_1^{n_1}}\frac{1}{D_2^*}\text{De }C_1^{*n_1}\exp\left(\frac{-E_1^*}{T_s^*}\right) - \frac{1}{m_2^{n_2}}\text{Df }C_2^{*n_2}\exp\left(\frac{-E_2^*}{T_s^*}\right)\right]$$

$$\psi_3 = -m_3(1-\gamma_3)\left[\frac{2}{3m_2^{n_2}}\frac{D_2^*}{D_3^*}\text{Df }C_2^{*n_2}\exp\left(\frac{-E_2^*}{T_s^*}\right) - \frac{1}{m_3^{n_3}}\text{Dg }C_3^{*n_3}\exp\left(\frac{-E_3^*}{T_s^*}\right)\right]$$

$$\psi_4 = -m_4(1-\gamma_4)\frac{3}{4m_3^{n_3}}\frac{D_3^*}{D_4^*}\text{Dg }C_3^{*n_3}\exp\left(\frac{-E_3^*}{T_s^*}\right)$$

$$\psi_5 = -m_5(1-\gamma_5)\frac{1}{m_1^{n_1}}\frac{1}{D_5^*}\text{De }C_1^{*n_1}\exp\left(\frac{-E_1^*}{T_s^*}\right)$$

### 4.4.4. Results and Discussion

Equation (4.34) is used for carrying out the mass transfer analysis during LCVD. Various parameters such as the wavelength of the laser, the laser power, the shape of the laser beam, the thermophysical and optical properties of the substrate and the depositing film, and the partial pressure of the species inside the LCVD chamber, can affect the LCVD process. Some typical results are presented in Figs. 4.16 through 4.28. For all these results, the laser beam radius is taken to be 1.537 mm at the top surface of the substrate and $L$, $W$, and $H$ are, respectively, 1 cm, 1 cm, and 14.605 cm; $d_{x0}$, $d_{xL}$, $d_{y0}$, and $d_{yW}$ are taken to be equal to 10.93 cm. The center of the laser beam is located at $x = 0.5$ cm, $y = 0.5$ cm, and $z = 0$, that is, at the center of the top surface of the substrate, and results are obtained by varying $x$ and keeping $y$ and $z$ fixed at 0.5 cm and 0, respectively. The Arrhenius constants $k_{0j}, j = 1, 2, 3$ for reactions E, F, and G are taken to be $6.5 \times 10^3$, $8.7 \times 10^3$, and $1.1 \times 10^4$ cm/s, respectively, which are calculated by using an equation from Johnston (1966). Although the solution of the mass transfer problem has been obtained for any order of the chemical reactions E, F, and G, results are presented for $n_1 = n_2 = n_3 = 1$. The Damkohler numbers (see Section 4.4.3) De, Df, and Dg for the chemical reactions E, F, and

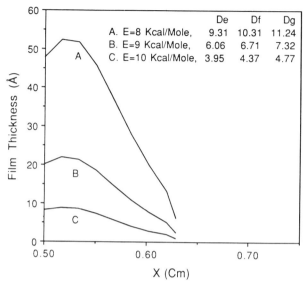

Figure 4.16. Spatial variations of the thickness profile and the morphology of Ti deposits with activation energies: laser power = 850 W, $P = 2$ Torr, $p = 0.058$ Torr, and $t = 4$ s.

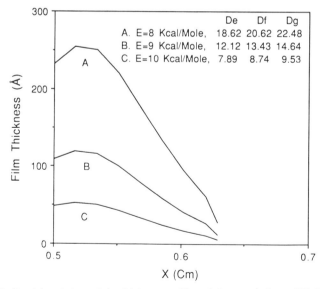

Figure 4.17. Spatial variations of the thickness profile and the morphology of Ti deposits with activation energies: laser power = 840 W, $P = 4$ Torr, $p = 0.115$ Torr, and $t = 4$ s.

G, respectively, have been calculated at the thermal decomposition temperature (1173 K) of reaction G. Also, when $t_0 = 0$, the initial distribution of various species is taken to be $f_1(x, y, z) = C_{01}$; $f_i(x, y, z) = 0$ for $i = 2, 3, 4$, and 5; and $C_{i\infty}$ is taken to be $C_{01}$ for $i = 1$, and zero for all other values of $i$.

Figures 4.16 through 4.18 represent the variation of the deposited Ti film thickness with $x$ for various parameters as shown in the figures. The thickness of the Ti film $T_h$ is calculated from the expression $T_h = \dot{r}_4 \gamma_4 \tau \Delta t^*/\rho$, where $\rho$ is the density of Ti. Also, it should be noted that the isotherms on the top surface of the substrate are circular due to the symmetric formulation of the heat transfer problem, which results in a circular chemical reaction zone. Due to the symmetry of the mass transfer problem, the complete morphology of the deposits can be obtained by rotating the film thickness profile around the vertical axis in Figs. 4.16 through 4.21.

The effects of the activation energy $E$ are examined in Figs. 4.16 through 4.18. These results are obtained by varying the activation energy, which is taken to be the same for all three reactions E, F, and G and by varying the total pressure $P$ and the partial pressure of TiBr$_4$, $p$, in the LCVD chamber. It can be seen from these figures that the film thickness decreases as the activation energy increases. For a given value of the total pressure and the partial pressure of TiBr$_4$, the Damkohler number, which is defined as the ratio of the chemical reaction rate to the diffusion rate,

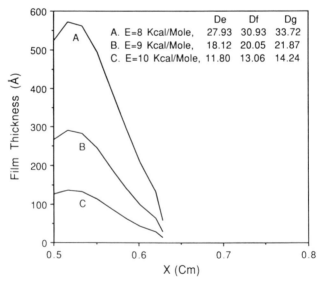

Figure 4.18. Spatial variations of the thickness profile and the morphology of Ti deposits with activation energies: laser power = 850 W, $P = 6$ Torr, $p = 0.173$ Torr, and $t = 4$ s.

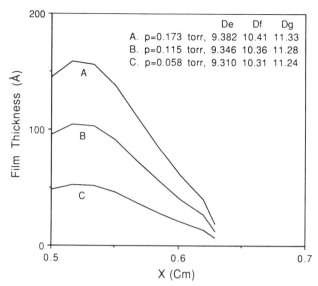

Figure 4.19. Spatial variations of the thickness profile and the morphology of Ti deposits with the partial pressure of $TiBr_4$: laser power = 850W, $P = 2$ Torr, $E = 8$ kcal/mol, $t = 4$ s.

decreases as the activation energy increases. We can, therefore, say that the film thickness decreases as the Damkohler number decreases. Also by comparing the results of Figs. 4.16 through 4.18, we find that the film thickness increases for the same value of $E$ as the total pressure and the partial pressure are increased. This is so because the increases in $P$ and $p$ decrease the diffusion coefficient and thereby increase the Damkohler number for the same value of $E$ resulting in thicker deposits. The thickness profiles are found to have depressions at $x = 0.5$ cm, that is, at the center of the deposits. This kind of volcanic morphology of the deposits has been reported in many studies as noted earlier. Here again, we find that the depression decreases as the Damkohler number decreases. For large Damkohler numbers, the reaction rate is much faster than the diffusion rate and hence there will be depletion of $TiBr_4$ near the center of the deposits as the diffusion-limited chemical reaction progresses, which will cause volcano-like deposition.

Figures 4.19 through 4.21 show the effects of the partial pressure of $TiBr_4$, $p$, and the total pressure $P$ on the morphology and the thickness profile of the deposits at time $t = 4$ s for a given laser power and activation energy $E$, which is taken to be the same for all three reactions E, F, and G. In each of these three figures, it can be seen that when $p$ decreases for a given value of $P$, the film thickness and the depression of the center of the deposits

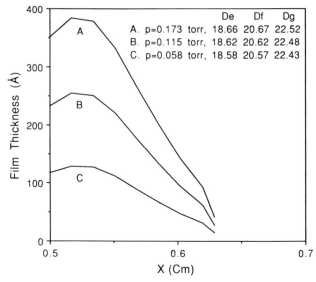

Figure 4.20. Spatial variations of the thickness profile and the morphology of Ti deposits with the partial pressure of TiBr$_4$: laser power = 850 W, $P = 4$ Torr, $E = 8$ kcal/mol, and $t = 4$ s.

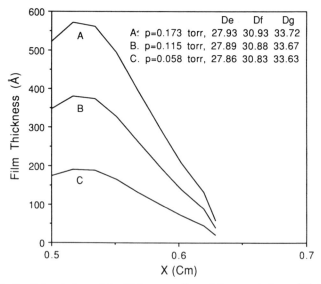

Figure 4.21. Spatial variations of the thickness profile and the morphology of Ti deposits with the partial pressure of TiBr$_4$: laser power = 850 W, $P = 6$ Torr, $E = 8$ kcal/mol, and $t = 4$ s.

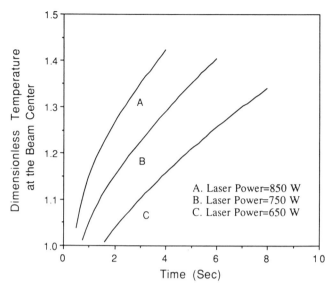

Figure 4.22. Time history of the dimensionless temperature at the beam center for various laser powers.

decrease. This can be explained by noting that the diffusion coefficient increases as $p$ decreases for a given value of $P$, which decreases the Damkohler number causing reduction in the film thickness as well as the depression of the center of the deposits. The results of Figs. 4.19 through 4.21 show that the film thickness and the depression increase when $p$ (or $P$) is increased for a given value of $P$ (or $p$).

Figure 4.22 shows the dimensionless temperature of the substrate surface at the center of the laser beam, that is, at $x = 0.5$ cm, $y = 0.5$ cm, and $z = 0$. The temperature field is made dimensionless by using the thermal decomposition temperature (1173 K) of the Ti film-forming chemical reaction G. As expected, the temperature rises faster with higher laser power. The melting point of the SS 304 substrate is 1670 K, and we find that the temperature at the beam center reaches the melting point of SS 304 at about 4.05, 6.5, and 10.4 s for laser powers of 850, 750, and 650 W, respectively. As the melting of the substrate is undesirable in LCVD processes, time histories of the film thickness are obtained up to 4, 6, and 8 s, respectively, for these laser powers.

Figures 4.23 through 4.25 show the evolution of the film thickness at the center of the laser beam for different laser powers. Here the activation

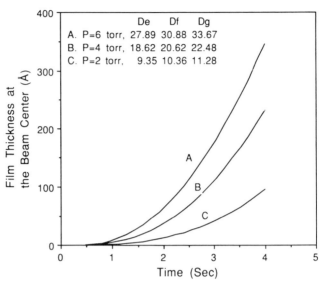

Figure 4.23. Temporal variations of the film thickness at the beam center with the total pressure inside the LCVD chamber: laser power = 850 W, $E = 8\,\text{kcal/mol}$, and $p = 0.115\,\text{Torr}$.

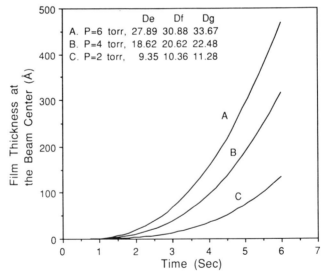

Figure 4.24. Temporal variations of the film thickness at the beam center with the total pressure inside the LCVD chamber: laser power = 750 W, $E = 8\,\text{kcal/mol}$, and $p = 0.115\,\text{Torr}$.

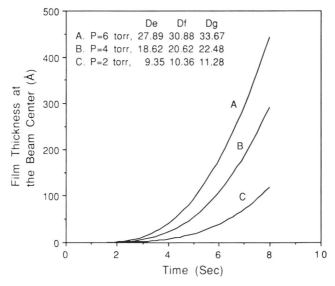

Figure 4.25. Temporal variations of the film thickness at the beam center with the total pressure inside the LCVD chamber: laser power = 650 W, $E = 8\,\text{kcal/mol}$, and $p = 0.115\,\text{Torr}$.

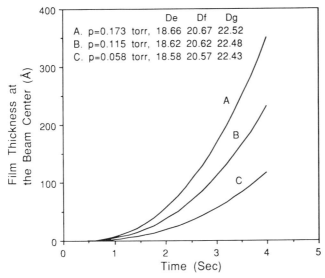

Figure 4.26. Temporal variations of the film thickness at the beam center with the partial pressure of $TiBr_4$ inside the LCVD chamber: laser power = 850 W, $E = 8\,\text{kcal/mol}$, and $P = 4\,\text{Torr}$.

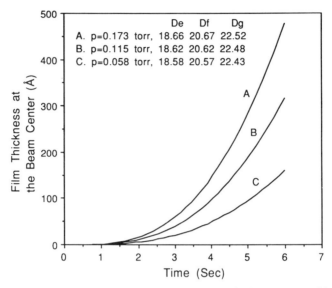

Figure 4.27. Temporal variations of the film thickness at the beam center with the partial pressure of titanium tetrabromide inside the LCVD chamber: laser power = 750 W, $E = 8\,\text{kcal/mol}$, $P = 4\,\text{Torr}$.

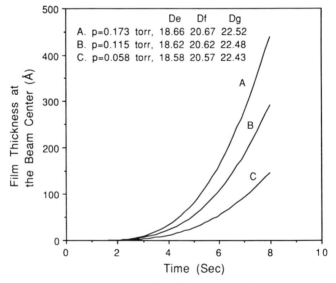

Figure 4.28. Temporal variations of the film thickness at the beam center with the partial pressure of $TiBr_4$ inside the LCVD chamber: laser power = 650 W, $E = 8\,\text{kcal/mol}$, $P = 4\,\text{Torr}$.

energy $E$ is taken to be the same for all three reactions E, F, and G. It can be seen from these three figures that the deposition starts earlier as the laser power increases, and this is because the substrate surface temperature as shown in Fig. 4.22 reaches the thermal decomposition temperature faster for a higher laser power. Each of these three figures shows the effect of the total pressure on the film thickness. As $P$ decreases, the diffusion coefficient increases, which decreases the Damkohler number causing reduction in the film thickness.

Figures 4.26 through 4.28 show the effect of the partial pressure of $TiBr_4$, $p$, on the film deposition for different laser powers and for a given value of the activation energy $E$, which is taken to be the same for all three reactions E, F, and G. Similar to the results of Figs. 4.23 through 4.25, here also we find that the deposition begins earlier as the laser power increases for reasons explained earlier. These three figures show that the film thickness decreases as $p$ decreases because the decrease in $p$ increases the diffusion coefficient and thereby decreases the Damkohler number. Another reason for the decrease in film thickness as $p$ decreases is that the concentration of the reactant decreases as $p$ decreases inside the LCV chamber.

### 4.4.5. Summary

The finite-medium Fourier integral transform has been used to solve the three-dimensional, transient, and nonlinear mass transfer problem in Cartesian coordinates in order to model the deposition of Ti film. The film is produced from the pyrolytic decomposition of $TiBr_4$ at the surface of SS 304 which is heated with a 10.6-mm CW $CO_2$ laser beam. The deposited films are found to have a volcanic shape for higher Damkohler numbers, that is, when the chemical reaction rate is much faster than the diffusion rate of various species inside the LCVD chamber. This implies that volcanic deposits will be formed for diffusion-limited chemical reactions in which depletion of the reactant occurs due to the rapid chemical reaction rate and the slow diffusion rate. The film thickness is found to increase with the increase of the partial pressure of $TiBr_4$ and with the increase of the total pressure inside the LCVD chamber. Also film deposition begins earlier as the laser power increases.

### 4.5. MODELING OF CERAMIC (TiN) FILM DEPOSITION

Laser-induced *c*hemical *p*rocessing (LCP) of materials involves unique processing characteristics which are of fundamental importance in many areas of technology (Bäuerle (1986)]. Among the LCP methods LCVD has been one of the most extensively studied, both experimentally and theoretically. The extremely high deposition rates [Allen et al. (1983)] achieved in LCVD together with the strongly localized heat and chemical treatments make LCVD an attractive technique for the deposition of a wide variety of materials useful in micromechanics, microelectronics, etc.

## 4.5.1. Introduction

The morphology of the deposited material plays an important role for many applications, and usually flat-topped deposits with sharp edges are desired. However, when the energy distribution in the laser beam used to induce film deposition is Gaussian, volcano-like shapes [Allen (1981), Chou et al. (1989), Moylan et al. (1986), Conde et al. (1990)] have very often been observed. Explanations given to account for these volcanoes have been based on high chemical reaction rates, convection from the surface, low sticking coefficients and melting, and even evaporation at the center of the heated spots. By studying the effects of convection on the deposition rate of Ni from $Ni(CO)_4$, Allen et al. (1984) explained the volcano-shaped profiles observed experimentally under certain operating conditions.

In Section 4.4, we discussed a three-dimensional transient mass transfer model for LCVD of Ti and showed volcano-like Ti deposits for high Damkohler numbers, that is, when the chemical reaction rate is much faster than the diffusion rate of various species inside the LCVD reactor. Similarly, Zeiri et al. (1991) obtained volcano-shape deposited patterns when the rate of decomposition was large compared to the rates at which the precursor molecules diffused into the irradiated region by using a Monte Carlo method to simulate the photolytic decomposition of $Ni(CO)_4$. The role played by the sticking coefficient, which depends on the surface coverage and temperature, has been taken into account in the model presented in Konstantinov (1990) and considered by Moylan et al. (1986) as an important parameter in determining deposit shapes.

In this section, we discuss a model, which has been presented in Conde et al. (1992), and by considering given chemical reaction kinetics for the deposition of TiN and a temperature-dependent sticking coefficient, we determine the profiles of TiN films formed during LCVD and compare them with experimental data [Conde et al. (1992)]. TiN possesses many interesting properties, such as high thermal and electrical conductivities, a high melting point, good thermodynamic stability, and low mass diffusivity. TiN thin films have several applications such as gold reflective coatings (in the jewelry industry) or as a diffusion barrier or in metallization (in integrated circuit technology) and for wear or corrosion resistant coatings (in the tool industry). In microelectronics, TiN thin films have frequently been used as a diffusion barrier between Si and Al for very large-scale integrated circuits. This broad range of applications has resulted in the development and understanding of the LCVD technique to deposit TiN films.

### 4.5.2. Experimental Procedure

The deposition system and the procedure used for LCVD of TiN has been described in detail in Conde *et al.* (1990). A CW $TEM_{00}$ $CO_2$ laser beam was focused 27 mm above the substrate surface, at normal incidence to the substrate. The diameter of the laser spot on the substrate surface was about 1.6 mm, as measured on a polymethylmethacrylate (PMMA, Perspex) plate.

Dots and lines of TiN were deposited in a closed chamber containing $TiCl_4$, $N_2$, and $H_2$. Prior to any deposition experiment, the reaction chamber was evacuated to a base pressure of $2 \times 10^{-6}$ Torr. The partial pressures of the reactants were measured with capacitive manometers. The partial pressure of $TiCl_4$ was kept at 7 Torr and the total pressure of the reactants was 207 Torr with $p_{N_2}/p_{H_2} \sim 1$, where $p_{H_2}$ and $p_{H_2}$ refer to the partial pressures of $N_2$ and $H_2$, respectively. Incoloy 800H plates (5 mm thick) were used as substrates; they were ground with 1200 SiC metallographic paper and cleaned with acetone in an ultrasonic bath before being placed inside the reactor.

The laser output power was varied between 300 and 600 W. For depositing TiN dots, the substrate was maintained in a fixed position for a given interaction time in the range from 2 to 8 s. The films were examined by scanning electron microscopy and their thicknesses were measured with a Rodenstock noncontact profilometer. A discussion of the various observed microstructures and cross-sectional profiles has been given in Conde *et al.* (1992).

### 4.5.3. Governing Equations and Boundary Conditions

Although a considerable amount of work has been reported on the deposition of TiN films by various CVD systems [Kato and Tamari (1975), Sadahiro *et al.* (1977), Kim and Chun (1983), Motojima and Kohno (1986), Itoh *et al.* (1986), Motojima and Mizutani (1988), Rong *et al.* (1989), Michalski and Wierzchon (1989)], the dominant mechanisms and rate-limiting processes are still not fully known. In a conventional CVD reactor, TiN is produced according to the following overall chemical reaction:

$$TiCl_4(g) + 2H_2(g) + \tfrac{1}{2}N_2(g) \rightarrow TiN(s) + 4HCl(g) \qquad (4.35)$$

The CVD of TiN films has been investigated as a function of deposition parameters such as the substrate temperature and the composition and flow rates of the reactants. For instance, Sadahiro et al. (1977) found that at constant temperature, the amount of TiN deposited increases linearly with both TiCl$_4$ content and the product $p_{H_2} \times p_{N_2}^{1/2}$. However, the order of the chemical reaction and the reaction pathways yielding the final TiN solid product still remain controversial [Rong et al. (1989)]. However, it is generally accepted [Kato and Tamari (1975), Sadahiro et al. (1977), Rong et al. (1989), Michalski and Wierzchon (1989)] that above a threshold temperature of about 900°C, TiCl$_4$ is reduced by H$_2$ to TiCl$_2$ and that the adsorption of the reactant species TiCl$_2$ controls the process. This same kinetic mechanism was assumed by Jang and Paik (1983) while studying the CVD of Ti from a TiCl$_4$ and H$_2$ gas mixture. They also found that the amount of deposited Ti varies linearly with the partial pressure of TiCl$_4$.

In the present model, we consider that TiN is produced in pyrolytic LCVD through the overall chemical reaction given by the expression (4.35), and by assuming the Sadahiro (1977) kinetics, the rate of production of TiN can be written as

$$R_{TiN} = k(T) C_{H_2} C_{N_2}^{1/2} C_{TiCl_4} \tag{4.36}$$

where rate constant $k(T)$ follows the Arrhenius equation

$$k(T) = k'_0 \exp[-E/RT(x, y, 0, t)]$$

Here $k'_0$ and $E$ are, respectively, the Arrhenius constant and activation energy of the chemical reaction (4.35).

Since the mole fractions of H$_2$ and N$_2$ inside the LCVD reactor were at least one order of magnitude higher than the TiCl$_4$ mole fraction in the experimental study [Conde et al. (1990)], we can neglect any variation of $C_{H_2}$ and $C_{N_2}$ during the deposition process and replace them in Eq. (4.36) by their initial values $^0C_{H_2}$ and $^0C_{N_2}$, respectively. Thus Eq. (4.36) can be written as

$$R_{TiN} = A k'_0 \exp[-E/RT(x, y, 0, t)] C \tag{4.37}$$

where $A = {^0C_{H_2}} {^0C_{N_2}^{1/2}}$ is a constant, and we have discarded the index TiCl$_4$, that is, $C \equiv C_{TiCl_4}$.

Equation (4.37) shows that the rate of production of TiN is determined by the concentration of TiCl$_4$ and, therefore, the thickness of the deposited TiN film may be obtained by solving the following mass diffusion equation:

$$\frac{\partial C}{\partial t} = D \left( \frac{\partial^2 C}{\partial x^2} + \frac{\partial^2 C}{\partial y^2} + \frac{\partial^2 C}{\partial z^2} \right) \tag{4.38}$$

In order to define the boundary conditions that have to be satisfied in order to solve the mass transfer equation (4.38), we consider a geometrical configuration as shown in Fig. 4.15, which shows the experimental LCVD chamber used for depositing TiN spots [Conde et al. (1990)], the substrate, and the Cartesian coordinate system under consideration. OLO'W represents the top surface of the substrate, and the point O the origin of the chosen coordinate system. In Fig. 4.15, the LCVD chamber is divided into two regions. The top surface of the substrate forms the base of region I, which extends up to the top inner surface of the LCVD chamber, and region II constitutes the rest of the chamber. Thus, the length, width, and height of the region I are $L$, $W$, and $H$, respectively. The boundary conditions can be written in the same way as in Section 4.4.2.

### 4.5.4. Method of Solution

As in Section 4.4.3, Eq. (4.38) is solved by applying a double-integral Fourier transform in the $x$ and $y$ directions, yielding

$$\frac{1}{\text{Fo}_m}\frac{\partial \bar{C}'^*}{\partial t^*} = -\lambda_{lx}^2 \bar{C}'^* - \lambda_{my}^2 W^{*2} \bar{C}'^* + H^{*2}\frac{\partial^2 \bar{C}'^*}{\partial z^{*2}} \quad (4.39a)$$

subject to the following boundary and initial conditions:

$$\frac{\partial \bar{C}'^*}{\partial z^*} = \bar{\psi}(\lambda_{lx}, \lambda_{my}, 0, t^*) \quad \text{at} \quad z^* = 0 \quad (4.39b)$$

$$\frac{\partial \bar{C}'^*}{\partial z^*} = 0 \quad \text{at} \quad z^* = 1 \quad (4.39c)$$

and

$$\bar{C}'^*(\lambda f_{lx}, \lambda_{my}, z^*, 0) = 0 \quad (4.39d)$$

Here the dimensionless variables are defined as follows:

$$C^* = C/C_0, \quad x^* = x/L, \quad y^* = y/W, \quad z^* = z/H, \quad t^* = t/\tau$$

while the mass transfer Fourier number $\text{Fo}_m = D\tau/L^2$, the aspect ratio $W^* = L/W$, and the slenderness ratio $H^* = L/H$. Also, we define $C'^* = C^* - C_\infty^*$, where $C_\infty^* = C_\infty/C_0$. In this model, $C_\infty$ is considered to be equal to $C_0$.

The Fourier transform variable $\bar{C}'^*$, that is, $\bar{C}'^*(\lambda_{lx}, \lambda_{my}, z^*, t^*)$, is defined as

$$\bar{C}'^*(\lambda_{lx}, \lambda_{my}, z^*, t^*) = \int_0^1 \int_0^1 dx^* \, dy^* \, C'^*(x^*, y^*, z^*, t^*) K_{lx}(x^*) K_{my}(y^*)$$

which satisfies the following inversion formula

$$C'^*(x^*, y^*, z^*, t^*) = \sum_{l=0}^{\infty} \sum_{m=0}^{\infty} \frac{\bar{C}'^*(\lambda_{lx}, \lambda_{my}, z^*, t^*)}{N_{lx} N_{my}} K_{lx}(x^*) K_{my}(y^*)$$

where the kernels $K_{lx}(x^*)$, $K_{my}(y^*)$, the characteristic equations for the eigenvalues $\lambda_{lx}$, $\lambda_{my}$, and the normalization constants $N_{lx}$, $N_{my}$ for the above integral transform in the $x$ and $y$ directions are given, respectively, by

$$K_{lx}(x^*) = X_0 \sin(\lambda_{lx} x^*) + \lambda_{lx} \cos(\lambda_{lx} x^*)$$

$$\tan \lambda_{lx} = \frac{\lambda_{lx}(X_0 - X_L)}{\lambda_{lx}^2 + X_0 X_L}$$

$$N_{lx} = \frac{1}{2}\left[X_0 + (\lambda_{lx}^0 + X_0^2)\left(1 - \frac{X_L}{\lambda_{lx}^2 + X_L^2}\right)\right]$$

and

$$K_{my}(y^*) = Y_0 \sin(\lambda_{my} y^*) + \lambda_{my} \cos(\lambda_{my} y^*)$$

$$\tan \lambda_{my} = [\lambda_{my}(Y_0 - Y_W)]/(\lambda_{my}^2 + Y_0 Y_W)$$

$$N_{my} = \frac{1}{2}\left[Y_0 + (\lambda_{my}^2 + Y_0^2)\left(1 - \frac{Y_W}{\lambda_{my}^2 + Y_W^2}\right)\right]$$

Here the dimensionless mass transfer coefficients $X_0$, $X_L$, $Y_0$, and $Y_W$ are $\beta_{x0} L/D$, $-\beta_{xL} L/D$, $\beta_{y0} W/D$, and $-\beta_{yW} W/D$, respectively, where $\beta_{x0} = D_{II}/d_{x0}$, $\beta_{xL} = D_{II}/d_{xL}$, $\beta_{y0} = D_{II}/d_{y0}$, $\beta_{yW} = D_{II}/d_{yW}$, $D_{II}$ is the diffusion coefficient of $TiCl_4$ in region II, and the distances $d_{x0}$, $d_{xL}$, $d_{y0}$, and $d_{yW}$ are as shown in Fig. 4.15.

The surface source term $\bar{\psi}(\lambda_{lx}, \lambda_{my}, 0, t^*)$ in the boundary condition (4.39b) is given by

$$\bar{\psi}(\lambda_{lx}, \lambda_{my}, 0, t^*) = \int_0^1 \int_0^1 dx^* \, dy^* K_{lx}(x^*) K_{my}(y^*) \psi(x^*, y^*, 0, t^*)$$

where

$$\psi(x^*, y^*, 0, t^*) = \mathrm{Dm}(1-\gamma)\frac{m_1}{m_2} C^* \exp\left\{\frac{-E^*}{T_s^*}\right\}$$

and the Damkohler number, $\mathrm{Dm} = k'_0 A H/D$, the dimensionless activation energy $E^* = E/(RT_D)$, and the dimensionless temperature of the substrate surface $T_s^* = T(x^*, y^*, 0, t^*)/T_D$.

In Section 4.4, Eq. (4.39a) was solved by applying a Fourier transform under the assumption that the surface source term $\bar{\psi}(\lambda_{lx}, \lambda_{my}, 0, t^*)$ remains unchanged over a small time interval $\Delta t^*$. This assumption holds good for slow chemical reactions in which the reaction rate varies slowly with time. However, one will need an extremely small time interval and, therefore, a long computational time to obtain results for fast chemical reactions if one uses the solution technique of Section 4.4. For this reason, the problem given by the expressions (4.39a–d) is solved in this model by using the Laplace transform technique, which yields the following expression for the concentration:

$$C^*(x^*, y^*, z^*, t^*) = C_\infty^* - b^2 \sum_{l=0}^{\infty} \sum_{m=0}^{\infty} \frac{K_{lx}(x^*) K_{my}(y^*)}{N_{lx} N_{my}} \int_0^{t^*} dt \bar{\psi}[\lambda_{lx}, \lambda_{my}, 0, (t^* - \tau)]$$

$$\times e^{-\lambda \tau} \left[ 1 + 2 \sum_{n=0}^{\infty} (-1)^n \exp(-n^2 \pi^2 b^2 \tau) \cos n\pi(1 - z^*) \right]$$

(4.40)

where

$$b^2 = \mathrm{Fo}_m H^{*2}$$

and

$$\lambda = \mathrm{Fo}_m (\lambda_{lx}^2 + W^{*2} \lambda_{my}^2)$$

Equation (4.40) is used to obtain results at $z^* = 0$. In order to determine the concentration at any time $t^*$, we divide $t^*$ into $i$ time intervals of step size $\Delta t^*$ such that $t^* = i\Delta t^*$, where $i$ is an integer, and then the concentration is evaluated at $\Delta t^*, 2\Delta t^*, 3\Delta t^*$, and so on. In Eq. (4.40), the integral on the right-hand side and $\bar{\psi}$ are evaluated at each time step by using the trapezoidal formula to obtain a set of linear algebraic equations for the concentrations at various locations which is solved by the method of

Gaussian elimination with partial pivoting. To obtain the substrate surface temperature $T_s^*$, we take into account the temperature-dependent thermophysical and optical properties of the substrate material, and use the mathematical technique of Section 4.3 for a stationary laser beam irradiating the center of the top surface of the substrate to solve the heat conduction equation for the substrate.

### 4.5.5. Data Estimation

In order to carry out a mass transfer analysis during LCVD, we first evaluate the temperature distribution at the top surface of the substrate $T_s^*(x^*, y^*, 0, t^*)$ at any time $t^*$ by solving the heat transfer problem for the substrate as in Section 4.3. Various parameters such as the laser beam profile and thermophysical and optical properties of the substrate and the depositing film can affect $T_s^*$. Since the temperature-dependent thermal conductivity, the specific heat, and the density of Incoloy 800 H are known (INCO Alloys Int.), only the absorptivity has to be estimated. For all the results of this model, the absorptivity is taken to be 0.23, which, although it appears to be large, can be due to multiple reflections of the laser beam in the microcavities in the microstructures created [Conde et al. (1992)] during LCVD. Also, we consider a Gaussian laser beam characterized by a laser spot radius at the substrate surfce, $r_0$, which changes from 1.0 to 1.5 mm as the laser power and the interaction time increase from 300 W and 2 s to 600 W and 4 s, respectively.

The geometrical parameters $L$, $W$, and $H$ (see Fig. 4.15) are, respectively, 1, 1, and 12 cm, and $d_{x0}$, $d_{xL}$, $d_{y0}$, and $d_{yW}$ are taken as 24.5 cm each. The center of the laser beam is located at $x = 0.5$ cm, $y = 0.5$ cm, and $z = 0$, that is, at the center of the top surface of the substrate, and results are obtained by varying $x$ and keeping $y$ and $z$ fixed at 0.5 cm and 0, respectively. The diffusion coefficient of $TiCl_4$ in the gas mixture has been determined by first calculating the binary diffusion coefficients of $TiCl_4$ with respect to each of the other species in the hard-sphere approximation and then by using the Stefan–Maxwell equation of Bird et al. 1960, pp. 511, 271). In region I, the diffusion coefficient, $D_I$ has been calculated at the threshold temperature $T_D = 1173$ K for the chemical reaction (4.35), whereas, in region II, the diffusion coefficient $D_{II}$ has been evaluated at 338 K, which is the reaction chamber wall temperature [Conde et al. (1990)].

In this model, the activation energy is taken to be $E = 11.0 \pm 1.2$ kcal/mol, which is very close to the value 12.2 kcal/mol given by Sadahiro et al. (1977) for the deposition of TiN by conventional CVD at a substrate temperature greater than or equal to 1173 K. Moreover, Jang and Paik

(1983) have also reported a value equal to 12.7 kcal/mol for the thermal decomposition of TiCl$_4$ above 1340 K in the presence of H$_2$. It should be noted that all the curves presented in this section have been obtained for an Arrhenius constant $k'_0 = (5.2 \pm 1.0) \times 10^9$ (cm/s)(cm$^3$/g)$^{3/2}$ and an activation energy $E = 11.0 \pm 1.2$ kcal/mol, that is, $k'_0$ and $E$ are determined with an error lower than 20% and 11%, respectively. The Arrhenius constant given above is the one determined by using Eq. (4.37). If we consider that the chemical reaction is first order with respect to the TiCl$_4$ concentration, the deposition rate of TiN may be written as

$$R_{\text{TiN}} = k_0 \exp(-E/RT)C$$

where the new value of $k_0$, which is equal to $Ak'_0$, is determined to be $(8.4 \pm 1.6) \times 10^2$ cm/s. By using trimolecular surface reactions based on the theory discussed in Leidler (1987), we calculated $k_0 = 1.2 \times 10^2$ cm/s, which is smaller than the value of $k_0$ used in this model to fit all of the experimental results. It should be noted that although the theoretical value of $k_0$ is usually higher than the experimentally determined value, it is found to be lower than the actual value in many cases [Leidler (1987)].

As has been previously discussed, LCVD often yields deposited films with volcano-like shapes. Among the various factors that can cause this type of profile, the low sticking coefficient at the center of the laser-heated spot is of particular importance. We therefore consider a temperature-dependent sticking coefficient for TiN, $\gamma_{\text{TiN}}(T)$, defined in such a way that $\gamma_{\text{TiN}}(T) = 1$ below a temperature $T_m$, $\gamma_{\text{TiN}}(T) = 0$ above a temperature $T_M$ ($T_m < T_M$), and between these two values, the sticking coefficient decreases linearly as the temperature increases, that is,

$$\gamma_{\text{TiN}}(T) = \begin{cases} 1 & \text{for } T < T_m \\ 1 + (T_m - T)/(T_M - T_m) & \text{for } T_m \leqslant T \leqslant T_M \\ 0 & \text{for } T > T_M \end{cases} \quad (4.41)$$

### 4.5.6. Results and Discussion

In this study, we investigate the effect of the sticking coefficient of TiN on the thickness and the morphology of the TiN deposits using the temperature-dependent sticking coefficient discussed in the previous section [Eq. (4.41)]. The results presented in Figs. 4.29 through 4.38 have been obtained for $T_m = 1473$ K and $T_M = 1640$ K. $T_m$ is chosen to be 1473 K because it is the one usually quoted as the maximum substrate temperature for which TiN films with good properties can be deposited in CVD reactors.

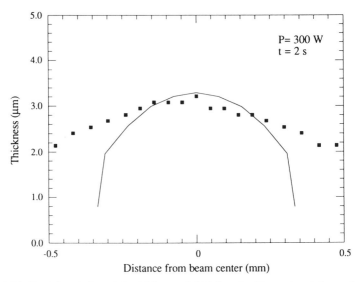

Figure 4.29. Spatial variation of the thickness of TiN deposits: ■ experimental, — calculated, $r_0 = 1.0$ mm, $T_m = 1473$ K, $T_M = 1640$ K, $E = 12.2$ kcal/mol, $k_0 = 7.4 \times 10^2$ cm/s.

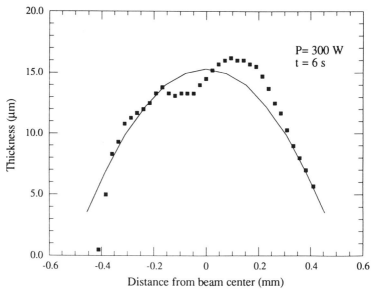

Figure 4.30. Spatial variation of the thickness of TiN deposits: ■ experimental, — calculated, $r_0 = 1.0$ mm, $T_m = 1473$ K, $T_M = 1640$ K, $E = 12.2$ kcal/mol, $k_0 = 9.0 \times 10^2$ cm/s.

# PYROLYTIC LCVD MODELING

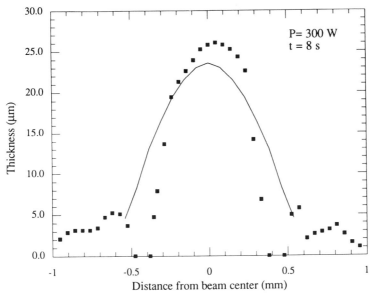

Figure 4.31. Spatial variation of the thickness of TiN deposits: ■ experimental, — calculated, $r_0 = 1.0$ mm, $T_m = 1473$ K, $T_M = 1640$ K, $E = 12.2$ kcal/mol, $k_0 = 9.0 \times 10^2$ cm/s.

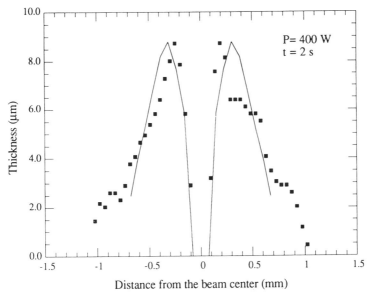

Figure 4.32. Spatial variation of the thickness of TiN deposits: ■ experimental, — calculated, $r_0 = 1.0$ mm, $T_m = 1473$ K, $T_M = 1640$ K, $E = 11.0$ kcal/mol, $k_0 = 8.4 \times 10^2$ cm/s.

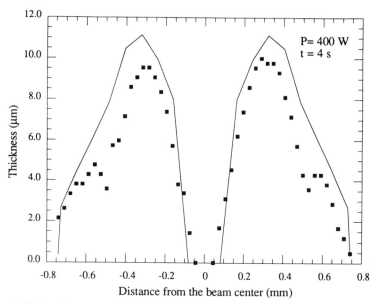

Figure 4.33. Spatial variation of the thickness of TiN deposits: ■ experimental, — calculated, $r_0 = 1.1$ mm, $T_m = 1473$ K, $T_M = 1640$ K, $E = 12.2$ kcal/mol, $k_0 = 6.9 \times 10^2$ cm/s.

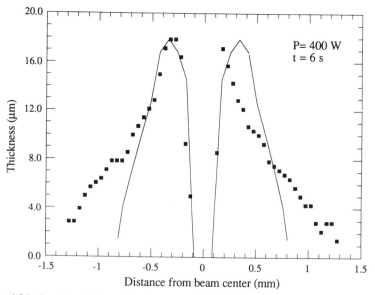

Figure 4.34. Spatial variation of the thickness of TiN deposits: ■ experimental, — calculated, $r_0 = 1.1$ mm, $T_m = 1473$ K, $T_M = 1640$ K, $E = 12.2$ kcal/mol, $k_0 = 7.9 \times 10^2$ cm/s.

# PYROLYTIC LCVD MODELING

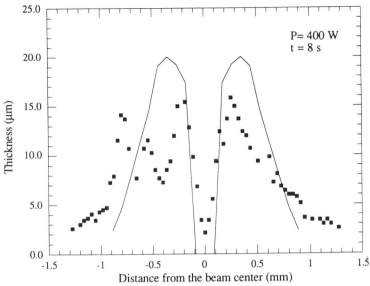

Figure 4.35. Spatial variation of the thickness of TiN deposits: ■ experimental, — calculated, $r_0 = 1.1$ mm, $T_m = 1473$ K, $T_M = 1640$ K, $E = 12.2$ kcal/mol, $k_0 = 6.9 \times 10^2$ cm/s.

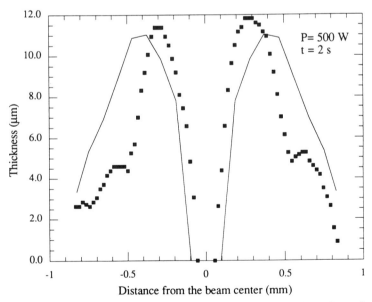

Figure 4.36. Spatial variation of the thickness of TiN deposits: ■ experimental, — calculated, $r_0 = 1.2$ mm, $T_m = 1473$ K, $T_M = 1640$ K, $E = 9.8$ kcal/mol, $k_0 = 10.0 \times 10^2$ cm/s.

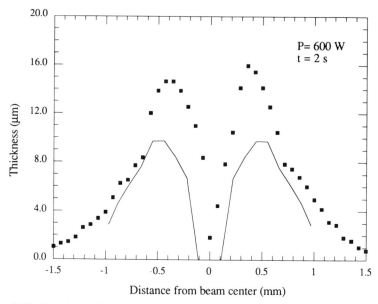

Figure 4.37. Spatial variation of the thickness of TiN deposits: ■experimental, —calculated, $r_0 = 1.4$ mm, $T_m = 1473$ K, $T_M = 1640$ K, $E = 9.8$ kcal/mol, $k_0 = 10.0 \times 10^2$ cm/s.

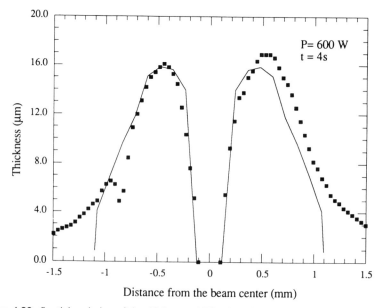

Figure 4.38. Spatial variation of the thickness of TiN deposits: ■experimental, —calculated, $r_0 = 1.5$ mm, $T_m = 1473$ K, $T_M = 1640$ K, $E = 10.0$ kcal/mol, $k_0 = 8.5 \times 10^2$ cm/s.

$T_M$ represents a temperature slightly under the substrate melting temperature (1658 K).

Figures 4.29 through 4.38 show the variation of the deposited TiN film thickness with $x$ for various parameters. The thickness of the TiN film $T_h$ is calculated from the expression $T_h = R_{TiN}\gamma_{TiN}\tau \Delta t^*/\rho$, where $\rho$ is the density of TiN. It should be noted that the TiN, which is produced from the chemical reaction in the central regions of the volcanoes shown in Figs. 4.32 through 4.41, but does not deposit there due to a fully nonsticking condition $[\gamma_{TiN}(T) = 0]$, is considered to deposit entirely on the top of the inner wall of the volcanoes. This assumption is due to the fact that deposition is not observed outside the heated spots. In general, excellent qualitative agreement is found between the theoretical and the experimental results in Figs. 4.29 through 4.38 over a wide range of laser output powers and irradiation times.

The calculated thickness profiles are very sensitive to the Arrhenius constant. To illustrate this point, in Fig. 4.39 we plot the experimental results obtained for a laser power of 400 W and an irradiation time of 4 s together with the theoretical predictions using three different values for the Arrhenius constant. It can be seen from this figure that the thickness profile changes considerably with increasing $k_0$.

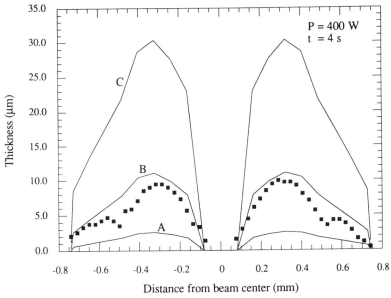

Figure 4.39. Effect of the Arrhenius constant on the TiN deposition profile: ■ experimental, — calculated, $r_0 = 1.1$ mm, $T_m = 1473$ K, $T_M = 1640$ K, $E = 12.2$ kcal/mol, (A) $k_0 = 1.4 \times 10^2$ cm/s, (B) $k_0 = 6.9 \times 10^2$ cm/s, (C) $k_0 = 3.4 \times 10^3$ cm/s.

The effect of the temperature-dependent TiN sticking coefficient on the deposited profile is also examined. Since we have assumed that $\gamma_{\text{TiN}}(T)$ varies linearly with temperature in the range $T_m \leqslant T \leqslant T_M$, the slope of the line and, therefore, the spatial variation of the TiN deposit will depend on the chosen limits for the temperature range. In Fig. 4.40, we study the effect of lowering $T_m$ and increasing $T_M$. Curve B of Fig. 4.40 is plotted by using the same values of $T_m$ and $T_M$ used in Figs. 4.29 through 4.39, while curve A is for $T_m = 1173\,K$, which is the threshold temperature for the chemical reaction, and the curve C is for $T_M = 2000\,K$, which is lower than the TiN melting point. We find that the deposited profile is strongly influenced by the chosen temperature limits. Curve C ($T_m = 1473\,K$, $T_M = 2000\,K$) does not reproduce the volcanic shape of the deposit, while curve A ($T_m = 1173\,K$, $T_M = 1640\,K$) shows a decreased mass deposition rate in relation to the experimental result. However, Fig. 4.40 suggests that curve A may fit the experimental results if it is plotted by taking a value of $k_0$ higher than $6.9 \times 10^2$ cm/s and keeping its other parameters unchanged, which could lead to two sets of values for $T_m$, $T_M$, and $k_0$, such that $T_m = 1473\,K$, $T_M = 1640\,K$, and $k_0 = 6.9 \times 10^2$ cm/s as for curve B, and $T_m = 1173\,K$, $T_M = 1640\,K$, and $k_0 > 6.9 \times 10^2$ cm/s for the same set of experi-

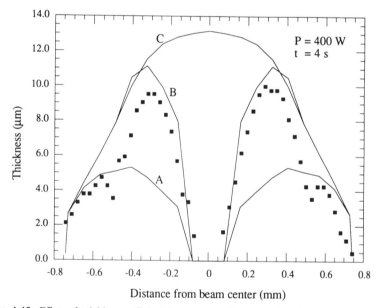

Figure 4.40. Effect of sticking coefficient temperature limits on the TiN deposition profile: ■experimental, —calculated, $r_0 = 1.1$ mm, $E = 12.2$ kcal/mol, $k_0 = 6.9 \times 10^2$ cm/s, (A) $T_m = 1173\,K$, $T_M = 1640\,K$, (B) $T_m = 1473\,K$, $T_M = 1640\,K$, (C) $T_m = 1473\,K$, $T_M = 2000\,K$.

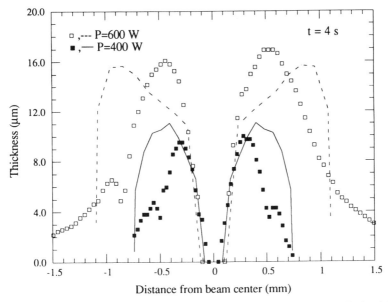

Figure 4.41. Effect of the Arrhenius constant on the TiN deposition profile calculated for a sticking coefficient defined by $T_m = 1173$ K, $T_M = 1640$ K, □, ■ experimental, --- calculated with $k_0 = 4.2 \times 10^3$ cm/s, — calculated with $k_0 = 1.9 \times 10^3$ cm/s. Other theoretical parameters are as indicated in Figs. 6 and 11 for 400 W and 600 W, respectively.

mental data. To resolve this, Fig. 4.41 is plotted by taking $T_m = 1173$ K, $T_M = 1640$ K, and $k_0 = 4.2 \times 10^3$ cm/s and $1.9 \times 10^3$ cm/s, which are higher than $6.9 \times 10^2$ cm/s. It can be seen from Fig. 4.41 that the agreement between the calculated and experimental results is very poor, which indicates that the parameters of Fig. 4.41 are unsuitable for explaining the observed TiN deposition profiles.

### 4.5.7. Summary

The deposition of TiN dots by pyrolytic LCVD with a CW $TEM_{00}$ $CO_2$ laser beam has been modeled by solving the three-dimensional, transient mass diffusion problem using an integral transform technique. Due to the lack of information on the chemical reaction mechanisms and the chemical kinetics, we have assumed that the rate of production of TiN is first order with respect to the $TiCl_4$ concentration. In general, excellent agreement has been found between the model predictions and the experimental results. The volcanic shape observed under some conditions has been reproduced by assuming a temperature-dependent sticking coefficient for TiN which decreases linearly from one to zero when the temperature increases from 1473 K to 1640 K. Best results are obtained with an activation energy $E = 11.0 \pm 1.2$ kcal/mol and an Arrhenius constant, $k_0 = (8.4 \pm 1.6) \times 10^2$ cm/s. The results indicate that the mass transfer model can be used to determine important chemical parameters with a relatively high accuracy.

## 4.6. CONVECTION DURING LCVD

In Sections 4.3 through 4.5, we have discussed the modeling of pyrolytic LCVD for the configuration of Fig. 4.1a by considering the chemical reactions that occur at the substrate surface, neglecting the gas-phase reactions. At low pressures, when the reactant concentration is very small, the reactions in the gas phase can be neglected [Wise (1958)]. Also, the effects of convection have not been taken into account in the models of the Sections 4.3 through 4.5. In the absence of forced convection, free convection can become important when the total pressure inside the LCVD chamber is high. At low pressures, the effect of free convection may be neglected. There can also be convection during LCVD due to the Stephan flow [Frank-Kamenetskii (1969)], which occurs when a chemical reaction is accompanied by a change in volume. However, the effect of the Stephan flow amounts to a second-order correction to the fluid flow in most of the chemically reacting systems so it can be neglected for most cases.

Allen et al. (1986) have pointed out that convection during pyrolytic LCVD is negligible for low deposition rates and/or small spot sizes. Earlier Allen et al. (1984) pointed out that convection during pyrolytic LCVD is negligible for low deposition rates and/or small spot sizes. They suggested that the diffusion process dominates if $\xi r_0 \leqslant 1.0$, where

$$\xi = \frac{RTj_f S'}{DP}$$

$$S' = -S'_{q+1} + \sum_{i=1}^{q+p} S'_i$$

and

$$S'_i = \pm S_i$$

The plus and minus signs in $\pm S_i$ correspond to the products and reactants, respectively. However, Copley (1988) has shown that the reactant pressure obtained by considering the transport of the species resulting from both diffusion and convection differs from that predicted by the diffusion analysis by 10% of the total pressure if $\xi r_0 = 0.1$.

Boyd and Vest (1975) have analyzed the onset of convection due to the heating of a gas phase by a horizontal laser beam, where the geometric configuration is essentially similar to the arrangement of Fig. 4.1b, by considering the following heat conduction problem in the gas phase:

GOVERNING EQUATION:

$$\frac{1}{\alpha}\frac{\partial T}{\partial t} = \frac{1}{r}\frac{\partial}{\partial r}\left(r\frac{\partial T}{\partial r}\right) + \frac{A}{k}I_0\exp\left(\frac{-r^2}{r_1^2}\right) \tag{4.42a}$$

BOUNDARY CONDITIONS:

$$T(\infty, t) = 0 \tag{4.42b}$$

$$\lim_{r \to 0}\left(r\frac{\partial T}{\partial r}\right) = 0 \tag{4.42c}$$

INITIAL CONDITION:

$$T(r, 0) = 0 \tag{4.42d}$$

Here, $T(r, t)$ represents the excess temperature above ambient, that is, $T(r, t) = T'(r, t) - T_\infty$, where $T'(r, t)$ is the actual temperature of the gas phase at any location $r$ at any time $t$, and $T_\infty$ is the ambient temperature. The axial variations of the temperature and laser intensity have been neglected by assuming that $A \ll 1$. The solution of the above problem can be written as [Boyd and Vest (1975), Gordon et al. (1965)]:

$$T(r^*, t^*) = -\frac{r_1^2 A I_0}{4k}\left[E_i\left(\frac{-r^{*2}}{4t^* + 1}\right) - E_i(-r^{*2})\right] \tag{4.43}$$

where $r^* = r/r_1$, $t^* = \alpha t/r_1^2$, and $E_i$ is the exponential integral function, that is,

$$E_i(-x) = -\int_1^\infty \frac{e^{-x\theta}}{\theta}d\theta \quad \text{for} \quad x > 0 \tag{4.44}$$

Noting that the derivative of $E_i(-x)$ is

$$\frac{d}{dx}E_i(-x) = \frac{e^{-x}}{x} \tag{4.45}$$

the temperature gradient can be written as

$$\frac{dT'(r, t)}{dr} = -\frac{r_1 A I_0}{2kr^*}\left[\exp\left(-\frac{r^{*2}}{4t^* + 1}\right) - \exp(-r^{*2})\right] \tag{4.46}$$

The temperature is a maximum along the center line of the laser beam and decreases radially outward. Therefore, during the heating of the gas phase by a horizontal laser beam, the density configuration can be approximated to be locally similar to that of a fluid layer heated from below. The stability analysis for an infinite fluid layer of thickness $\delta$ heated from below is known as the Benard problem [Chandrasekhar (1961)]. When the vertical temperature gradient is constant, and the shear stresses at the upper and lower boundaries are negligible, a horizontal layer of fluid becomes linearly unstable at a critical Rayleigh number $\mathrm{Ra}_c$, which is given by

$$\mathrm{Ra}_c = \frac{g\gamma |dT'(r,\ t)/dr|\delta^4}{\alpha \nu} = 658 \qquad (4.47)$$

where $g$ is the acceleration due to gravity, and $\gamma$ is the volumetric thermal expansion coefficient defined by

$$\gamma = -\frac{1}{\rho}\left(\frac{\partial \rho}{\partial T'}\right)_P$$

Here $\rho$ and $P$ are the density and pressure of the medium, respectively. If the Rayleigh number is slightly higher than $\mathrm{Ra}_c$, convection causes multicellular motion with adjacent cells rotating in opposite directions. Hauf and Grigull (1970) have observed similar cellular motion during the development of convection above suddenly heated horizontal wires. Expression (4.47) provides a means of determining the relative importance of diffusion and convection in the gas phase heated by a horizontal laser beam.

## NOMENCLATURE

| | |
|---|---|
| $A$ | Absorptivity (in Sections 4.3 and 4.6) |
| $C$ | Concentration of $TiCl_4$ in Region I (see Fig. 4.15). |
| $C_0$ | Initial concentration of $TiCl_4$ |
| $C_{01}$ | Distribution of $TiBr_4$ concentration at time $t_0 = 0$ |
| $C_i$ | Concentration of the $i$th species in units of mass per unit volume; $i = 1, 2, 3, 4,$ and 5 represent the species $TiBr_4$, $TiBr_2$, $TiBr$, $Ti$, and $Br_2$, respectively (in Section 4.4) |
| $C_i$ | Concentration of the $i$th species in units of mass per unit volume; $i = H_2, N_2,$ and $TiCl_4$ represent the species $H_2$, $N_2$, and $TiCl_4$, respectively (in Section 4.5) |
| $C_{iII}$ | Concentration of the $i$th species in region II (see Fig. 4.15) |

| | |
|---|---|
| $C_{ir}$ | Concentration of the $i$th species in the $r$th region. $C_{ir}$ represents the mass per unit area for surface reactions (Fig. 4.1a), and the mass per unit volume for volumetric reactions (Fig. 4.1b) |
| $C_{i\infty}$ | Concentration of the $i$th species at a location far away from region I (see Fig. 4.15) |
| $C_p(T)$ | Temperature-dependent heat capacity of the substrate |
| $C_\infty$ | Concentration of $TiCl_4$ at a location far away from region I (see Fig. 4.15) |
| $^0C_{H_2}$ | Initial concentration of hydrogen |
| $^0C_{N_2}$ | Initial concentration of nitrogen |
| $D$ | Diffusion coefficient (in Section 4.6) |
| $D$ | Diffusion coefficient of $TiCl_4$ in region I (see Fig. 4.15) |
| $D$ | $Y$-coordinate of the center of the laser beam on the top surface of the substrate (see Fig. 4.2) |
| $D_i$ | Diffusion coefficient of the $i$th species in region I (Fig. 4.15 in Section 4.4) |
| $D_{iII}$ | Diffusion coefficient of the $i$th species in region II (see Fig. 4.15) |
| $D_{im}$ | Diffusion coefficient of the $i$th species with respect to the mixture of the other gases in the $r$th region |
| $D_{iT}$ | Thermal diffusion coefficient of the $i$th species in the mixture in the $r$th region |
| De | Damkohler number for the reaction E, $De = C_{01}^{n_1-1} k_{01} H / D_1$ (see Section 4.4.2) |
| Df | Damkohler number for the reaction F, $Df = C_{01}^{n_2-1} k_{02} H / D_2$, (see Section 4.4.2) |
| Dg | Damkohler number for the reaction G, $Dg = C_{01}^{n_3-1} k_{03} H / D_3$, (see Section 4.4.2) |
| $E$ | Activation energy |
| $E_j$ | Activation energy of the $j$th reaction. $j = 1, 2,$ and $3$ refer to the reactions E, F, and G, respectively (see Section 4.4.2) |
| $f_i(x, y, z)$ | Concentration distribution of the $i$th species in region I (see Fig. 4.15) at the initial time $t_0$ |
| $g$ | Acceleration due to gravity |
| $h_{ij}$ | Heat transfer coefficient at the $i = j$ plane of the substrate. For example, $h_{xL}$ represents the heat transfer coefficient at the $x = L$ plane of the substrate |
| $h_{x0}$ | Heat transfer coefficient at $x = 0$ |
| $h_{xL}$ | Heat transfer coefficient at $x = L$ |
| $h_{y0}$ | Heat transfer coefficient at $y = 0$ |
| $h_{yW}$ | Heat transfer coefficient at $y = W$ |
| $h_{z0}$ | Heat transfer coefficient at $z = 0$ |
| $h_{zH}$ | Heat transfer coefficient at $z = H$ |

| | |
|---|---|
| $H$ | Enthalpy of the mixture (in Section 4.2) |
| $H$ | Thickness of the substrate |
| $\bar{H}_i$ | Partial molal enthalpy of the $i$th species |
| $i^*$ | Total number of species in the mixture |
| $\hat{i}$ | Unit vector along the $x$-direction in Cartesian coordinates |
| $I_0$ | Laser intensity at the center of the beam |
| $\hat{j}$ | Unit vector along the $y$-direction in Cartesian coordinates |
| $j_f$ | Film deposition rate in units of mass per unit time per unit area of the substrate surface |
| $\mathbf{j}_i$ | Molar flux of the $i$th species with respect to stationary coordinates (moles of the $i$th species flowing per unit time per unit area) |
| $\mathbf{J}_{ir}$ | Mass flux of the $i$th species in the $r$th region (mass of the $i$th species flowing per unit time per unit area in the $r$th region) |
| $k$ | Thermal conductivity of the mixture or the gas phase (in Sections 4.2 and 4.6) |
| $\hat{k}$ | Unit vector along the $z$-direction in Cartesian coordinates. |
| $k(T)$ | Reaction rate constant (in Section 4.5) |
| $k(T)$ | Temperature-dependent thermal conductivity of the substrate |
| $k_0$ | Modified Arrhenius constant, $k_0 = k_0'\,{}^0C_{H_2}\,{}^0C_{N_2}^{1/2}$ |
| $k_0'$ | Arrhenius constant |
| $k_{0j}$ | Arrhenius constant for the $j$th reaction; $j = 1, 2,$ and 3 refer to the reactions E, F, and G, respectievly (see Section 4.4.2) |
| $k_i$ | Surface reaction rate constant with respect to the concentration of the $i$th species: $i = 1, 2,$ and 3 represent the species $TiBr_4$, $TiBr_2$, and $TiBr$, respectively |
| $L$ | Length of the substrate |
| $m_i$ | Molecular weight of the $i$th species; $i = 1, 2, 3, 4,$ and 5 represent the species $TiBr_4$, $TiBr_2$, $TiBr$, $Ti$, and $Br_2$, respectively |
| $m_i$ | Molecular weight of the $i$th species; $i = 1$ and 2 represent $TiCl_4$ and $TiN$, respectively (in Section 4.5) |
| $n_i$ | Number of moles of the $i$th species in the mixture (in Section 4.2) |
| $n_j$ | Order of the $j$th reaction with respect to the $j$th species; $j = 1, 2,$ and 3 refer to the reactions E, F, and G, respectively (see Section 4.4.2); also $j = 1, 2,$ and 3 represent the species $TiBr_4$, $TiBr_2$, and $TiBr$, respectively |
| $p$ | Partial pressure of $TiBr_4$ inside the deposition chamber at time $t_0 = 0$ |
| $p_{H_2}$ | Partial pressure of hydrogen inside the deposition chamber |
| $p_{N_2}$ | Partial pressure of nitrogen inside the deposition chamber |
| $P$ | Power of the incident laser beam |

| | |
|---|---|
| $P$ | Total pressure inside the LCVD chamber (in Sections 4.2, 4.4) |
| $q$ | Heat flux (amount of thermal energy flowing per unit time per unit area) |
| $r$ | Radial distance from the center of the laser beam on the top surface of the substrate |
| $r$ | Radial variable (in Section 4.6) |
| $r_0$ | Laser spot radius at the substrate surface at $1/e^2$ point (in Sections 4.5 and 4.6) |
| $r_0$ | Radius of the laser beam at its waist at $1/e^2$-point (in Section 4.3) |
| $r_1$ | Characteristic radius of the laser beam at $1/e$-point |
| $r_i$ | Sticking probability of the $i$th species at the substrate surface |
| $R$ | Reflectivity (in Section 4.3) |
| $R$ | Universal gas constant |
| $R_{TiN}$ | Mass of TiN produced per unit area of the substrate surface per unit time |
| $S_i$ | Reaction coefficient of the $i$th species (mole number for the $i$th species in the chemical reaction); $1 \leq i \leq q$, $q+1 \leq i \leq q+p$, and $q+p+1 \leq i \leq q+p+n$ represent the reactant, product, and inert species, respectively, that is, $\Sigma_{i=1}^{q} S_i X_i = \Sigma_{i=1}^{p} S_{q+1} X_{q+1}$ |
| $t$ | Time variable |
| $t_0$ | Initial time (in Section 4.4) |
| $T$ | Temperature |
| $T(x, y, 0, t)$ | Temperature at the top surface of the substrate at any point $(x, y)$ at time $t$ |
| $T_a$ | Ambient temperature inside the LCVD chamber |
| $\bar{T}_a$ | Temperature around which the Taylor series expansion is carried out to linearize the boundary conditions |
| $T_D$ | Threshold temperature for the chemical reaction (4.35) (in Section 4.5) |
| $T_d$ | Dissociation temperature for pyrolytic LCVD (K) |
| $T_i$ | Initial temperature of the substrate |
| $T_M$ | Upper limit of temperature for the linear variation of $\gamma_{TiN}(T)$ [see Eq. (4.41)] |
| $T_m$ | Lower limit of temperature for the linear variation of $\gamma_{TiN}(T)$ [see Eq. (4.41)] |
| $u$ | Velocity in the $x$-direction |
| $\mathbf{u}$ | Velocity vector, $\mathbf{u} = u\hat{i} + v\hat{j} + w\hat{k}$ |
| $\nabla \mathbf{u}$ | A dyadic product |
| $(\nabla \mathbf{u})^t$ | Transpose of the dyadic $\nabla \mathbf{u}$ |
| $u_0$ | Magnitude of the local fluid velocity ($u_0 = \sqrt{\mathbf{u} \cdot \mathbf{u}} = \sqrt{u^2 + v^2 + w^2}$) |
| $U$ | Velocity of the laser beam with respect to the substrate |

| | |
|---|---|
| $\hat{U}$ | Internal energy per unit mass of the mixture |
| $v$ | Velocity in the $y$-direction |
| $W$ | Width of the substrate |
| $w$ | Velocity in the $z$ direction |
| $x$ | Distance along the $x$-axis in Cartesian coordinates |
| $y$ | Distance along the $y$-axis in Cartesian coordinates |
| $z$ | Distance along the $z$-axis in Cartesian coordinates |

GREEK SYMBOLS

| | |
|---|---|
| $\alpha$ | Thermal diffusivity |
| $\beta_{ijk}$ | Mass transfer coefficient of the $i$th species at the $j = k$ plane For example, $\beta_{ixL}$ represents the mass transfer coefficient of the $i$th species at the plane $x = L$ |
| $\gamma$ | Sticking coefficient of TiCl$_4$ at the substrate surface |
| $\gamma_{\text{TiN}}$ | Sticking coefficient of TiN at the substrate surface |
| $\delta$ | Unit tensor |
| $\nabla$ | The del operator, $\nabla \equiv \hat{i}\partial/\partial x + \hat{j}\partial/\partial y + \hat{k}\partial/\partial z$ in Cartesian coordinates |
| $\varepsilon$ | Thermal emissivity of the substrate |
| $\kappa$ | Bulk viscosity |
| $\mu$ | Viscosity |
| $\rho$ | Density of the mixture (in Section 4.2) |
| $\rho(T)$ | Temperature-dependent density of the substrate |
| $\sigma$ | Stefan–Boltzmann constant |
| $\tau$ | Shear stress tensor |
| $\tau$ | Characteristic time ($\tau = L/U$) (in Section 4.3) |
| $\tau$ | Irradiation time (in Sections 4.4 and 4.5) |

# REFERENCES

Abraham, E., and Halley, J. M. (1987), Some Calculations of Temperature Profiles in Thin Films with Laser Heating, *Appl. Phys.* **A42**, 279.

Allen, S. D. (1981), *J. Appl. Phys.* **52**, 6501.

Allen, S. D., Goldstone, J. A., Stope, J. P., and Jan, R. Y. (1986), *J. Appl. Phys.* **59**, 1653.

Allen, S. D., Jan, R. Y., Edwards, R. H., Mazuk, S. M., and Vernon, S. D. (1984), Optical and Thermal Effects in Laser Chemical Vapor Deposition, *Proc., SPIE on Laser Assisted Deposition, Etching, and Doping,* S. D. Allen, Ed., SPIE-The Int. Soc. for Optical Engineering, Washington, Vol. 459, pp. 42–48.

Allen, S. D., Trigubo, A. B., and Jan, R. Y. (1983), *Mater. Res. Soc. Symp. Proc.* **17**, 207.

Bäuerle, D. (1986), *Chemical Processing with Lasers,* Springer Series in Material Sciences, Vol. 1, Springer, Berlin.

Bell, A. E. (1979), *RCA Review* **40**, 295.

Bilenchi, R., Gianinoni, I., and Musci, M. (1982), *J. Appl. Phys.* **53**, 6479.

Bird, R. B., Stewart, W. E., and Lightfoot, E. N. (1960), *Transport Phenomena*, Wiley, New York.
Bloom, J., and Giling, L. J. (1978), in: *Current Topics in Materials Science*, E. Kaldis, ed., North-Holland, Amsterdam, Chapter 4.
Boyd, R. D., and Vest, C. M. (1975), Onset of Convection Due to Horizontal Laser Beams, *Appl. Phys. Lett.* **26**, 287.
Breinan, E. M., and Kear, B. H. (1983), in: *Proc., Laser Material Processing*, M. Bass, ed., North-Holland, Amsterdam, pp. 235–296.
Burgener, M. L., and Reedy, R. E. (1982). Temperature Distributions Produced in a Two-Layer Structure by a Scanning CW Laser or Electron Beam, *J. Appl. Phys.* **53**, 4357.
Calder, I. D., and Sue, R. (1982), Modeling of CW Laser Annealing of Multilayer Structures, *J. Appl. Phys.* **53**, 7545.
Carslaw, H. S., and Jaeger, J. C. (1986), Conduction of Heat in Solids, 2nd Ed., Clarendon, Oxford, pp. 10–11.
Chandrasekhar, S. (1961), *Hydrodynamic and Hydromagnetic Stability*, Oxford University Press, London, pp. 35–36.
Choi, M., Baum, H. R., and Greif, R. (1987), *J. Heat Trans.* **109**, 642.
Chou, W. B., Azer, M. N., and Mazumder, J. (1989), Laser Chemical Vapor Deposition of Ti from $TiBr_4$, *J. Appl. Phys.* **66**, 191.
Cline, H. E., and Anthony, T. R. (1977), *J. Appl. Phys.* **48**, 3895.
Coltrin, M. E., Kee, R. J., and Miller, J. A. (1984), *J. Electrochem. Soc.* **131**, 425.
Coltrin, M. E., Kee, R. J., and Miller, J. A. (1986), *J. Electrochem. Soc.* **133**, 1206.
Conde, O., Ferreira, M. L. G., Hochholdinger, P., Silvestre, A. J., and Vilar, R. (1992), $CO_2$ Laser Induced CVD of TiN, *Appl. Surf. Sci.* **54**, 130.
Conde, O., Kar, A., and Mazumder, J. (1992), Laser Chemical Vapor Deposition of TiN Dots: A Comparison of Theoretical and Experimental Results, *J. Appl. Phys.* **72**, 754.
Conde, O., Mariano, J., Silvestre, A. J., and Vilar, R. (1990), *Proc. 3rd European Conf. on Laser Treatment of Materials*, H. W. Bergmann and R. Kupfer, eds., Sprechsal, Coburg, p. 145.
Copley, S. M. (1988), Mass Transport During Laser Chemical Vapor Deposition, *J. Appl. Phys.* **64**, 2064.
de Heer, J. (1986), *Phenomenological Thermodynamics with Applications to Chemistry*, Prentice Hall, Englewood Cliffs, NJ, p. 60.
El-Adawi, M. K., and Elshehawey, E. F., 1986, Heating a Slab Induced by a Time-Dependent Laser Irradiance—An Exact Solution, *J. Appl. Phys.* **60**, 2250.
Ferrieu, F., and Auvert G. (1983), Temperature Evolutions in Silicon Induced by a Scanned CW Laser, Pulsed Laser, or an Electron Beam, *J. Appl. Phys.* **54**, 2646.
Flint, J. H., Meunier, M., Adler, D., and Haggerty, J. S. (1984), A-Si:H Films Produced from Laser Heated Gases: Process Characteristics and Film Properties, *Proc., SPIE on Laser Assisted Deposition, Etching, and Doping*, S. D. Allen, ed., SPIE–The International Society for Optical Engineering, Washington, Vol. 459, pp. 66–70.
Frank-Kamenetskii, D. A. (1969), *Diffusion and Heat Transfer in Chemical Kinetics*, Second enlarged and revised edition, J. P. Appleton, English Translation Editor, Plenum Press, New York, pp. 158–191.
Funaki, K., Uchimura, K., and Kuniya, Y. (1961), Kogyo Kagaku Zasshi, Vol. 64, p. 1914.
Gattuso, T. R., Meunier, M., Adler, D., and Haggerty J. S. (1983), IR Laser-Induced Deposition of Silicon Thin Films, in: *Laser Diagnostics and Photochemical Processing for Semiconductor Devices*, Mat. Res. Soc. Symp. Proc., R. M. Osgood, S. R. J. Brueck, and H. R. Schlossberg, eds., North-Holland, Amsterdam, Vol. 17, pp. 215–222.
Gibbons, J. F., and Sigmon, T. W. (1982), Solid Phase Regrowth, in: *Laser Annealing of Semiconductors*, J. M. Poate and J. W. Mayer, eds., Academic Press, New York, Ch. 10, pp. 325–381.
Gordon, J. P., Leite, R. C. C., Moore, R. S., Porto, S. P. S., and Whinnery, J. R. (1965), Long Transient Effects in Lasers with Inserted Liquid Samples, *J. Appl. Phys.* **36**, 3.

Hauf, W., and Grigull, U. (1970), in: *Advances in Heat Transfer*, J. P. Hartnett and T. F. Irvine, Jr., eds., Academic Press, New York, Vol. 6, pp. 134–366.
Hermann, I. P., Hyde, R. A., McWilliams, B. M., Weisberg, A. H., and Wood, L. L. (1983), in: *Materials Research Symposium*, R. M. Osgood, S. R. J. Brueck, and H. R. Schlossberg, eds., North-Holland, New York, Vol. 17, pp. 9–18.
Hess, L. D., Forber, R. A., Kokorowski, S. A., and Olson, G. L. (1980), in: *Proc. Society of Photo-Optical Instrumentation Engineers*, J. F. Ready, ed., Soc. of Photo-Optical Instrumentation Engineers, Washington, Vol. 198, pp. 31–34.
Hirschfelder, J. O., Curtiss, C. F., and Bird, R. B. (1954), *The Molecular Theory of Gases and Liquids*, Wiley, New York.
Holstein, W. L. (1992), Design and Modeling of Chemical Vapor Deposition Reactors, *Prog. Crystal Growth and Charact.* **24**, 111–211.
INCO Alloys International, Huntington, WV 25720, data supplied.
Incropera, F. P., and Dewitt, D. P. (1985), *Fundamentals of Heat and Mass Transfer*, 2nd ed., Wiley, New York, p. 8.
Itoh, H., Kato, K., and Sugiyama, K. (1986), *J. Mater. Sci.* **21**, 751.
Jang, H., and Paik, Y. H. (1983), *J. Kor. Inst. Met.* **21**, 38.
Jensen, K. F. (1987), *Chem. Eng. Sci.* **42**, 923.
Johnston, H. S. (1966), *Gas Phase Reaction Rate Theory*, Ronald, New York, p. 138.
Kant, R. (1988), *J. Appl. Mech.* **55**, 93.
Kar, A., and Mazumder, J. (1987), *J. Appl. Phys.* **61**, 2645.
Kar, A., and Mazumder, J. (1988), *Acta Metal.* **36**, 702.
Kar, A., and Mazumder, J. (1989), Three-Dimensional Transient Thermal Analysis for Laser Chemical Vapor Deposition on Uniformly Moving Finite Slabs, *J. Appl. Phys.* **65**, 2923.
Kar, A., Azer, M. N., and Mazumder, J. (1991), Three-Dimensional Transient Mass Transfer Model for Laser Chemical Vapor Deposition of Titanium on Stationary Finite Slabs, *J. Appl. Phys.* **69**, 757.
Kato, A., and Tamari, N. (1975), *J. Cryst. Growth* **29**, 55.
Kim, D. M., Kwong, D. L., Shah, R. R., and Crosthwait, D. L. (1981), Laser Heating of Semiconductors—Effect of Carrier Diffusion in Nonlinear Dynamic Heat Transport Process, *J. Appl. Phys.* **52**, 4995.
Kim, M. S., and Chun, J. S. (1983), *Thin Solid Films* **107**, 129.
Kokorowski, S. A., Olson, G. L., and Hess, L. D. (1955), in: *Proc. of Laser and Electron-Beam Solid Interactions and Materials Processing*, J. F. Gibbons, L. D. Hess, and T. W. Sigmon, eds., Elsevier-North-Holland, New York, pp. 139–146.
Konstantinov, L., Kowak, R., and Hess, P. (1990), *Appl. Surf. Sci.* **46**, 102.
Kwong, D. L., and Kim, D. M. (1982), Pulsed Laser Heating of Silicon: The Coupling of Optical Absorption and Thermal Conduction During Irradiation, *J. Appl. Phys.* **53**, 366.
Lax, M. (1977), *J. Appl. Phys.* **48**, 3919.
Lax, M. (1978), *J. Appl. Phys. Lett.* **33**, 786.
Lax, M. (1979), in: *Proc. Laser–Solid Interactions and Laser Processing 1978*, S. D. Ferris, H. J. Leamy, and J. M. Poate, eds., American Institute of Physics, New York, pp. 149–154.
Leidler, K. J. (1987), *Chemical Kinetics*, Harper & Row, New York, pp. 262–265.
Liarokapis, E., and Raptis, Y. S. (1985), Temperature Rise Induced by a CW Laser Beam Revisited, *J. Appl. Phys.* **57**, 5123.
Mazumder, J. (1987), in: *Proc., Interdisciplinary Issues in Materials Processing and Manufacturing*, S. K. Samanta, R. Komanduri, R. McMeeking, M. M. Chen, and A. Tseng, eds., ASME, New York, pp. 599–630.
Mazumder, J., and Allen, S. D. (1980), in: *Proc., Society of Photo-Optical Instrumentation Engineers*, J. F. Ready, ed., Soc. of Photo-Optical Instr. Eng., Washington, Vol. 198, pp. 73–80.

McWilliams, B. M., Chin, H. W., Herman, I. P., Hyde, R. A., Mitlitsky, F., Whitehead, J. C., and Wood, L. L. (1984), Wafer-Scale Laser Pantography: VI. Direct Write Interconnection of VLSI Gate Arrays, *Proc. SPIE on Laser Assisted Deposition, Etching, and Doping*, S. D. Allen, ed., SPIE—The Int. Soc. for Optical Engineering, Washington, Vol. 459, pp. 49–54.

Meunier, M., Gattuso, T. R., Adler, D., and Haggerty, J. S. (1983), *Appl. Phys. Lett.* **43**, 273.

Meyer, J. R., Bartoli, F. J., and Kruer, M. R. (1980a), Optical Heating in Semiconductors, *Phys. Rev.* **B21**, 1559.

Meyer, J. R., Kruer, M. R., and Bartoli, F. J. (1980b), Optical Heating in Semiconductors: Laser Damage in Ge, Si, InSb, and GaAs, *J. Appl. Phys.* **51**, 5513.

Michalski, J., and Wierzchon, T. (1989), *J. Mater. Sci. Lett.* **8**, 779.

Moody, J. E., and Hendel, R. H. (1982), *J. Appl. Phys.* **53**, 4364.

Motojima, S., and Kohno, M. (1986), *Thin Solid Films*, **137**, 59.

Motojima, S., and Mizutani, H. (1988), *J. Mater. Sci.* **23**, 3435.

Moylan, C. R., Baum, T. H., and Jones, C. R. (1986), *Appl. Phys.* **A40**, 1.

Nissim, Y. I., A. Lietoila, R. B. Gold, and J. F. Gibbons (1980), *J. Appl. Phys.* **51**, 274.

Piglmayer, K., Doppelbauer, J., and Bäuerle, D. (1984), *Materials Research Society Symposium*, A. W. Johnson, D. J. Ehrlich, and H. R. Schlossberg, eds., Elsevier, New York, Vol. 29, pp. 47–54.

Rong, C.Z., Sheng, Du Y., and Fang, M. H. (1989), *Surf. Eng.* **5**, 315.

Sadahiro, T., Cho, T., and Yamaya, S. (1977), *J. Japan Inst. Met* **41**, 542.

Sanders, D. J. (1984), Temperature Distributions Produced by Scanning Gaussian Laser Beam, *Appl. Opt.* **23**, 30.

Singh, J., and Mazumder, J. (1987), *Acta Metal.* **35**, 1995.

Sitrl, E., Hunt, L. P., and Sawyer, D. H. (1974), *J. Electrochem. Soc.* **121**, 919.

Skouby, D. C., and Jensen, K. F. (1988), *J. Appl. Phys.* **63**, 198.

Sneddon, I. N. (1972), *The Use of Integral Transforms*, McGraw-Hill, New York.

West, G. A., and Gupta, A. (1984), Laser-Induced Chemical Vapor Deposition of Silicon Nitride Films in Laser-Controlled Chemical Processing of Surfaces, *Proc. Materials Research Society Symposium*, A. W. Johnson, D. J. Ehrlich, and H. R. Schlossberg, eds., North-Holland, Amsterdam, Vol. 29, pp. 61–66.

Wise, H. (1958), *J. Chem. Phys.* **29**, 634.

Yarmoff, J. A., and McFeely, F. R. (1988), *J. Appl. Phys.* **63**, 5213.

Zeiger, H. J., and Ehrlich, D. J. (1989), Lateral Confinement of Microchemical Surface Reactions: Effects on Mass Diffusion and Kinetics, *J. Vac. Sci. Technol.* **B7**, 466.

Zeiri, Y., Atzmony, U., Bloch, J., and Lucchese, R. R. (1991), *J. Appl. Phys.* **69**, 4110.

# FIVE

# PHOTOLYTIC LCVD MODELING

## 5.1. INTRODUCTION

As noted in Chapter 4, mathematical models are useful for understanding the mechanism of the chemical reaction and deposition process; the relative importance of various process parameters such as the laser irradiance, speed of the substrate with respect to the laser beam, laser pulselength, and wavelength; to analyze the LCVD experimental data; and to design and control the LCVD system in an optimum way. The mathematical modeling of photolytic LCVD differs from pyrolytic LCVD modeling in that the former involves photochemical reactions, whereas, the latter relies on the thermal decomposition of the reactant molecules. However, the transport mechanisms for the distribution of various species inside the deposition chamber are similar in the pyrolytic and photolytic processes and, for this reason, the transport equations are identical for the two processes. The source term, which represents the rate of production of the film material, is different for these two processes because it involves the Arrhenius rate expression and the laser intensity (or photon flux) for the pyrolytic and photolytic LCVD processes, respectively.

To model the photolytic LCVD process, we have to know the geometrical arrangement of the system, such as the direction of the laser beam propagation and the location of the substrate with respect to the laser beam, as well as the process conditions, such as the reactant flow rate and the pressure inside the chamber. As in the case of pyrolytic LCVD, photolytic LCVD can also be carried out by arranging the laser beam and substrate perpendicular (see Fig. 4.1a) or parallel (see Fig. 4.1b) to one another. The configuration of Fig. 4.1b is usually used to deposit a film over a large area, and in this configuration the film growth rate depends on the decomposition of the reactant molecules in the gas phase; the film qualities, such as density and microstructure, are controlled by the substrate temperature.

The mode of operation of the LCVD system is a factor in the relative importance of the transport mechanisms, which are diffusion and forced and natural convection. In the batch operation mode, that is, when there is no flow of the reactant gas due to an external agent, such as a pump, during the deposition process, there is no forced convection. In the continuous operation mode, that is, when the reactant gas flows continuously into and out of the LCVD chamber during the deposition process, the effect of forced convection can be important. At low pressures inside the chamber, natural convection is usually not important, and diffusion is the dominant transport mechanism in the absence of forced convection.

For the purpose of modeling the photolytic LCVD process, the deposition chambers shown in Figs. 4.1a and b can be divided into two regions, where the first region, which will be denoted region 1, refers to the volume illuminated by the laser beam, and the second region, which will be denoted region 2, refers to the rest of the volume inside the chamber. In light of this classification, for purposes of photolytic LCVD modeling, regions 1 and 2 can be referred to as the active or illuminated or bright region and the passive or dark region, respectively.

## 5.2. GOVERNING EQUATIONS AND BOUNDARY CONDITIONS

The transport equations that have to be solved to model photolytic LCVD are the conservation of mass, momentum, and energy equations [Bird et al. (1960)]. In this section, we will discuss the governing equations and boundary conditions for the LCVD system shown in Fig. 4.1a, where the laser beam is perpendicular to the substrate. Studies for the configuration of Fig. 4.1b can be found in Flint et al. (1984), Gattuso et al. (1983), Meunier et al. (1983), Bilenchi et al. (1982), West and Gupta (1984).

The mass conservation equation for the $i$th species (where $i = 1, 2, \ldots, i^*$ and $i^*$ is the total number of species inside the deposition chamber in the $r$th region ($r = 1, 2$) is given by

$$\frac{\partial N_{ir}}{\partial t} + \nabla \cdot (N_{ir}\mathbf{u}) = -\nabla \cdot \mathbf{J}_{ir} + S_{pir} \tag{5.1}$$

where **u** is the same as defined in Chapter 4.

Assuming that the flux $\mathbf{J}_{ir}$ is given by Fick's law, we can write

$$\mathbf{J}_{ir} = -D_{ir}(N_{1r},\ldots,N_{i*r}, T)\nabla N_{ir} \qquad (5.2)$$

The source term $S_{pir}$, that is, the rate of production of the $i$th species per unit volume in the $r$th region is given by (see Appendix C)

$$S_{pir} = -\nabla \cdot \psi \qquad (5.3)$$

We will obtain an explicit expression for $S_{pir}$ from Eq. (5.3) under the following assumptions:

1. Only photochemical reactions occur during the deposition process:

$$A + h\nu \rightarrow X + Z \qquad (5.4)$$

2. Only single-photon absorption occurs to induce the photochemical reactions, and one molecule of $A$ produces one atom or molecule of the desired material $X$, as shown in Eq. (5.4).
3. The laser beam is cylindrical, which is a good approximation if the beam divergence is very small and we consider a portion of the beam around its waist. It should be noted that a Gaussian beam has hyperbolic caustics which make the beam hyperboloid instead of cylindrical.
4. The Beer–Lambert law is applicable for the propagation of the beam.

As a result of these assumptions, Eq. (5.3) can be written as

$$S_{pir} = N_{Ar}\sigma_A\phi_0(r,t)\exp[-N_{Ar}\sigma_A(H-z)] \qquad (5.5)$$

where the quantum yield of $x$ is taken to be unity.

Since the photochemical reactions occur only in region 1 (see Fig. 4.1a), that is, in the bright region, we have $S_{pi2} = 0$. The momentum conservation equation in region 1 is given by

$$\rho\left[\frac{\partial \mathbf{u}}{\partial t} + (\mathbf{u}\cdot\nabla)\mathbf{u}\right] = -\nabla\cdot\tau - \nabla\cdot P\delta + \rho(P,T)\mathbf{g} \qquad (5.6)$$

where the shear stress tensor

$$\tau = -\mu(\nabla\mathbf{u} + (\nabla\mathbf{u})^t) + (\tfrac{2}{3}\mu - \kappa)(\nabla\cdot\mathbf{u})\delta$$

and $\rho$, $P$, $\delta$, $\mathbf{g}$, $\mu$, $\nabla \mathbf{u}$, $(\nabla \mathbf{u})^t$, and $\kappa$ are as defined in Chapter 4. The momentum conservation equation for region 2 is similar to Eq. (5.6). The energy conservation equation in region 1 is given by

$$\rho \left[ \frac{\partial}{\partial t}(\hat{U} + \tfrac{1}{2}u^2) + \mathbf{u} \cdot \nabla(\hat{U} + \tfrac{1}{2}u^2) \right] = -\nabla \cdot \mathbf{q} - \nabla \cdot [(\tau + P\delta) \cdot \mathbf{u}] + \rho(P, T)\mathbf{u} \cdot \mathbf{g}$$

(5.7)

where $\hat{U}$, $u$, and $\mathbf{q}$ are as defined in Chapter 4. The energy conservation equation for region 2 is similar to Eq. (5.7).

To write the boundary conditions needed to solve the above conservation equations, we consider regions 1 and 2 to be cylindrical with radii $r_1$ and $r_2$, respectively. We also consider that the substrate is located in such a way that its top surface, that is, the bottom surface of region 1 and the bottom surface of region 2 are at the $Z = 0$ plane in Fig. 4.1a. Similarly, the inner surface of the window and the top surface of region 2 are considered to be at the $Z = H$ plane in Fig. 4.1a. It should be noted that the bottom surface of region 1 is at the $z = 0$ plane at time $t = 0$, and that its height increases as the film is deposited. Assuming that the height of the surface of the film increases at a spatially uniform speed $dZ(t)/dt$, we let $Z(t)$ be the location of the top surface of the film at any instant of time $t$. Also, let us consider that photochemical reactions occur at the film surface due to the interactions of laser with the adsorbed species, which produce $\dot{r}_{irs}$ number of the $i$th species per unit area per unit time at the film surface $Z = Z(t)$ in the $r$th region.

Now the boundary conditions for the Eq. (5.1) can be written. To write the boundary condition at the film surface, that is, at $Z = Z(t)$, we will use the following mass balance across the surface for $r = 1$:

$$\begin{pmatrix} \text{Net diffusive flux} \\ \text{in the } z\text{-direction} \end{pmatrix} = \begin{pmatrix} \text{Net convective flux} \\ \text{in the } z\text{-direction} \end{pmatrix}$$

$$\left[ \begin{pmatrix} \text{Diffusive flux} \\ \text{out of the surface} \end{pmatrix} - \begin{pmatrix} \text{Diffusive flux} \\ \text{into the surface} \end{pmatrix} \right]_{\text{Along the normal to the surface } \hat{n}}$$

$$= \left[ \begin{pmatrix} \text{Convective flux} \\ \text{out of the surface} \end{pmatrix} - \begin{pmatrix} \text{Convective flux} \\ \text{into the surface} \end{pmatrix} \right]_{\text{Along the normal to the surface } \hat{n}}$$

or

$$-D_{i1}\frac{\partial N_{i1}}{\partial z} - \left( -D_{i1s}\frac{\partial N_{i1s}}{\partial z} \right) = \dot{r}_{i1s} + N_{i1}\frac{dZ(t)}{dt} - N_{i1f}\frac{dZ(t)}{dt} \quad (5.8)$$

Assuming that there is no diffusion in the solid phase, that is, $D_{i1s} = 0$ in the film and substrate, and defining the sticking coefficient $\gamma_{ir0}$ by the following expression:

$$\gamma_{ir0} = \frac{N_{irf}/dZ(t)/dt}{\dot{r}_{irs} + N_{ir}/dZ(t)/dt} \tag{5.9}$$

for $r = 1, 2$, we can rewrite the boundary condition (5.8) as follows:

$$D_{i1} \frac{\partial N_{i1}}{\partial Z} = -(1 - \gamma_{i10}) \left( \dot{r}_{i1s} + N_{i1} \frac{dZ(t)}{dt} \right) \quad \text{at} \quad Z = Z(t) \tag{5.10}$$

It should be noted that $Z(0) = 0$, and that $Z(t)$ can be determined from the energy conservation equation by following the procedure, which is usually used to track the solid–liquid interface during melting or solidification [Kar and Mazumder (1988)]. However, if the film thickness is very small and its effect on the process is negligible, we can take $Z(t) \approx 0$.

The boundary conditions at $z = H$ for regions 1 and 2 and at $z = 0$ and $r = r_2$ for region 2 depend on the physical situation that exists inside the chamber during the LCVD process. Three possible situations and their associated boundary conditions are discussed below.

SITUATION I: There are no photochemical reactions and film deposition at $z = H$ in regions 1 and 2, and at $r = r_2$ and $z = 0$ in region 2. This situation leads to the Neumann boundary conditions, where the fluxes are zero, that is,

$$\frac{\partial N_{i2}}{\partial z} = 0 \quad \text{at} \quad z = 0 \tag{5.11a}$$

$$\frac{\partial N_{i1}}{\partial z} = 0 \quad \text{at} \quad z = H \tag{5.11b}$$

$$\frac{\partial N_{i2}}{\partial z} = 0 \quad \text{at} \quad z = H \tag{5.11c}$$

and

$$\frac{\partial N_{i2}}{\partial r} = 0 \quad \text{at} \quad r = r_2 \tag{5.11d}$$

SITUATION II: The concentrations of various species do not change at $z = H$ in regions 1 and 2, and at $r = r_2$ and $z = 0$ in region 2. This situation leads to the Dirichlet boundary conditions, where the concentrations are specified, that is,

$$N_{i2} = N_{i20} \quad \text{at} \quad z = 0 \tag{5.12a}$$

$$N_{i1} = N_{i1H} \quad \text{at} \quad z = H \tag{5.12b}$$

$$N_{i2} = N_{i2H} \quad \text{at} \quad z = H \tag{5.12c}$$

and

$$N_{i2} = N_{i2r} \quad \text{at} \quad r = r_2 \tag{5.12d}$$

SITUATION III: There are no photochemical reactions in region 2, that is, $\dot{r}_{i2s} = 0$, but film deposition occurs at $z = H$ for regions 1 and 2 and at $r = r_2$ and $z = 0$ for region 2. This situation leads to the Cauchy boundary conditions, where the net diffusive and convective fluxes are equal to each other, as in the boundary condition (5.10), which can be utilized to write the following expressions:

$$D_{i2} \frac{\partial N_{i2}}{\partial z} = -(1 - \gamma_{i20})N_{i2} \frac{dZ_0(t)}{dt} \quad \text{at} \quad z = Z_0(t) \tag{5.13a}$$

$$D_{i1} \frac{\partial N_{i1}}{\partial z} = -(1 - \gamma_{i1H})\left(-\dot{r}_{i1sH} + N_{i1} \frac{dH_1(t)}{dt}\right) \quad \text{at} \quad z = H_1(t) \tag{5.13b}$$

$$D_{i2} \frac{\partial N_{i2}}{\partial z} = -(1 - \gamma_{i2H})N_{i2} \frac{dH_2(t)}{dt} \quad \text{at} \quad z = H_2(t) \tag{5.13c}$$

and

$$D_{i2} \frac{\partial N_{i2}}{\partial r} = -(1 - \gamma_{i2r})N_{i2} \frac{dR_2(t)}{dt} \quad \text{at} \quad r = R_2(t) \tag{5.13d}$$

As in the case of boundary condition (5.10), the locations of the film–gas interfaces $Z_0(t)$, $H_1(t)$, $H_2(t)$, and $R_2(t)$ are such that $Z_0(0) = 0$, $H_1(0) = H_2(0) = H$, and $R_2(0) = r_2$, and that the method of determining these locations is analogous to the method used in Kar and Mazumder (1988). However, if the film thickness is very small with negligible effects on the LCVD process, we can assume $Z_0(t) \approx 0$, $H_1(t) \approx H_2(t) \approx H$, and $R_2(t) \approx r_2$. Also, any combination of the boundary conditions (5.11a–d), (5.12a–d), and (5.13a–d) can be used to solve the mass conservation Eq. (5.1).

To complete the boundary conditions for Eq. (5.1), we have to satisfy the continuity of mass and mass flux at the interface of regions 1 and 2 and at $\theta = 0$ and at $\theta = 2\pi$, that is,

$$N_{i1} = N_{i2} \quad \text{at} \quad r = r_1 \tag{5.14a}$$

and

$$D_{i1}\frac{\partial N_{i1}}{\partial r} = D_{i2}\frac{\partial N_{i2}}{\partial r} \quad \text{at} \quad r = r_1 \tag{5.14b}$$

$$N_{ir}(r, 0, z, t) = N_{ir}(r, 2\pi, z, t) \quad \text{for} \quad r = 1 \text{ and } 2 \tag{5.14c}$$

$$\left.\frac{\partial N_{ir}}{\partial \theta}\right|_{\theta=0} = \left.\frac{\partial N_{ir}}{\partial \theta}\right|_{\theta=2\pi} \quad \text{for} \quad r = 1 \text{ and } 2 \tag{5.14d}$$

The initial condition is given by

$$N_{ir}(r, \theta, z, 0) = N_{ir}^* \tag{5.15}$$

The boundary conditions for the momentum and energy conservation equations have been discussed in Chapter 4. When the boundary conditions are known, Eqs. (5.1), (5.6), and (5.7) can be solved to determine the distributions of various species inside the deposition chamber and the velocity and temperature fields during the LCVD process. However, these equations are difficult to solve in practice because they are coupled non-linear partial differential equations. For this reason, the mass, momentum, and energy conservation equations and the boundary conditions are simplified by using appropriate approximations to model the LCVD process.

A review of several LCVD models can be found in Zeiger et al. (1989). The chemical reactions that occur during photolytic LCVD can be classified into those of the gas or liquid phase and the substrate surface. Deposition due to the surface reactions involves the following processes:

1. Adsorption of reactant molecules at the substrate surface.
2. Interactions between the laser beam and the adsorbed molecules leading to photochemical reactions to produce the atoms of the film material.
3. Adherence of the desired atoms to the substrate surface depending on the sticking coefficient.

On the other hand, deposition due to reactions of the gas or liquid phase involves the following processes:

1. Interactions between the laser beam and the reactant molecules along the path of the beam, leading to photochemical reactions to produce the desired atoms.
2. Transport of these atoms from the reaction zone to the substrate surface.

3. Adherence of the desired atoms to the substrate surface depending on the sticking coefficient.

However, as pointed out in Chapter 3, photolytic LCVD can be considered to occur mainly due to decomposition of the gas or liquid phase of the reactant molecules, although the surface reactions are considered to take place initially to form critical nuclei.

## 5.3. UNIFORM SOURCE AND FLUX MODEL

This model is based on the following assumptions:

1. The atoms of the film material are produced at a constant rate per unit volume of the gas or liquid column illuminated by the laser beam, which is referred to as the bright region, such as region 1 in Fig. 4.1. This assumption holds good when the concentration of the reactant molecules and the intensity of the laser beam are uniform in the bright region. For a Gaussian laser beam, the intensity varies radially as well as along the direction of beam propagation, that is, longitudinally. The radial variation of the intensity is usually expressed as $I(r) = I_0 \exp(-2r^2/w_0^2)$, which implies that $I(r) = I_0/e$ at $r = w_0/\sqrt{2}$. Within this $1/e$ point, that is, within the diameter $\sqrt{2}w_0$, the beam is assumed to have a uniform intensity $\bar{I}$ in this model.
2. The flux of the atoms of interest is constant at the surface of the bright region.
3. The bright region is under the steady state condition; that is, the rates of production and loss of the atoms of interest are equal. The loss of atoms is assumed to be due only to the diffusive flux. Loss of atoms due to recombination or other effects is not considered in this model.
4. There is no convection inside the LCVD chamber.
5. The reason zone is considered to be a semi-infinitely long cylinder of diameter $\sqrt{2}w_0$, which is a good approximation when the Rayleigh range is much larger than the focal spot diameter.

The governing equation for this model can be obtained from Eq. (5.1) by noting that its left-hand side is zero due to assumptions (3) and (4), taking the volume integral of its right-hand side, applying the divergence theorem, and using assumptions (1) and (2), which leads to the following result.

$$(\text{Flux}) \begin{pmatrix} \text{Area of a} \\ \text{surface } S \end{pmatrix} = \begin{pmatrix} \text{Volumetric} \\ \text{production rate} \end{pmatrix} \begin{pmatrix} \text{Volume bounded} \\ \text{by the surface } S \end{pmatrix} \quad (5.16)$$

The volumetric rate of production of the atoms of interest is given by [Zeiger et al. (1989)]

$$j_p = \bar{I}\sigma(n_0 - n) \quad (5.17)$$

and the diffusive flux at the surface of the reaction zone is given by

$$j_d = (\text{Diffusive speed})(\text{Number density})$$
$$= (D/\sqrt{2}w_0)n \quad (5.18)$$

where the characteristic length is taken to be $\sqrt{2}w_0$. For any length of the reaction zone, we obtain the following expression from Eqs. (5.16)–(5.18):

$$n = \frac{\bar{I}\sigma n_0 w_0^2/2}{D + \bar{I}\sigma w_0^2/2} \quad (5.19)$$

Eliminating $n$ from Eq. (5.17) by using Eq. (5.19), we can write

$$j_p = Dn_0 \frac{\bar{I}\sigma/D}{1 + \bar{I}\sigma w_0^2/2D} \quad (5.20)$$

The volumetric source, such as $j_p$, which is given by Eq. (5.20), can be converted [Rockwell (1956)] into a surface source by using the following relationship:

$$j_s = j_p \frac{(\pi/4)(\sqrt{2}w_0)^2}{\pi(\sqrt{2}w_0)}$$
$$= \left(\frac{Dn_0}{\sqrt{2}w_0}\right) \frac{\bar{I}\sigma(w_0^2/2D)}{1 + \bar{I}\sigma(w_0^2/2D)} \quad (5.21)$$

Based on the solid angle subtended by the semi-infinitely long cylindrical column of the reaction zone at a point on the substrate surface, the normal component of the number of the atoms of interest impinging per unit time per unit area of the substrate surface, which will be referred to as the particle flux, has been given as [Zeiger *et al.* (1989), Tsao and Ehrlich (1984a,b)]

$$j_f = \frac{j_s}{4\pi} \tan^{-1}\left(\frac{\sqrt{2}w_0^2}{b}\right) \qquad (5.22)$$

for unit sticking probability.

At very low laser intensities, $\bar{I}\sigma(w_0^2/2D) \ll 1$ in Eq. (5.21) and, therefore,

$$j_s = \tfrac{1}{2}\bar{I}\sigma n_0(w_0/\sqrt{2}) \qquad (5.23)$$

which is independent of the diffusion coefficient, and indicates that the deposition process is in the reaction-rate-controlled regime. The surface source term $j_s$ is proportional to the laser intensity and the laser beam diameter in Eq. (5.23) and, so the film growth rate will increase in this regime as the spot size increases for fixed $\bar{I}$.

At very high laser intensities, $\bar{I}\sigma(w_0^2/2D) \gg 1$ in Eq. (5.21) and, therefore,

$$j_s = \frac{Dn_0}{\sqrt{2}w_0} \qquad (5.24)$$

which is independent of the laser intensity, and indicates that the deposition process is in the diffusion-controlled regime. Here $j_s$ depends linearly on the diffusion coefficient and inversely on the laser beam diameter, so that the film growth rate may be enhanced by reducing the spot size for fixed $D$.

DERIVATION OF EQ. (5.22)

Equation (5.22) can be derived by using the concept of view factor which is used for studying radiation heat transfer [Siegel and Howell (1981)]. In Fig. 5.1, the cylinder represents the reaction zone whose base is at the substrate surface and which extends up to infinity in the $z$-direction. We want to consider a portion of the cylinder that can be viewed from the point $A$ which is located on the substrate surface with Cartesian coordinates $(b, 0, 0)$. $AB$ and $AC$ are two tangents to the base circle of the cylinder at the substrate surface and, therefore, $BCDE$ will be the portion of the cylinder that can be seen from the point $A$. However, this front portion $BCDE$ as well as the rear portion of the cylinder contribute equally to the particle flux at $A$, since they subtend the same solid angle at point $A$. Consider an infinitesimal surface $F$ on the front portion $BCDE$, which is located at a distance $R_1$ from point $A$. Assuming that the angular distribution of the atoms of interest is isotropic at the surface of the reaction zone, that is, $j_s$ is isotropic at the

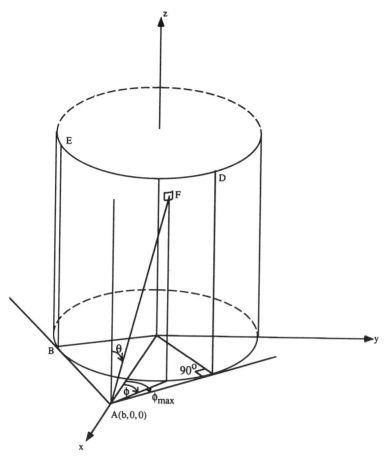

Figure 5.1. Geometry of the reaction zone for calculating the particle flux at the substrate surface.

surface of the cylinder in Fig. 5.1, we can write the particle flux at $A$ as

$$\begin{aligned}
j_f &= \begin{pmatrix} \text{Flux due to the front} \\ \text{portion } BCDE \end{pmatrix} + \begin{pmatrix} \text{Flux due to the rear} \\ \text{portion of the cylinder} \end{pmatrix} \\
&= 2\frac{j_s}{4\pi} \int \begin{pmatrix} \text{Solid angle subtended by the} \\ \text{differential area } F \text{ at the point } A \end{pmatrix} \cos\theta \\
&= 2\frac{j_s}{4\pi} \int_{\theta=0}^{\pi/2} \int_{-\phi_{\max}}^{\phi_{\max}} \frac{(R_1 d\theta)(R_1 \sin\theta d\phi)}{R_1^2} \cos\theta \\
&= \frac{j_s}{4\pi}(2\phi_{\max})
\end{aligned} \qquad (5.25)$$

From Fig. 5.1, we find that

$$2\phi_{max} = 2\sin^{-1}\left(\frac{w_0}{\sqrt{2b}}\right) = \tan^{-1}\left[\frac{(\sqrt{2}w_0/b)\sqrt{1-(w_0/\sqrt{2b})^2}}{1-2(w_0/\sqrt{2b})^2}\right]$$

$$\approx \tan^{-1}\left(\frac{\sqrt{2}w_0}{b}\right) \quad \text{for} \quad b \gg w_0 \qquad (5.26)$$

From Eqs. (5.25) and (5.26), we can write

$$j_f \approx \frac{j_s}{4\pi}\tan^{-1}\left(\frac{\sqrt{2}w_0}{b}\right) \quad \text{for} \quad b \gg w_0 \qquad (5.27)$$

It should be noted that $\phi_{max} = \pi/2$ when point $A$ is inside the cylinder, and in that case $j_f = j_s/4$. We also note that the solid angle subtended by the semi-infinitely long cylinder at the point $A$ is given by [Guest (1961), Masket (1957)]

$$2\phi_{max} \approx \tan^{-1}\left(\frac{\sqrt{2}w_0}{b}\right) \quad \text{for} \quad b \gg w_0$$

Although Eq. (5.27), which is identical to Eq. (5.22), is derived for $b \gg w_0$, an expression

$$j_f = \frac{j_s}{2\pi}\tan^{-1}\left(\frac{\sqrt{2}w_0}{b}\right)$$

which is twice the right-hand side of Eq. (5.22), has been used by Zeiger *et al.* (1989) to obtain results for $b \leqslant w_0$ as well as $b > w_0$.

## 5.4. IDENTIFICATION OF GAS AND ADSORBED PHASE REACTIONS

In the last section, the volumetric source term $j_p$ was converted into a surface source term $j_s$ to obtain an expression for the film growth rate. Wood *et al.* (1983a, b) have developed a model to determine the film growth rate and find out whether adsorbed or gas-phase decomposition takes place during LCVD by considering the volumetric source term directly. We consider Fig. 5.2, which shows the cylindrical geometry used for this model,

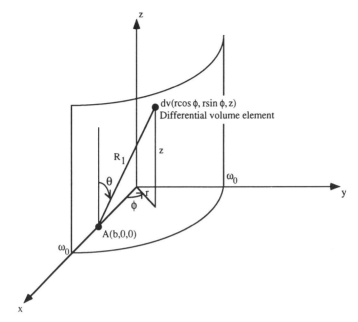

Figure 5.2. Geometry for the calculations of flux due to a semi-infinitely long cylindrical volumetric source.

point $A$, where we want to calculate the film growth rate, and a differential volume element. The Cartesian coordinates of the point of interest, $A$, and the differential volume element $dV$ are $(b, 0, 0)$ and $(r\cos\phi, r\sin\phi, z)$, respectively. The cylindrical reaction zone is considered to be semi-infinitely long, which is a good approximation when the confocal beam parameter (Rayleigh range) is much larger than the focal spot diameter (beam waist size), and its radius is taken to be $w_0$, which is equal to the radius of the beam waist. Also, we assume that a Gaussian laser beam is used for photolytic LCVD.

The film growth rate, that is, the number of atoms deposited on the substrate per unit area per unit time due to the dissociation of adsorbed molecules can be written as

$$j_{fa}(b, w_0) = n_a \sigma_a I(b) = n_a \sigma_a \frac{2P_t}{\pi w_0^2 E} \exp\left(\frac{-2b^2}{w_0^2}\right) \qquad (5.28)$$

It should be noted that $n_a$ is of the order of one monolayer density [Ehrlich and Osgood (1981)].

To model the deposition rate due to gas phase dissociation. Wood *et al.* (1983a, b) have assumed that the volumetric source is isotropic and used the following expression for the normal component of the number of the atoms of interest impinging per unit time per unit area of the substrate surface:

$$j_f(b, w_0) = \gamma \int \frac{1}{4\pi R_1^2} \left[ n_0 \sigma \frac{2P_t}{\pi w_0^2 E} \exp\left(\frac{-2r^2}{w_0^2}\right) \right] (dV)(\cos \theta) \exp\left(\frac{-R_1}{1}\right) \quad (5.29)$$

It should be noted that $j_f(b, w_0)$ represents the particle flux that sticks to the substrate surface.

The derivation of Eq. (5.29) is similar to the determination of neutron flux [Morgan and Emerson (1967)] from a neutron source. However, it should be noted that the neutron flux is a scalar quantity, but the mass or particle flux $j_f$ is considered to be directed along the normal to the surface [Morgan and Emerson (1967)]. The exponential term, $\exp(-R_1/l)$ represents the attenuation of the particles moving toward point $A$. Ehrlich *et al.* (1980) have found by studying the gas phase population decay that $l$ is about 10 cm under typical photolytic LCVD conditions, which means that $l \gg w_0$ and, therefore, $\exp(-R_1/l) \approx 1$. Also, we find that

$$\cos \theta = \hat{k} \cdot \hat{R}_1 = \frac{\hat{k} \cdot [r\cos\phi - b)\hat{i} + r\sin\phi\hat{j} + z\hat{k}]}{R_1} \quad (5.30)$$

Thus, Eq. (5.29) can be written as

$$j_f(b, w_0) = \frac{\gamma n_0 \sigma P_t}{2\pi^2 w_0^2 E} \int_0^{w_0} r \exp\left(\frac{-2r^2}{w_0^2}\right)$$

$$\times dr \left\{ \int_0^{2\pi} d\phi \int_0^{\infty} \frac{z \, dz}{[r\cos\phi - b)^2 + (r\sin\phi)^2 + z^2]^{3/2}} \right\}$$

$$= \frac{\gamma n_0 \sigma P_t}{2\pi^2 w_0^2 E} \int_0^{w_0} r \exp\left(\frac{-2r^2}{w_0^2}\right) dr \left[ 2 \int_0^{\pi} \frac{d\phi}{(r^2 + b^2 - 2br\cos\phi)^{1/2}} \right] \quad (5.31)$$

$$= \frac{\gamma n_0 \sigma P_t}{\pi^2 w_0^2 E} \int_0^{w_0} r \exp\left(\frac{-2r^2}{w_0^2}\right) \frac{2}{b+r} K(k) \, dr \quad (5.32)$$

for $r^2 + b^2 > 2br > 0$, where $k^2 = 4br/(r+b)^2$ and $K(k)$ is the complete elliptic integral of the first kind (Byrd and Friedman, 1954). From Byrd and

Friedman (1954), we know that

$$K(k) = \frac{\pi}{2}\left(1 + \frac{1}{4}k^2 + \frac{9}{64}k^4 + \frac{25}{256}k^6 + \frac{1225}{16384}k^8 + \frac{3969}{65536}k^{10} + \cdots\right)$$

$$= \frac{\pi}{2}\sum_{i=0}^{\infty} A_i k^{2i}$$

where

$$A_i = \frac{[(\tfrac{1}{2})(\tfrac{1}{2}+1)\cdots(\tfrac{1}{2}+i-1)]^2}{(i!)^2} \quad \text{for} \quad i = 1, 2, 3, \ldots, \quad \text{and} \quad A_0 = 1$$

Noting that

$$\exp\left(\frac{-2r^2}{w_0^2}\right) = \sum_{j=0}^{\infty} B_j (r^{-2})^j$$

where $B_j = (1/j!)(-2/w_0^2)^j$, and assuming that the integrand is such that the integration and summation can be interchanged in Eq. (5.32), we can write

$$j_f(b, w_0) = \frac{\gamma n_0 \sigma P_t}{\pi w_0^2 E} \sum_{i=0}^{\infty} A_i (4b)^i \sum_{j=0}^{\infty} B_j \int_0^{w_0} \frac{r^{i+2j+1}}{(b+r)^{2i+1}} dr \quad (5.33)$$

where the integration with respect to $r$ can be carried out by using the following formula [Gradshteyn and Ryzhik (1980)]:

$$\int \frac{r^p}{(b+r)^q} dr = -\frac{1}{q-1}\frac{r^p}{(b+r)^{q-1}} + \frac{p}{q-1}\int \frac{r^{p-1}}{(b+r)^{q-1}} dr$$

It can be shown [Wood et al. (1983a, b)] from Eq. (5.29) that for an arbitrary scaling factor $L$ the deposition rate $j_f$ can be expressed as

$$j_f(Lb, Lw_0) = \frac{1}{L} j_f(b, w_0) \quad (5.34)$$

that is,

$$j_f(b, w_0) = \frac{1}{w_0} f_f(b/w_0) \quad (5.35)$$

From Eq. (5.28), we find that

$$j_{f_a}(b, w_0) = \frac{1}{w_0^2} f_{f_a}(b/w_0) \tag{5.36}$$

In general, the deposition rate $j$ can be expressed as a function of $b/w_0$, $[f(b/w_0)]$ as follows:

$$j = \frac{1}{w_0^m} f(b/w_0) \tag{5.37}$$

where $m = 1$ and $2$ represent the deposition due to the gas and adsorbed phase decomposition, respectively. Therefore, by examining the deposition rate data as a function of $b/w_0$, the value of $m$ can be determined to differentiate between the gas and adsorbed phase decomposition. The deposition rates at the laser beam center can be obtained by setting $b = 0$ in Eqs. (5.28) and (5.31), which leads to the following expression:

$$j_{f_a}(0, w_0) = \frac{2 n_a \sigma_a P_t}{\pi E} \frac{1}{w_0^2} \tag{5.38}$$

and

$$j_f(0, w_0) = \frac{\gamma n_0 \sigma P_t}{\pi w_0^2 E} \int_0^{w_0} \exp\left(\frac{-2r^2}{w_0^2}\right) dr$$

$$= \frac{\gamma n_0 \sigma P_t}{\pi w_0^2 E} \int_0^{\infty} \exp\left(\frac{-2r^2}{w_0^2}\right) dr \tag{5.39}$$

since the contribution to $j_f$ by the term

$$\frac{\gamma n_0 \sigma P_t}{\pi w_0^2 E} \int_0^{\infty} \exp\left(\frac{-2r^2}{w_0^2}\right) dr$$

is very small because the laser intensity is extremely small at $r \geqslant w_0$. Evaluating the integral in Eq. (5.39), $j_f(0, w_0)$ is found to be

$$j_f(0, w_0) = \frac{\gamma n_0 \sigma P_t}{\sqrt{8\pi} E} \frac{1}{w_0} \tag{5.40}$$

Equations (5.38) and (5.40) show that $j_{f_a}(0, w_0) \propto 1/w_0^2$ and $j_f(0, w_0) \propto 1/w_0$, respectively.

## 5.5. HEMISPHERIC MODEL

Besides considering the surface and volumetric sources in the cylindrical coordinates as discussed in the last two sections, the spherical coordinate system has also been used to analyze the distribution of the reactant molecules by considering a localized hemispherical region at the laser-irradiated substrate surface as shown in Fig. 5.3.

### 5.5.1. Single-Region ($w_0 \leq r < \infty$) Model with Hemispheric Surface Reaction

Ehrlich and Tsao (1983) have presented a model to study the effect of gas phase diffusion on surface reactions during photolytic LCVD under the following assumptions:

1. The reaction is assumed to occur at the surface of a hemisphere of radius $\omega_0$ as shown in Fig. 5.3. It should be noted that the actual photolytic LCVD usually involves a plane substrate surface, and that the reactant and product molecules move toward and away from the reaction zone, respectively.
2. There is no convection inside the deposition chamber. The partial pressure of the reactant molecules is usually very low during photolytic LCVD, and the deposition chamber is filled mainly with an

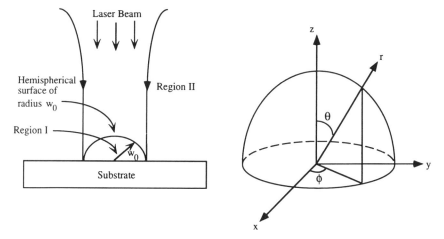

Figure 5.3. Geometry for the hemispheric model. $0 \leq r \leq w_0$ and $w_0 \leq r \leq \infty$ define regions I and II, respectively.

inert carrier gas which assists in maintaining the pressure equilibrium during the LCVD process.
3. The radial diffusive flux of the reactant molecules at the surface of the hemisphere (see Fig. 5.3) is assumed to be equal to the radial flux of the reactant molecules that take part in the chemical reaction at $r = w_0$.

The governing equation and the boundary and initial conditions for this model can be written [Ehrlich and Tsao (1983)] as follows: It should be noted that the problem is symmetric in the azimuthal $\phi$ direction, and that the flux in the polar $\theta$ direction is zero at the substrate surface, that is, $\partial n/\partial \theta = 0$ at $\theta = \pm \pi/2$, which is the symmetry condition for a full sphere. Therefore, $n$ can be taken to be independent of $\theta$ and $\phi$:

GOVERNING EQUATION:

$$D \frac{1}{r^2} \frac{\partial}{\partial r}\left(r^2 \frac{\partial n}{\partial r}\right) = \frac{\partial n}{\partial t} \tag{5.41a}$$

BOUNDARY CONDITIONS:

$$D \frac{\partial n}{\partial r} = \tfrac{1}{2}\eta u n \quad \text{at} \quad r = w_0 \tag{5.41b}$$

$$n(r, t) = n_0 \quad \text{as} \quad r \to \infty \tag{5.41c}$$

INITIAL CONDITION:

$$n(r, t) = n_0 \quad \text{at} \quad t = 0 \tag{5.41d}$$

It should be noted that the collisional reaction efficiency $\eta$ and the root mean square velocity $u$ usually depend on temperature. If the variations of $\eta$ and $u$ with temperature are taken into account, the mass and energy conservation equations have to be solved simultaneously to model the LCVD process properly. In the present model, $\eta$ and $u$ are taken to be independent of temperature.

To solve the mass transfer problem, let us define a new variable,

$$N = rn \tag{5.42}$$

which transforms the problem given by expressions (5.41a–d) into the

# PHOTOLYTIC LCVD MODELING

following form:

$$D \frac{\partial^2 N}{\partial r^2} = \frac{\partial N}{\partial t} \tag{5.43a}$$

$$D \frac{\partial N}{\partial r} = (\tfrac{1}{2}\eta u + D/r)N \quad \text{at} \quad r = w_0 \tag{5.43b}$$

$$N(r, t) = rn_0 \quad \text{as} \quad r \to \infty \tag{5.43c}$$

$$N(r, t) = rn_0 \quad \text{at} \quad t = 0 \tag{5.43d}$$

We will solve the problem given by expressions (5.43a–d) by using the Laplace transform, which is defined as

$$\bar{N}(r, s) = \int_0^\infty N(r, t) e^{-st} \, dt \tag{5.44a}$$

and the associated inversion formula is given by

$$N(r, t) = \frac{1}{2\pi i} \int_{\beta - i\infty}^{\beta + i\infty} \bar{N}(r, s) e^{st} \, ds \tag{5.44b}$$

where $\beta$ is a real number which is so large that all the singularities of $\bar{N}(s, t)$ lie to the left of the line that joins the points $\beta - i\infty$ and $\beta + i\infty$. The Laplace transform of Eqs. (5.43a–c) leads to the following expressions:

$$D \frac{d^2 \bar{N}(r, s)}{dr^2} = s\bar{N}(r, s) - rn \tag{5.45a}$$

$$D \frac{d\bar{N}(r, s)}{dr} = (\tfrac{1}{2}\eta u + D/r)\bar{N}(r, s) \quad \text{at} \quad r = w_0 \tag{5.45b}$$

$$\bar{N}(r, s) = rn_0/s \quad \text{as} \quad r \to \infty \tag{5.45c}$$

Note that the initial condition (5.43d) has been utilized in Eq. (5.45a). The solution of Eq. (5.45a) is given by

$$\bar{N}(r, s) = A_s \exp[-\sqrt{(s/D)}\, r] + B_s \exp[\sqrt{(s/D)}\, r] + rn_0/s \tag{5.46}$$

where $A_s$ and $B_s$ are two constants of integration that have to be determined by satisfying the boundary conditions.

The solution (5.46) and the boundary condition (5.45c) imply that $B_s = 0$. Now applying the boundary condition (5.45b) to Eq. (5.46), we obtain

$$A_s = -\frac{n_0 w_0}{r_0} \frac{\exp[\sqrt{(s/D)}w_0]}{s(\sqrt{s/D} + 1/r_0 + 1/w_0)}$$

where $r_0 = 2D/(\eta u)$ and, finally, Eq. (5.46) can be written as

$$\bar{N}(r, s) = \frac{r n_0}{s} - \frac{n_0 w_0}{r_0} \frac{\exp[-\sqrt{(s/D)}(r - w_0)]}{s(\sqrt{s/D} + 1/r_0 + 1/w_0}\quad (5.47)$$

Taking the inverse Laplace transform [Carslaw and Jaeger (1959)] of Eq. (5.47) and utilizing Eq. (5.42), we obtain

$$\frac{n(r,t)}{n_0} = 1 - \frac{w_0^2}{r(r_0 + w_0)} \left\{ \mathrm{erfc}\left(\frac{r - w_0}{2\sqrt{Dt}}\right) - \left\{\exp\left[\left(\frac{1}{r_0} + \frac{1}{w_0}\right)(r - w_0)\right.\right.\right.$$
$$\left.\left.\left. + \left(\frac{1}{r_0} + \frac{1}{w_0}\right)^2 Dt\right]\right\} \left\{\mathrm{erfc}\left[\frac{r - w_0}{2\sqrt{Dt}} + \left(\frac{1}{r_0} + \frac{1}{w_0}\right)\sqrt{Dt}\right]\right\}\right\} \quad (5.48)$$

The flux, which is directed in the negative direction of $r$ at the surface of the hemisphere is given by

$$j(w_0, t) = D \left.\frac{\partial n(r, t)}{\partial r}\right|_{r = w_0}$$
$$= \frac{D n_0}{r_0 + w_0} \left\{ 1 + \frac{w_0}{r_0} \left\{\exp\left[\left(\frac{1}{r_0} + \frac{1}{w_0}\right)^2 Dt\right]\right\}\right.$$
$$\left. \times \left\{\mathrm{erfc}\left[\left(\frac{1}{r_0} + \frac{1}{w_0}\right)\sqrt{Dt}\right]\right\}\right\} \quad (5.49)$$

At steady state, the concentration and the flux at the surface of the hemisphere are given by

$$n(w_0, \infty) = \frac{n_0 r_0}{r_0 + w_0} \quad (5.50a)$$

and

$$j(w_0, \infty) = \frac{Dn_0}{r_0 + w_0} \qquad (5.50b)$$

Based on the relative values of $r_0$ and $w_0$ the reaction at the surface of the hemisphere can be divided into the following two regimes:

REACTION-CONTROLLED REGIME: This regime is attained when $r_0 \gg w_0$. Due to the large value of $r_0$ compared to $w_0$, Eq. (5.50b) becomes

$$j(w_0, \infty) \approx Dn_0/r_0 = \tfrac{1}{2} \eta u n_0 \qquad (5.50c)$$

Equation (5.50c) shows that the flux depends on the reaction rate, but it does not depend on the diffusion coefficient $D$. However, $j(w_0, \infty)$ is proportional to pressure, since it is proportional to $n_0$, which, in turn, is proportional to pressure. Therefore, the deposition rate can be increased by increasing $n_0$ in this regime. Since $j(w_0, \infty)$ does not depend on $w_0$, according to Eq. (5.50c), the deposition rate cannot be affected by altering the spot size for a given laser intensity

DIFFUSION CONTROLLED REGIME: This regime is attained when $r_0 \ll w_0$, and in this regime Eq. (5.50b) becomes

$$j(w_0, \infty) \approx Dn_0/w_0 \qquad (5.50d)$$

Equation (5.50d) shows that the flux depends on the diffusion coefficient, but it does not depend on the reaction rate. Since the diffusion coefficient is inversely proportional to the pressure and $n_0$ is directly proportional to the pressure, $j(w_0, \infty)$ is independent of pressure in this regime. However, $j(w_0, \infty)$ is inversely proportional to $w_0$, which implies that the deposition rate can be enhanced by decreasing the laser focal spot size for a given laser intensity.

Although the reaction-controlled and diffusion-controlled regimes have been discussed above for steady state conditions, it should be noted from (Eq. 5.49) that the flux $j(w_0, \infty)$ attains its steady state value within a very short time. For a pressure and a temperature of 100 Torr and 100°C, respectively, $D = 2.8 \, \text{cm}^2/\text{s}$, and $\eta = 0.1$, the reaction flux $j(w_0, \infty)$ has been plotted by Zeiger et al. (1989) and Ehrlich and Tsao (1983) as a function of time for $\omega_0 = 1, 10, 100,$ and $1000 \, \mu\text{m}$, which shows that the time required to reach the steady state increases as the value of $w_0$ increases, and that the steady state is reached at about $t = 10^{-7}$ and $10^{-5}$ s for $w_0 = 1$ and $10 \, \mu\text{m}$, respectively, and within a few milliseconds for $w_0 = 100$ and $1000 \, \mu\text{m}$.

### 5.5.2. Single-Region $(w_0 \leq r \leq R_i)$ Model with Hemispheric Surface Reaction: Steady State Analysis

In the previous section, we determined the reaction flux at $r = w_0$, $j(w_0, t)$, by considering an infinitely large deposition chamber. Now we will analyze the effect of the size of the deposition chamber on the reaction flux by formulating the problem in the following way:

GOVERNING EQUATION:

$$D \frac{1}{r^2} \frac{\partial}{\partial r} \left( r^2 \frac{\partial n}{\partial r} \right) = \frac{\partial n}{\partial t} \quad (5.51a)$$

BOUNDARY CONDITIONS:

$$D \frac{\partial n}{\partial r} = \tfrac{1}{2} \eta u n \quad \text{at} \quad r = w_0 \quad (5.51b)$$

$$n(r, t) = n_0 \quad \text{at} \quad r = R_i \quad (5.51c)$$

Boundary condition (5.51c) is based on the assumption that the reactant molecules are replenished at the inner wall of the deposition chamber without causing any convection in the system.

This boundary condition is difficult to create in practice. A natural boundary condition would be to assume that the wall of the deposition chamber is impervious to the reactant molecules and, therefore, the flux $D(\partial n/\partial r)$ is zero at $r = R_i$. This aspect will be discussed in the next section.

INITIAL CONDITION:

$$n(r, t) = n_0 \quad \text{at} \quad t = 0 \quad (5.51d)$$

This transient problem, where $n \equiv n(r, t)$, can be solved by using the Laplace transform in the same way as in Section 5.5.1. Since the steady state is reached within a very short time as discussed in Section 5.5.1, we will solve the following steady state problem where $n \equiv n(r, \infty) \equiv n(r)$:

GOVERNING EQUATION:

$$\frac{d}{dr} \left( r^2 \frac{dn}{dr} \right) = 0 \quad (5.52a)$$

BOUNDARY CONDITIONS:

$$dn/dr = n/r_0 \quad \text{at} \quad r = w_0 \tag{5.52b}$$

$$n(r) = n_0 \quad \text{at} \quad r = R_i \tag{5.52c}$$

The solution of Eq. (5.52a) is given by

$$n(r) = C_1/r + C_2 \tag{5.52d}$$

where $C_1$ and $C_2$ are two constants that have to be determined by satisfying the boundary conditions. Applying the boundary conditions (5.52b) and (5.52c) to the solution (5.52d), we obtain

$$(r_0 + w_0)C_1 + w_0^2 C_2 = 0 \tag{5.52e}$$

and

$$C_1 + R_i C_2 = n_0 R_i \tag{5.52f}$$

which can be solved to determine the following expressions for $C_1$ and $C_2$:

$$C_1 = \frac{n_0 w_0^2 R_i}{w_0^2 - R_i(r_0 + w_0)} \tag{5.52g}$$

and

$$C_2 = -\frac{n_0(r_0 + w_0)R_i}{w_0^2 - R_i(r_0 + w_0)} \tag{5.52h}$$

From Eqs. (5.52d), (5.52g), and (5.52h), we can write

$$n(r) = \frac{n_0}{r} \frac{w_0^2 R_i}{w_0^2 - R_i(r_0 + w_0)} - n_0 \frac{R_i(r_0 + w_0)}{w_0^2 - R_i(r_0 + w_0)} \tag{5.53a}$$

The reaction flux $j(r)$ is given by

$$j(r) = D\frac{dn}{dr} = -\frac{Dn_0}{r^2} \frac{w_0^2 R_i}{w_0^2 - R_i(r_0 + w_0)} \tag{5.53b}$$

The number density of the reactant molecules and the reaction flux at

$r = w_0$ can be obtained from Eqs. (5.53a) and (5.53b) as follows:

$$n(w_0) = \frac{n_0 R_i r_0}{R_i(r_0 + w_0) - w_0^2} \quad (5.53c)$$

and

$$j(w_0) = \frac{D n_0 R_i}{R_i(r_0 + w_0) - w_0^2} \quad (5.53d)$$

As $R_i \to \infty$, Eqs. (5.53c) and (5.53d) become identical to Eqs. (5.50a) and (5.50b), respectively. Based on the relative values of $r_0$ and $\omega_0$, the reaction at $r = \omega_0$ can be divided into the following two regimes:

REACTION-CONTROLLED REGIME: This regime is attained when $r_0 \gg w_0$. Since $R_i > w_0$, $R_i r_0$ will be much larger than $w_0^2$ when $r_0 \gg w_0$ and, therefore, Eq. (5.53d) can be written as

$$j(w_0) \approx D n_0 / r_0 = \tfrac{1}{2} \eta u n_0 \quad (5.53e)$$

which is identical to Eq. (5.50c).

DIFFUSION-CONTROLLED REGIME: This regime is attained when $r_0 \ll w_0$, and Eq. (5.53d) becomes

$$j(w_0) \approx \frac{D n_0}{w_0(1 - w_0/R_i)} \quad (5.53f)$$

which is a little different from Eq. (5.50d). Equation (5.53f) shows that the reaction flux $j(w_0)$ can be increased by decreasing the size of the deposition chamber.

### 5.5.3. Single-Region ($w_0 \leqslant r \leqslant R_i$) Model with Hemispheric Surface Reaction: Transient Analysis

The boundary condition (5.51c) used in Section (5.5.2) is difficult to achieve in practice. However, the inner wall of the deposition chamber can be considered to be impervious to the reactant molecules, so that the flux $D(\partial n/\partial r)$ of the reactant molecules penetrating through the wall is zero. We will use this condition to solve the following transient problem:

# PHOTOLYTIC LCVD MODELING

GOVERNING EQUATION:

$$D \frac{1}{r^2} \frac{\partial}{\partial r}\left(r^2 \frac{\partial n}{\partial r}\right) = \frac{\partial n}{\partial t} \qquad (5.54a)$$

BOUNDARY CONDITIONS:

$$D \frac{\partial N}{\partial R} = \frac{1}{2} \eta u n \quad \text{at} \quad r = w_0 \qquad (5.54b)$$

$$\frac{\partial n}{\partial r} = 0 \quad \text{at} \quad r = R_i \qquad (5.54c)$$

INITIAL CONDITION:

$$n(r, t) = n_0 \quad \text{at} \quad t = 0 \qquad (5.54d)$$

It should be noted that when $\partial n/\partial t = 0$, that is, at the steady state, the solution of this transient problem is given by $n(r, \infty) = 0$, which implies that after a sufficiently long time, the reactant molecules will be completely depleted inside the deposition chamber due to photochemical reactions. Also, due to the decrease in the reactant molecule population density as the reaction progresses, the deposition rate will decrease, which can affect the film thickness, as the deposition time increases. The solution of the above transient problem will allow us to study (a) the variation of the reactant molecule population density with time, (b) the variation of the reaction flux with time, and (c) the time required for the reactant molecules to be completely consumed by the chemical reaction, which can be utilized to optimize the LCVD process. We will solve Eq. (5.54a) by using the method of separation of variables, and for this purpose we let

$$n(r, t) = F(r)G(t) \qquad (5.55)$$

Substituting Eq. (5.55) into the expressions (5.54a–c), we obtain

$$\frac{1}{F(r)} \frac{1}{r^2} \frac{d}{dr}\left(r^2 \frac{dF(r)}{dr}\right) = \frac{1}{DG(t)} \frac{dG(t)}{dt} \qquad (5.56a)$$

$$\frac{dF(r)}{dr} = \frac{F(r)}{r_0} \quad \text{at} \quad r = w_0 \qquad (5.56b)$$

$$\frac{dF(r)}{dr} = 0 \quad \text{at} \quad r = R_i \qquad (5.56c)$$

In Eq. (5.56a), since the left-hand side is a function only of $r$ and the right-hand side is a function only of $t$, each side must be equal to a constant. Thus, we can write

$$\frac{1}{F(r)} \frac{1}{r^2} \frac{d}{dr}\left(r^2 \frac{dF(t)}{dr}\right) = \frac{1}{DG(t)} \frac{dG(t)}{dt} = -\lambda^2 \qquad (5.57a)$$

which leads to the following two equations:

$$\frac{d^2(rF)}{dr^2} = -\lambda^2(rF) \qquad (5.57b)$$

and

$$\frac{dG}{dt} = -\lambda^2 DG \qquad (5.57c)$$

where $\lambda$ is the separation constant, and is also known as the eigenvalue of the problem. The solution of Eq. (5.57b) is given by

$$F = A \frac{\sin(\lambda r)}{r} + B \frac{\cos(\lambda r)}{r} \qquad (5.57d)$$

Applying boundary conditions (5.56b) and (5.56c) to the solution (5.57d), we obtain

$$A[\lambda r_0 w_0 \cos(\lambda w_0) - (r_0 + w_0)\sin(\lambda w_0)]$$
$$- B[\lambda r_0 w_0 \sin(\lambda w_0) + (r_0 + w_0)\cos(\lambda w_0)] = 0 \qquad (5.57e)$$

and

$$A[\lambda R_i \cos(\lambda R_i) - \sin(\lambda R_i)] - B[\lambda R_i \sin(\lambda R_i) + \cos(\lambda R_i)] = 0 \qquad (5.57f)$$

For nonzero values of $A$ and $B$, we find from Eqs. (5.57e) and (5.57f) that

$$\frac{A}{B} = \frac{\lambda w_0 r_0 \sin(\lambda w_0) + (r_0 + w_0)\cos(\lambda w_0)}{\lambda w_0 r_0 \cos(\lambda w_0) - (r_0 + w_0)\sin(\lambda w_0)}$$
$$= \frac{\lambda R_i \sin(\lambda R_i) + \cos(\lambda R_i)}{\lambda R_i \cos(\lambda R_i) - \sin(\lambda R_i)} \qquad (5.57g)$$

which yields

$$\tan \lambda(R_i - w_0) = \frac{\lambda[R_i(r_0 + w_0) - r_0 w_0]}{r_0 w_0 R_i \lambda^2 + r_0 + w_0} \tag{5.57h}$$

and

$$A = B \frac{\lambda R_i \sin(\lambda R_i) + \cos(\lambda R_i)}{\lambda R_i \cos(\lambda R_i) - \sin(\lambda R_i)} \tag{5.57i}$$

Equation (5.57h) is a transcendental equation which has to be solved numerically to determine its various roots. If we denote these roots by $\lambda_l$, various values of the eigenvalue $\lambda$ are given by $\lambda = \lambda_l$ for $l = 1, 2, 3, \ldots, \infty$. Letting $A = A_l$, $B = B_l$, and $F = F_l$ when $\lambda = \lambda_l$, we can write from Eqs. (5.57d) and (5.57i) that

$$F_l = \bar{B}_l \psi_l(r)/r \tag{5.57j}$$

where

$$\psi_l(t) = \lambda_l R_i \cos \lambda_l (r - R_i) + \sin \lambda_l (r - R_i)$$

and

$$\bar{B}_l = \frac{B_l}{\lambda_l R_i \cos(\lambda_l R_i) - \sin(\lambda_l R_i)}$$

Here $\psi_l(r)$ is the eigenfunction corresponding to the eigenvalue $\lambda_l$. Similarly, letting $G = G_l$ when $\lambda = \lambda_l$, the solution of Eq. (5.57c) can be written as

$$G_l = C_l \exp(-\lambda_l^2 Dt) \tag{5.57k}$$

where $C_l$ is the constant of integration.

For all possible values of $\lambda_l$ [the positive roots of Eq. (5.57h)], the following expression for $n(r, t)$ is obtained from Eqs. (5.55), (5.57j), and (5.57k):

$$n(r, t) = \sum_{l=1}^{\infty} \frac{C_l}{r} \psi_l(r) \exp(-\lambda_l^2 Dt) \tag{5.58a}$$

where $C_l = \bar{B}_l \bar{C}_l$, which has to be determined by satisfying the initial condition (5.54d). Applying the initial condition (5.54d) to Eq. (5.58), multiplying both sides of the resulting expression by $r\psi_m(r)$, integrating with respect to $r$ from $w_0$ to $R_i$, and using the orthogonality relationship:

$$\int_{w_0}^{R_i} \psi_l(r)\psi_m(r)\,dr = 0 \quad \text{if } l \neq m$$

$$N_l = \frac{1}{2}\left\{(1 + \lambda_l^2 R_i^2)\left[(R_i - w_0) + \frac{1/r_0 + 1/w_0}{\lambda_l^2 + (1/r_0 + 1/w_0)^2}\right] - R_i\right\} \quad \text{if } l = m$$

we obtain

$$C_l = \frac{n_0}{N_l}\int_{w_0}^{R_i} r[\lambda_l R_i \cos \lambda_l(r - R_i) + \sin \lambda_l(r - R_i)]\,dr$$

$$= \left(\frac{1}{\lambda_l^2} + w_0 R_i\right) \sin \lambda_l(R_i - w_0) - \frac{R_i - w_0}{\lambda_l}\cos \lambda_l(R_i - w_0) \quad (5.58b)$$

Using this expression for $C_l$, we can determine the distribution of the reactant molecules, $n(r, t)$, from Eq. (5.58a). The reaction flux at $r = w_0$ is given by

$$j(r, t) = D\frac{\partial n}{\partial r}\bigg|_{r=w_0}$$

$$= \sum_{l=1}^{\infty}\frac{DC_l}{w_0^2}[(\lambda_l^2 R_i w_0 - 1)\sin \lambda_l(R_i - w_0)$$

$$- \lambda_l(R_i - w_0)\cos \lambda_l(R_i - \omega_0)]\exp(-\lambda_l^2 Dt) \quad (5.58c)$$

Mathematically, the reactant molecules will be depleted completely, that is, $n(r, t) = 0$, as $t \to \infty$, as indicated by Eq. (5.58a). However, to obtain uniform film thickness and to carry out the LCVD process in an optimized way, we would like to stop the deposition process when a certain fraction of $n_0$ is consumed by the chemical reaction. If we define the half-life $t_{1/2}$ of the reactant molecules to be the time that one-half of the initial concentration of the reactant molecules is consumed, $t_{1/2}$ at any location $r$ can be determined from the following expression:

$$\frac{n_0}{2} = \sum_{l=1}^{\infty}\frac{C_l}{r}\psi_l(r)\exp(-\lambda_l^2 Dt_{1/2}) \quad (5.58d)$$

## 5.5.4. Two-Region ($0 \leq r \leq w_0$ and $w_0 \leq r < \infty$) Model with Hemispheric Volumetric Reaction

In Sections 5.5.1–5.5.3, we have assumed that the chemical reaction occurs at the surface of a hemisphere of radius $w_0$ (see Fig. 5.3). We will now analyze the distribution of the reactant molecules by considering that the chemical reaction occurs within the hemisphere of radius $w_0$, and for this purpose we will divide the space inside the deposition chamber into two regions: $0 \leq r \leq w_0$ and $w_0 \leq r < \infty$ define regions I and II, respectively, as shown in Fig. 5.3. The governing equations and the boundary and initial conditions can be written as follows:

GOVERNING EQUATIONS:

$$D \frac{1}{r^2} \frac{\partial}{\partial r}\left(r^2 \frac{\partial n_1}{\partial r}\right) - n_1 \sigma I^* = \frac{\partial n_1}{\partial t} \quad \text{in region 1}, \quad 0 \leq r \leq w_0 \quad (5.59a)$$

The radial variations of the photon beam and its attenuation as it propagates have been ignored for this analysis; $n_1 \sigma I^*$ represents the consumption of the reactant molecules per unit time per unit volume and, therefore, $-n_1 \sigma I^*$ denotes the volumetric production rate:

$$D \frac{1}{r^2} \frac{\partial}{\partial r}\left(r^2 \frac{\partial n_2}{\partial r}\right) = \frac{\partial n_2}{\partial t} \quad \text{in region II}, \quad w_0 \leq r < \infty \quad (5.59b)$$

We have assumed that the diffusion coefficient $D$ is the same for both regions.

BOUNDARY CONDITIONS:

$$n_1(r, t) \text{ must be finite at } r = 0 \quad (5.59c)$$

$$D \frac{\partial n_1}{\partial r} = D \frac{\partial n_2}{\partial r} \quad \text{at} \quad r = w_0 \quad (5.59d)$$

$$n_1(r, t) = n_2(r, t) \quad \text{at} \quad r = w_0 \quad (5.59e)$$

$$n_2(r, t) = n_0 \quad \text{as} \quad r \to \infty \quad (5.59f)$$

Conditions (5.59d) and (5.59e) are sometimes referred to as the interface conditions, which represent the continuity of the mass flux and concentration, respectively, at the interface of the regions.

INITIAL CONDITIONS:

$$n_1(r, t) = n_0 \quad \text{at} \quad t = 0 \tag{5.59g}$$

$$n_2(r, t) = n_0 \quad \text{at} \quad t = 0 \tag{5.59h}$$

To solve this problem, let us define two new variables,

$$N_1 = rn_1 \tag{5.60a}$$

$$N_2 = rn_2 \tag{5.60b}$$

which transform the problem into the following form:

GOVERNING EQUATIONS:

$$\frac{\partial^2 N_1}{\partial r^2} - \frac{N_1}{w_0^2} \mathrm{Dm} = \frac{1}{D} \frac{\partial N_1}{\partial t} \quad \text{for} \quad 0 \leqslant r \leqslant w_0 \tag{5.61a}$$

$$\frac{\partial^2 N_2}{\partial r^2} = \frac{1}{D} \frac{\partial N_2}{\partial t} \quad \text{for} \quad w_0 \leqslant r < \infty \tag{5.61b}$$

BOUNDARY CONDITIONS:

$$N_1 = 0 \quad \text{at} \quad r = 0 \tag{5.61c}$$

$$\frac{\partial N_1}{\partial r} = \frac{\partial N_2}{\partial r} \quad \text{at} \quad r = w_0 \tag{5.61d}$$

To obtain the boundary condition (5.61d), we have applied the boundary condition (5.59e) to the condition (5.59d):

$$N_1(r, t) = N_2(r, t) \quad \text{at} \quad r = w_0 \tag{5.61e}$$

$$N_2(r, t) = rn_0 \quad \text{as} \quad r \to \infty \tag{5.61f}$$

INITIAL CONDITIONS:

$$N_1(r, t) = rn_0 \quad \text{at} \quad t = 0 \tag{5.61g}$$

$$N_2(r, t) = rn_0 \quad \text{at} \quad t = 0 \tag{5.61h}$$

We will solve this problem by using the Laplace transform defined by Eqs. (5.44a, b), which transforms the problem given by the expressions (5.61a–h) into the following form:

# PHOTOLYTIC LCVD MODELING

**GOVERNING EQUATIONS:**

$$\frac{d^2\bar{N}_1(r,s)}{dr^2} - \frac{D_m}{w_0^2}\bar{N}_1(r,s) = \frac{s}{D}\bar{N}_1(r,s) - \frac{rn_0}{D} \quad (5.62a)$$

$$\frac{d^2\bar{N}_2(r,s)}{dr^2} = \frac{s}{D}\bar{N}_2(r,s) - \frac{rn_0}{D} \quad (5.62b)$$

**BOUNDARY CONDITIONS:**

$$\bar{N}_1(r,s) = 0 \quad \text{at} \quad r = 0 \quad (5.62c)$$

$$\frac{d\bar{N}_1(r,s)}{dr} = \frac{d\bar{N}_2(r,s)}{dr} \quad \text{at} \quad r = w_0 \quad (5.62d)$$

$$\bar{N}_1(r,s) = \bar{N}_2(r,s) \quad \text{at} \quad r = w_0 \quad (5.62e)$$

$$\bar{N}_2(r,s) = rn_0/s \quad \text{as} \quad r \to \infty \quad (5.62f)$$

Note that the initial conditions (5.61g) and (5.61h) have been utilized in Eqs. (5.62a) and (5.62b), respectively. The solutions of Eqs. (5.62a) and (5.62b) are, respectively, given by

$$\bar{N}_1(r,s) = C_s \sinh\left(\sqrt{\frac{s}{D} + qr}\right) + D_s \cosh\left(\sqrt{\frac{s}{D} + qr}\right) + \frac{rn_0}{s + qD} \quad (5.63a)$$

and

$$\bar{N}_2(r,s) = A_s \exp\left(-\sqrt{\frac{s}{D}}r\right) + B_s \exp\left(\sqrt{\frac{s}{D}}r\right) + \frac{rn_0}{s} \quad (5.63b)$$

where $q = D_m/w_0^2$, and $C_s$, $D_s$, $A_s$, and $B_s$ are the constants of integration that have to be determined by satisfying the boundary conditions. The solution (5.63a) and the boundary condition (5.62c) imply that $D_s = 0$. Similarly, the solution (5.63b) and the boundary condition (5.62f) imply that $B_s = 0$. Applying the boundary conditions (5.62d) and (5.62e) to the solutions (5.63a) and (5.63b), we can write

$$C_s\sqrt{\frac{s}{D} + q}\cosh\left(\sqrt{\frac{s}{D} + q}w_0\right) + \frac{n_0}{s + qD} = -A_s\sqrt{\frac{s}{D}}\exp\left(-\sqrt{\frac{s}{D}}w_0\right) + \frac{n_0}{s}$$

(5.63c)

and

$$C_s \sinh\left(\sqrt{\frac{s}{D} + q}w_0\right) + \frac{n_0 w_0}{s + qD} = A_s \exp\left(-\sqrt{\frac{s}{D}}w_0\right) + \frac{n_0 w_0}{s} \quad (5.63d)$$

Solving Eqs. (5.63c) and (5.63d) for $A_s$ and $C_s$, we obtain the following results for $\bar{n}_1(r, s)$ and $\bar{n}_2(r, s)$ from Eqs. (5.60a), (5.60b), (5.63a), and (5.63b):

$$\bar{n}_1(r, s) = \frac{n_0}{s + qD}$$

$$+ \frac{n_0 qD(\sqrt{D} + \sqrt{s}w_0)}{s(s + qD)[\sqrt{s + qD}\cosh(\sqrt{s/D} + qw_0) + \sqrt{s}\sinh(\sqrt{s/D} + qw_0)]}$$

$$\times \frac{\sinh(\sqrt{s/D} + qr)}{r} \tag{5.63e}$$

$$\bar{n}_2(r, s) = \frac{n_0}{s} - \frac{n_0 qD}{s(s + qD)}$$

$$\times \frac{\sqrt{s + qD}w_0\cosh(\sqrt{s/D} + qw_0) - \sqrt{D}\sinh(\sqrt{s/D} + qw_0)}{\sqrt{s + qD}\cosh(\sqrt{s/D} + qw_0) + \sqrt{s}\sinh(\sqrt{s/D} + qw_0)}$$

$$\times \frac{1}{r}\exp\left[-\sqrt{\frac{s}{D}}(r - w_0)\right] \tag{5.63f}$$

Equations (5.63e) and (5.63f) can be inverted numerically (Bellman *et al.* (1966) to obtain $n_1(r, t)$ and $n_2(r, t)$. However, approximate expressions for $n_1(r, t)$ and $n_2(r, t)$ can be obtained for short and long times as discussed below. We will determine $n_1(0, t)$ by noting that

$$\lim_{r \to 0} \frac{\sinh(\sqrt{s/D} + qr)}{r} = \sqrt{\frac{s + qD}{D}} \tag{5.63g}$$

DETERMINATION OF $n_1(0, t)$ FOR SHORT TIMES: As $t \to 0$, $s \to \infty$, and for large values of $s$, $\cosh(\sqrt{s/D} + qw_0)$ and $\sinh(\sqrt{s/D} + qw_0)$ becomes infinitely large. Therefore, Eq. (5.63e) can be approximately written as

$$\bar{n}_1(0, s) \approx \frac{n_0}{s + qD} \quad \text{as} \quad s \to \infty$$

which can be inverted to obtain the following approximate expression for $n_1(0, t)$:

$$n_1(0, t) \approx n_0 \exp(-qDt) \quad \text{as} \quad t \to 0 \tag{5.63h}$$

DETERMINATION OF $n_1(0, t)$ FOR LONG TIMES: As $t \to \infty$, $s \to 0$, and for small

PHOTOLYTIC LCVD MODELING

values of $s$, we assume that $s \ll qD$ so that the following approximate relations are valid, that is,

$$s + qD \approx qD$$

$$\cosh\left(\sqrt{\frac{s+qD}{D}}\, w_0\right) \approx \cosh(\sqrt{q}w_0)$$

$$\sinh\left(\sqrt{\frac{s+qD}{D}}\, w_0\right) \approx \sinh(\sqrt{q}w_0)$$

Using these three expressions in Eq. (5.63e), we can write

$$\bar{n}_1(0, s) \approx \frac{n_0}{s+qD} + \frac{n_0 qD(\sqrt{D} + \sqrt{s}w_0)\sqrt{q}}{(s)(qD)[\sqrt{qD}\cosh(\sqrt{q}w_0) + \sqrt{s}\sinh(\sqrt{q}w_0)]}$$

$$= \frac{n_0}{s+qD} + \frac{n_0\sqrt{q}}{S} \frac{\sqrt{D} + w_0\sqrt{s}}{s(\sqrt{s}+C)},$$

where $S \equiv \sinh(\sqrt{q}w_0)$ and $C \equiv \sqrt{qD}\coth(\sqrt{q}w_0)$.

$$\bar{n}_i(0, s) = \frac{n_0}{s+qD} + \frac{n_0\sqrt{q}}{S}\left[\frac{\sqrt{D}}{C^2}\frac{C^2}{s(\sqrt{s}+C)} + w_0\frac{1}{\sqrt{s}(\sqrt{s}+C)}\right] \quad \text{as} \quad s \to 0$$

This can be inverted to obtain the following approximate expression for $n_1(0, t)$:

$$n_1(0, t) \approx n_0 \exp(-qDt) + \frac{n_0\sqrt{qD}}{SC^2}[C - C\exp(C^2 t)\operatorname{erfc}(C\sqrt{t})]$$

$$+ \frac{n_0 w_0\sqrt{q}}{S}\exp(C^2 t)\operatorname{erfc}(C\sqrt{t}) \quad \text{as} \quad t \to \infty \tag{5.63i}$$

At steady state, Eq. (5.63i) becomes

$$n_1(0, \infty) = \frac{n_0}{\cosh(\sqrt{Dm})} \tag{5.63j}$$

Equation (5.63j) shows that the distribution of the reactant molecules at the center of the hemisphere and at the steady state, that is, $n_1(0, \infty)$, depends on the Damkohler number Dm. As the Dm increases, $n_1(0, \infty)$ decreases, and so more of the reactant molecules are consumed before the steady state is achieved. However, Eq. (5.63j) has been derived by making several approximations. An extract derivation for the expression for $n_1(0, \infty)$ is presented in the next section.

### 5.5.5. Two-Region ($0 \leqslant r \leqslant w_0$ and $w_0 \leqslant r \leqslant R_i$) Model with Hemispheric Volumetric Reaction: Steady State Analysis

In Section 5.5.4, we analyzed the distribution of the reactant molecules by considering an infinitely large deposition chamber. Now we will analyze the effect of the size of the deposition chamber on the distribution of the reactant molecules by formulating the problem in the following way:

GOVERNING EQUATIONS:

$$D \frac{1}{r^2} \frac{\partial}{\partial r}\left(r^2 \frac{\partial n_1}{\partial r}\right) - n_1 \sigma I^* = \frac{\partial n_1}{\partial t} \quad \text{in region I}, \quad 0 \leqslant r \leqslant w_0 \quad (5.64a)$$

$$D \frac{1}{r^2} \frac{\partial}{\partial r}\left(r^2 \frac{\partial n_2}{\partial r}\right) = \frac{\partial n_2}{\partial t} \quad \text{in region II}, \quad w_0 \leqslant r \leqslant R_i \quad (5.64b)$$

BOUNDARY CONDITIONS:

$n_1(r, t)$ must be finite at $r = 0$ \hfill (5.64c)

$$D \frac{\partial n_1}{\partial r} = D \frac{\partial n_2}{\partial r} \quad \text{at} \quad r = w_0 \quad (5.64d)$$

$$n_1(r, t) = n_2(r, t) \quad \text{at} \quad r = w_0 \quad (5.64e)$$

$$n_2(r, t) = n_0 \quad \text{at} \quad r = R_i \quad (5.64f)$$

The boundary condition (5.64f) is difficult to create in practice. A natural boundary condition would be to assume that the reactant molecules do not penetrate through the inner wall of the deposition chamber and, therefore, the flux $D\partial n_2/\partial r$ is zero at $r = R_i$. This aspect is discussed in the next section.

INITIAL CONDITIONS:

$$n_1(r, t) = n_0 \quad \text{at} \quad t = 0 \quad (5.64g)$$

$$n_2(r, t) = n_0 \quad \text{at} \quad t = 0 \quad (5.64h)$$

# PHOTOLYTIC LCVD MODELING

This transient problem, where $n_i \equiv n_i(r, t)$ for $i = 1$ and 2, can be solved by using the Laplace transform in the same way as in Section 5.5.4. To analyze the distribution of the reactant molecules at the steady state, we will solve the following steady state problem where $n_i \equiv n_i(r, \infty) \equiv n_i(r)$ for $i = 1$ and 2:

GOVERNING EQUATIONS:

$$\frac{1}{r^2}\frac{d}{dr}\left(r^2 \frac{dn_1}{dr}\right) - qn_1 = 0 \quad \text{in region I}, \quad 0 \leqslant r \leqslant w_0 \quad (5.65a)$$

where $q = Dm/w_0^2$.

$$\frac{d}{dr}\left(r^2 \frac{dn_2}{dr}\right) = 0 \quad \text{in region II}, \quad w_0 \leqslant r \leqslant R_i \quad (5.65b)$$

BOUNDARY CONDITIONS:

$$n_1(r) \text{ must be finite at } r = 0 \quad (5.65c)$$

$$\frac{dn_1}{dr} = \frac{dn_2}{dr} \quad \text{at} \quad r = w_0 \quad (5.65d)$$

$$n_1(r) = n_2(r) \quad \text{at} \quad r = w_0 \quad (5.65e)$$

$$n_2(r) = n_0 \quad \text{at} \quad r = R_i \quad (5.65f)$$

The solutions of Eqs. (5.65a) and (5.65b) can be written as

$$n_1(r) = \frac{C_1}{r}\sinh(\sqrt{q}r) + \frac{D_1}{r}\cosh(\sqrt{q}r) \quad (5.66a)$$

and

$$n_2(r) = \frac{A_1}{r} + B_1 \quad (5.66b)$$

where $A_1$, $B_1$, $C_1$, and $D_1$ are the constants of integration that have to be determined by satisfying the boundary conditions. Applying the boundary conditions (5.65c) and (5.65f) to Eqs. (5.66a) and (5.66b), respectively, we obtain

$$D_1 = 0 \quad \text{and} \quad B_1 = n_0 - \frac{A_1}{R_i} \quad (5.66c)$$

Satisfying the boundary conditions (5.65d) and (5.65e), we obtain

$$\begin{bmatrix} (\sqrt{q}/w_0)C - S/w_0^2 & (1/w_0^2) \\ S/w_0 & -(1/w_0 - 1/R_i) \end{bmatrix} \begin{bmatrix} C_1 \\ A_1 \end{bmatrix} = \begin{bmatrix} 0 \\ n_0 \end{bmatrix} \quad (5.66d)$$

where $C \equiv \cosh(\sqrt{q}w_0)$ and $S \equiv \sinh(\sqrt{q}w_0)$. The solution of Eq. (5.66d) yields

$$C_1 = -\frac{1}{\Delta}\frac{n_0}{w_0^2} \quad \text{and} \quad A_1 = \frac{1}{\Delta} n_0 \left( \frac{\sqrt{q}}{w_0} C - \frac{S}{w_0^2} \right)$$

where the determinant

$$\Delta = -\left(\frac{1}{w_0} - \frac{1}{R_i}\right)\left(\frac{\sqrt{q}}{w_0} C - \frac{S}{w_0^2}\right) - \frac{S}{w_0^3}$$

and, therefore, Eqs. (5.66a) and (5.66b) can be written as

$$n_1(r) = -\frac{1}{\Delta}\frac{n_0}{w_0^2}\frac{\sinh(\sqrt{q}r)}{r} \quad (5.66e)$$

$$n_2(r) = n_0 + \frac{1}{\Delta} n_0 \left(\frac{\sqrt{q}}{w_0} C - \frac{S}{w_0^2}\right)\left(\frac{1}{r} - \frac{1}{R_i}\right) \quad (5.66f)$$

At $r = 0$, Eq. (5.66e) becomes

$$n_1(0) = -\frac{1}{\Delta}\frac{n_0\sqrt{q}}{w_0^2} \quad \text{for finite values of } R_i \quad (5.66g)$$

and for an infinitely large deposition chamber we have

$$n_1(0) = \frac{n_0}{\cosh(\sqrt{Dm})} \quad \text{when } R_i \to \infty \quad (5.66h)$$

which is identical to Eq. (5.63j). The effects of the size of the deposition chamber on $n_1(r)$, $n_2(r)$, and $n_1(0)$ are given by Eqs. (5.66e), (5.66f), and

(5.66g), respectively. Specifically, Eq. (5.66g) becomes

$$n_1(0) = \frac{n_0\sqrt{\mathrm{Dm}}}{(1 - w_0/R_i)(\sqrt{\mathrm{Dm}}\cosh(\sqrt{\mathrm{Dm}}) - \sinh(\sqrt{\mathrm{Dm}})) + \sinh(\sqrt{\mathrm{Dm}})} \quad (5.66i)$$

which shows that as the size of the deposition chamber, $R_i$, decreases, $n_1(0)$ increases, that is, fewer of the reactant molecules are consumed before the steady state is reached.

### 5.5.6. Two-Region ($0 \leq r \leq w_0$ and $w_0 \leq r \leq R_i$) Model with Hemispheric Volumetric Reaction: Transient Analysis

The boundary condition (5.64f) used in Section 5.5.5 is difficult to create in practice. However, the inner wall of the deposition chamber can be considered to be impervious to the reactant molecules so that the flux $D\,\partial n_2/\partial r$ of the reactant molecules penetrating through the wall is zero. We will use this condition to solve the following transient problem:

GOVERNING EQUATIONS:

$$D\frac{1}{r^2}\frac{\partial}{\partial r}\left(r^2\frac{\partial n_1}{\partial r}\right) - n_1\sigma I^* = \frac{\partial n_1}{\partial t} \quad \text{in region I,} \quad 0 \leq r \leq w_0 \quad (5.67a)$$

$$D\frac{1}{r^2}\frac{\partial}{\partial r}\left(r^2\frac{\partial n_2}{\partial r}\right) = \frac{\partial n_2}{\partial t} \quad \text{in region II,} \quad w_0 \leq r \leq R_i \quad (5.67b)$$

BOUNDARY CONDITIONS:

$$n_1(r, t) \text{ must be finite at } r = 0 \quad (5.67c)$$

$$D\frac{\partial n_1}{\partial r} = D\frac{\partial n_2}{\partial r} \quad \text{at} \quad r = w_0 \quad (5.67d)$$

$$n_1(r, t) = n_2(r, t) \quad \text{at} \quad r = w_0 \quad (5.67e)$$

$$D\frac{\partial n_2}{\partial r} = 0 \quad \text{at} \quad r = R_i \quad (5.67f)$$

Boundary conditions (5.67d) and (5.67e) are sometimes referred to as the interface conditions, which represent the continuity of flux and concentration, respectively, at the interface of the two regions.

INITIAL CONDITIONS:

$$n_1(r, t) = n_0 \quad \text{at} \quad t = 0 \tag{5.67g}$$

$$n_2(r, t) = n_0 \quad \text{at} \quad t = 0 \tag{5.67h}$$

At the steady state, that is, when $\partial n_i/\partial t = 0$ for $i = 1$ and 2, the solutions of this transient problem are given by $n_1(r, \infty) = 0$ and $n_2(r, \infty) = 0$. This means that after a sufficiently long time, the reactant molecules will be completely depleted inside the deposition chamber due to photochemical reactions. Also, due to the decrease in the reactant molecule population density as the reaction progresses, the film deposition rate will decrease, which can affect the film thickness, as the deposition time increases. The solutions of the above transient problem will allow us to determine the variations of the reactant molecule population density and the volumetric production rate of the atoms of interest with time, and the time required for the reactant molecules to be completely consumed by the chemical reaction, which are useful to optimize the LCVD process.

To solve this problem, let us define the new variables

$$N_1 = rn_1 \tag{5.68a}$$

$$N_2 = rn_2 \tag{5.68b}$$

which transform the problem into the following form:

$$\frac{\partial^2 N_1}{\partial r^2} - qN_1 = \frac{1}{D}\frac{\partial N_1}{\partial t} \quad \text{for} \quad 0 \leqslant r \leqslant w_0 \tag{5.69a}$$

where $q = Dm/w_0^2$, and

$$\frac{\partial^2 N_2}{\partial r^2} = \frac{1}{D}\frac{\partial N_2}{\partial t} \quad \text{for} \quad w_0 \leqslant r \leqslant R_i \tag{5.69b}$$

BOUNDARY CONDITIONS:

$$N_1(r, t) = 0 \quad \text{at} \quad r = w_0 \tag{5.69c}$$

$$\frac{\partial N_1}{\partial r} = \frac{\partial N_2}{\partial r} \quad \text{at} \quad r = w_0 \tag{5.69d}$$

To obtain the boundary condition (5.69d), we have applied the boundary condition (5.67e) to the condition (5.67d):

$$N_1(r, t) = N_2(r, t) \quad \text{at } r = w_0 \tag{5.69e}$$

$$\frac{\partial N_2}{\partial r} = \frac{N_2}{R_i} \quad \text{at} \quad r = R_i \tag{5.69f}$$

INITIAL CONDITIONS:

$$N_1(r, t) = rn_0 \quad \text{at} \quad t = 0 \tag{5.69g}$$

$$N_2(r, t) = rn_0 \quad \text{at} \quad t = 0 \tag{5.69h}$$

We will solve this problem by using the integral transform technique [Sneddon (1972)]. In particular, we will use the multiregion integral transform method [Kar and Mazumder (1988), Özisik and Murray (1974), Yener and Özisik (1974)] to solve this problem by defining the following transform variables:

INTEGRAL TRANSFORM:

$$\bar{N}_l(t) = \sum_{j=1}^{2} \int_{a_j}^{b_j} W_j N_j(r, t) K_{j,l}(r) \, dr \tag{5.70a}$$

INVERSION FORMULA:

$$N_j(r, t) = \sum_{l=1}^{\infty} \frac{1}{N_l^*} \bar{N}_l(t) K_{j,l}(r) \tag{5.70b}$$

NORMALIZATION CONSTANT:

$$N_l^* = \sum_{j=1}^{2} \int_{a_j}^{b_j} W_j K_{j,l}^2(r) \, dr \tag{5.70c}$$

Here, $a_j$ and $b_j$ are, respectively, the lower and upper values of $r$ in the $j$th region, that is, $a_1 = 0$, $b_1 = w_0$, $a_2 = w_0$, and $b_2 = R_i$. The kernels of the integral transform, $K_{j,l}(r)$ for $j = 1$ and 2, which are also referred to as the eigenfunctions, the associated eigenvalues $\lambda_l$ for $l = 1, 2, 3, \ldots, \infty$, and the weight function $W_j$ will be determined later in this section.

Multiplying Eqs. (5.69a) and (5.69b) by $\bar{W}_j K_{1,l}(r)$ and $\bar{W}_j K_{2,l}(r)$, respectively, integrating the resulting expressions with respect to $r$ from 0 to

$w_0$ and $w_0$ to $R_i$, respectively, and adding them together, we obtain

$$\sum_{j=1}^{2} \int_{a_j}^{b_j} \bar{W}_j \frac{\partial^2 N_j}{\partial r^2} K_{j,l}(r)\, dr - \sum_{j=1}^{2} q_j \int_{a_j}^{b_j} \bar{W}_j N_j(r, t) K_{j,l}(r)\, dr$$

$$= \frac{1}{D} \sum_{j=1}^{2} \int_{a_j}^{b_j} \frac{\partial N_j}{\partial t} \bar{W}_j K_{j,l}(r)\, dr \quad (5.71a)$$

where $q_1 = q$ and $q_2 = 0$.

Integrating the first term on the left-hand side of Eq. (5.71a) by parts twice, and applying the boundary conditions (5.69c–f), we obtain

$$\frac{1}{D} \frac{d\bar{N}_l}{dt} = -\lambda_l \bar{N}_l \quad (5.71b)$$

under the following conditions:

$$\bar{W}_j = W_j = 1$$

$$\frac{d^2 K_{j,l}}{dr^2} = -(\lambda_l - q_j) K_{j,l} \quad \text{where} \quad \lambda_l > q_j$$

$$= -\beta_{jl}^2 K_{j,l} \quad \text{where} \quad \beta_{jl}^2 = \lambda_l - q_j \quad (5.71c)$$

$$K_{1,l} = 0 \quad \text{at} \quad r = 0 \quad (5.71d)$$

$$K_{1,l} = K_{2,l} \quad \text{at} \quad r = w_0 \quad (5.71e)$$

$$\frac{dK_{1,l}}{dr} = \frac{dK_{2,l}}{dr} \quad \text{at} \quad r = w_0 \quad (5.71f)$$

$$\frac{dK_{2,l}}{dr} = \frac{1}{R_i} K_{2,l} \quad \text{at} \quad r = R_i \quad (5.71g)$$

The solution of Eq. (5.71b) is given by

$$\bar{N}_l(t) = \bar{N}_l(0) \exp(-\lambda_l D t) \quad (5.71h)$$

where $\bar{N}_l(0)$ is obtained from Eq. (5.70a); that is,

$$\bar{N}_l(0) = \sum_{j=1}^{2} \int_{a_j}^{b_j} N_j(r, 0) K_{j,l}(r)\, dr \quad (5.71i)$$

PHOTOLYTIC LCVD MODELING

and $K_{j,l}(r)$ for $j = 1$ and 2 are determined by solving Eq. (5.71c); that is,

$$K_{1,l}(r) = C_1 \sin(\beta_{1l} r) + D_1 \cos(\beta_{1l} r) \tag{5.71j}$$

$$K_{2,l}(r) = A_1 \sin(\beta_{2l} r) + B_1 \cos(\beta_{2l} r) \tag{5.71k}$$

Satisfying the boundary condition (5.71d), we obtain $D_1 = 0$ and, therefore, Eq. (5.71j) becomes

$$K_{1,l}(r) = C_1 \sin(\beta_{1l} r) \tag{5.71l}$$

Applying the boundary conditions (5.71e–g) to Eqs. (5.71k) and (5.71l), we obtain

$$\begin{bmatrix} S_{20} & C_{20} & S_{10} \\ \beta_{2l} C_{20} & -\beta_{2l} S_{20} & -\beta_{1l} C_{10} \\ \beta_{2l} C_i - S_i/R_i & -\beta_{2l} S_i - C_i/R_i & 0 \end{bmatrix} \begin{bmatrix} A_1 \\ B_1 \\ C_1 \end{bmatrix} = 0 \tag{5.71m}$$

where $S_{20} \equiv \sin(\beta_{2l} w_0)$, $C_{20} \equiv \cos(\beta_{2l} w_0)$, $S_i \equiv \cos(\beta_{2l} R_i)$, $C_i \equiv \cos(\beta_{2l} R_i)$, $S_{10} \equiv \sin(\beta_{1l} w_0)$, and $C_{10} \equiv \cos(\beta_{1l} w_0)$.

Since Eq. (5.71m) is a homogeneous equation, it will have a nonzero solution only if the determinant of the matrix is zero, that is,

$$\begin{vmatrix} S_{20} & C_{20} & S_{10} \\ \beta_{2l} C_{20} & -\beta_{2l} S_{20} & -\beta_{1l} C_{10} \\ \beta_{2l} C_i - S_i/R_i & -\beta_{2l} S_i - C_i/R_i & 0 \end{vmatrix} = 0 \tag{5.71n}$$

Equation (5.71m) can be solved numerically to determine the eigenvalues $\lambda_l$ for $l = 1, 2, 3, \ldots, \infty$, noting that $\lambda_l$ and $\beta_{jl}$, $j = 1$ and 2, are related to each other as given by Eq. (5.71c). Also, Eq. (5.71m) suggests that any two of the three unknowns can be expressed in terms of the third unknown. We express $A_1$ and $B_1$ in terms of $C_1$ in the following:

$$A_1 = f_a(\lambda_l) C_1 \tag{5.71o}$$

$$B_1 = f_b(\lambda_l) C_1 \tag{5.71p}$$

where the expressions for $f_a(\lambda_l)$ and $f_b(\lambda_l)$ can be obtained by solving Eq. (5.71m). It should be noted that $C_1$ can be set equal to unity without any loss of generality. However, we will retain $C_1$ as it is. From Eqs. (5.70b),

(5.71h), (5.71k), (5.71l), (5.71o), and (5.71p), we can write

$$N_1(r, t) = \sum_{l=1}^{\infty} \frac{C_1}{N_l^*} \bar{N}_l(0) \sin(\beta_{1l} r) \exp(-\lambda_l D t) \tag{5.71q}$$

and

$$N_2(r, t) = \sum_{l=1}^{\infty} \frac{C_l}{N_l^*} \bar{N}_l(0)[f_a(\lambda_l) \sin(\beta_{2l} r) + f_b(\lambda_l) \cos(\beta_{2l} r)] \exp(-\lambda_l D t) \tag{5.71r}$$

where $N_l^*$ and $\bar{N}_l(0)$ are given by Eqs. (5.70c) and (5.71i), respectively, and $C_1$ is eliminated when the expressions for $N_l^*$ and $\bar{N}_l(0)$ are substituted in Eqs. (5.71q) and (5.71r). Equations (5.68a), 5.68b), (5.71q), and (5.71r) provide the following expressions for the distributions of the reactant molecules, that is,

$$n_1(r, t) = \sum_{l=1}^{\infty} \frac{C_1}{N_l^*} \bar{N}_l(0) \frac{\sin(\beta_{1l} r)}{r} \exp(-\lambda_l D t) \tag{5.71s}$$

$$n_2(r, t) = \sum_{l=1}^{\infty} \frac{C_1}{N_l^*} \bar{N}_l(0) \frac{1}{r} [f_a(\lambda_l) \sin(\beta_{2l} r) + f_b(\lambda_l) \cos(\beta_{2l} r)] \exp(-\lambda_l D t)$$

$$\tag{5.71t}$$

Let us analyze the change in the reactant molecule population density with time at $r = 0$, which is given by

$$n_1(0, t) = \sum_{l=1}^{\infty} \frac{C_1 \beta_{1l}}{N_l^*} \bar{N}_l(0) \exp(-\lambda_l D t) \tag{5.71u}$$

Mathematically, the reactant molecules will be depleted completely; that is, the values of $n_1(r, t)$ and $n_2(r, t)$ will be zero when $t \to \infty$, as indicated by Eqs. (5.71s) and (5.71t). However, to obtain uniform film thickness and to carry out the LCVD process in an optimum way, we would like to stop the deposition process when a certain fraction of $n_0$ is consumed in the chemical reaction. If we define the half-life $t_{1/2}$ of the reactant molecules to be the time when one-half of the initial concentration of the reactant molecules is consumed, $t_{1/2}$ at $r = 0$ can be determined from the following expression:

$$\frac{n_0}{2} = \sum_{l=1}^{\infty} \frac{C_1 \beta_{1l}}{N_l^*} \bar{N}_l(0) \exp(-\lambda_l D t_{1/2}) \tag{5.71v}$$

## 5.6. DEPOSITION RATE EXPRESSIONS

In the last section, we determined the concentration of the reactant molecules near the substrate surface at any time $t$, such as $n_1(0, t)$, and at the steady state, such as, $n_1(0, \infty)$ or $n_1(0)$, which can be utilized to determine the film deposition rate. Assuming that each reaction produces one atom or molecules of the film material, that is, one atom or molecule of the desired product species, the number of the product species produced at the substrate surface per unit time per unit volume of the gas phase is given by $n_1(0, t)\sigma I^*$ at any time $t$ and $n_1(0)\sigma I^*$ at the steady state. The deposition of the product species on the substrate depends on the sticking coefficient, which can be defined in terms of the flux or the concentration of the product species as follows:

Sticking coefficient $\gamma_s$

$$= \frac{\text{Number of product species deposited per unit time per unit deposition area}}{\text{Flux of product species toward the substrate at its surface}}$$
(5.72a)

Sticking coefficient $\gamma_v$

$$= \frac{\text{Number of product species deposited per unit time per unit volume of the deposited film}}{\text{Number of product species produced at the substrate surface per unit time per unit volume of the gas phase}}$$
(5.72b)

The flux of the product species toward the substrate at its surface is given by $n_1(0)\sigma I^* u_0$, where $u_0$ represents the velocity of the product species toward the substrate along the normal to the substrate surface. The deposition rate can be expressed in terms of the number of product species deposited or the film thickness, as discussed below.

DEPOSITION RATE IN TERMS OF THE NUMBER OF PRODUCT SPECIES: If we define the deposition rate $D_s$ to be the number of the product species deposited per unit time per unit deposition area, it can be expressed as

$$D_s = \tfrac{1}{2} r_s n_1(0) \sigma I^* u_0$$
(5.72c)

On the other hand, if we define the deposition rate $D_V$ to be the number of the product species deposited per unit time per unit volume of the deposited film, it can be expressed as

$$D_v = \gamma_v n_1(0, t)\sigma I^* \qquad (5.72d)$$

$D_s$ and $D_v$ can be multiplied by $M/A_v$ to express them in units of mass per unit area per unit time and mass per unit volume per unit time, respectively.

DEPOSITION RATE IN TERMS OF THE FILM THICKNESS: If we define the deposition rate $D^*$ to be the change in the film thickness per unit time, it can be expressed as

$$dh/dt = D^* = D_s M/A_v \rho \qquad (5.72e)$$

where $h$ is the film thickness at any time $t$; $dh/dt$ can also be related to $D_v$ by the following expression:

$$\frac{dh}{dt} = D^* = \frac{D_v M}{A_v} A_0 h \frac{1}{A_0} \frac{1}{\rho} = \frac{D_v M h}{A_v \rho} \qquad (5.72f)$$

Here $A_0$ is the substrate surface area where the film is deposited. In Eq. (5.72f), $(D_v M/A_v)A_0 h$ and $(D_v M/A_v)A_0 h(1/A_0)$ represent the deposition rates in units of mass per unit time and mass per unit area per unit time, respectively. Equation (5.72f) is written by assuming that the film thickness $h$ is uniform across the area $A_0$.

## 5.7. KINETIC THEORY MODEL

In Section 5.5, we divided the deposition process into two regimes based on the geometrical configurations of the laser beam and the deposition chamber. Chen (1987) has presented a model for laser photochemical deposition by dividing the process into two regimes based on the pressure inside the deposition chamber. When the pressure is low, the mean free path of the product species is much larger than the laser beam focal spot diameter, and for this situation Chen (1987) used the kinetic theory approach to model the transport of the product particles, which is transferred to as ballistic deposition. On the other hand, when the pressure is high, the mean free path of the product species is much smaller than the laser beam focal spot diameter, and for this situation he has used the diffusion theory approach to model the movement of the product particles, which is referred to as the diffusive deposition. He has developed the model by considering the reactant (precursor) molecules in three different phases,

# PHOTOLYTIC LCVD MODELING

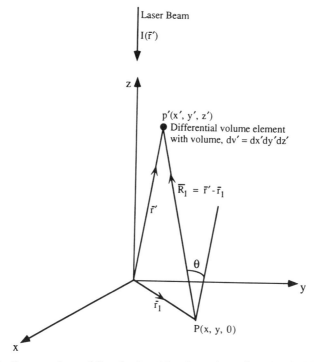

Figure 5.4. Geometry for modeling the deposition due to laser photochemical decomposition in the gas phase.

which are the gas, physisorbed, and chemisorbed phases. The diffusion of the product and precursor species, which is important when the pressure inside the deposition chamber is high, has also been taken into account in the model.

The coordinate system Chen used (1987) for modeling the laser photochemical deposition is shown in Fig. 5.4, where the differential volume element is located at an arbitrary point $P'$ with coordinates $(x', y', z')$. The volume of this differential element is given by $dV' = dx'dy'dz'$. In Fig. 5.4, $P$ is the point of interest on the substrate surface, where we want to determine the deposition rate by considering the volumetric rate of generation of the product species at all of the points in the gas phase, that is, in the region bounded by $-\infty < x < \infty$, $-\infty < y < \infty$, and of $0 \leqslant z < \infty$. The volumetric generation rate $S(\mathbf{r}', t)$, that is, the number of the product species produced per unit time per unit volume at any location $\mathbf{r}'$ and time $t$, is given by

$$S(\mathbf{r}', t) = n(\mathbf{r}')\sigma I(r') \tag{5.73}$$

It should be noted that $\sigma = \phi_r \sigma_a$, and that $\sigma$ and $\sigma_a$ depend on the wavelength of the incident laser. Chen (1987) has compared his model predictions with the experimental data given in Wood et al. (1983a,b) and Krchnavek et al. (1987).

### 5.7.1. Gas Phase Decomposition: Ballistic Deposition Model

This model has been developed [Chen (1987)] for the case when the total pressure inside the deposition chamber is so low that $\bar{\lambda} \gg 2w_0$. Under this condition, the deposition is considered to occur as a result of the ballistic motion of the product species directly from the location where they are produced. If the concentration of the precursor molecules does not change appreciably during a given LCVD process, or if the precursor molecules are properly replenished inside the deposition chamber during the LCVD process, the number density of the precursor molecules, $n(\mathbf{r}')$, can be considered to be constant. Also, assuming that the ideal gas law holds good for the precursor molecules at the operating pressure inside the LCVD chamber, we can write

$$n(\mathbf{r}') = p_r/k_B T = n = \text{constant} \qquad (5.74)$$

We also assume that the regular distribution of the product species is isotropic. Under these assumptions, the deposition rate, that is, the particle flux at the substrate surface due to ballistic motion, $j_{f_{\text{ball}}}$, can be written as

$$j_{f_{\text{ball}}} = \gamma \int \frac{1}{4\pi R_1^2} [n\sigma I(\mathbf{r}')] dV' \cos\theta \qquad (5.75a)$$

where $R_1 = [(x' - x)^2 + (y' - y)^2 + z'^2]^{1/2}$. Equation (5.75a) is similar to Eq. (5.29) except that the exponential attenuation term $\exp(-R_1/l)$ which appears in Eq. (5.29) is not considered in Eq. (5.75a). $J_{f_{\text{ball}}}$ represents the particle flux that sticks to the substrate surface. The deviation of Eq. (5.75a) is similar to the determination of neutron flux [Morgan and Emerson (1967)] due to a neutron source. However, the neutron flux is a scalar quantity, but the mass or particle flux $J_{f_{\text{ball}}}$ is considered to be directed along the normal to the substrate surface. In Eq. (5.75a), $\cos\theta$ can be replaced by noting that

$$\mathbf{R}_1 = \mathbf{r}' - \mathbf{r}_1 = (x' - x)\hat{i} + (y' - y)\hat{j} + z'\hat{k}$$

and

$$\frac{\cos\theta}{R_1^2} = \frac{1}{R_1^2}\left(\hat{k}\cdot\frac{\mathbf{R}_1}{R_1}\right)$$

$$= \frac{z'}{[(x'-x)^2 + (y'-y)^2 + z'^2]^{3/2}}$$

$$= \left[\frac{z'-z}{[(x'-x)^2 + (y'-y)^2 + (z'-z)^2]^{3/2}}\right]_{z=0}$$

$$= \left[\frac{\partial}{\partial z}\left(\frac{1}{[(x'-x)^2 + (y'-y)^2 + (z'-z)^2]^{1/2}}\right)\right]_{z=0}$$

$$= \frac{\partial}{\partial z}\left(\frac{1}{R}\right) = -\left[\frac{\partial}{\partial z'}\left(\frac{1}{R}\right)\right]_{z=0}$$

where $R = [(x'-x)^2 + (y'-y)^2 + (z'-z)^2]^{1/2}$. Using these expressions for $(\cos\theta)/R_1^2$ in Eq. (5.75a), we can write

$$j_{f_{\text{ball}}} = \frac{\gamma n \sigma}{4\pi} \int_{-\infty}^{\infty} dx' \int_{-\infty}^{\infty} dy' \int_{0}^{\infty} I(x', y', z') \frac{z' dz'}{[(x'-x)^2 + (y'-y)^2 + z'^2]^{3/2}} \quad (5.75b)$$

$$= \gamma \left[\frac{\partial F(\mathbf{r})}{\partial z}\right]_{z=0} \quad (5.75c)$$

where

$$F(\mathbf{r}) = n\sigma \int_{-\infty}^{\infty} dx' \int_{-\infty}^{\infty} dy' \int_{0}^{\infty} I(x', y', z') \frac{1}{4\pi R} dz' \quad (5.75d)$$

with

$$\mathbf{r} = x\hat{i} + y\hat{j} + z\hat{k}$$

Equation (5.75b) will be used to determine the deposition rate, and Eqs. (5.75c, d) will be used to show the similarity between the ballistic and diffusive deposition models later in this section. To determine the deposition rate, Chen (1987) has considered that the radius of the laser beam varies slowly with $z$, and that the photochemical decomposition occurring at a distance far away from the focal spot has a negligible effect on the deposition rate. Under these assumptions, the laser intensity $I(x', y', z')$ can be expressed as

$$I(x', y', z') = \frac{2P_t}{\pi w_0^2 E} \exp\left[-\frac{2(x'^2 + y'^2)}{w_0^2}\right] \quad (5.75e)$$

Substituting Eq. (5.75e) into Eq. (5.75b), and carrying out the integration with respect to $z'$, we obtain

$$j_{f_{\text{ball}}} = \frac{\gamma n \sigma P_t}{2\pi^2 w_0^2 E} \int_{-\infty}^{\infty} dx' \int_{-\infty}^{\infty} \frac{\exp[-2(x'^2 + y'^2)/w_0^2]}{[(x' - x)^2 + (y' - y)^2]^{1/2}} dy' \quad (5.75f)$$

We will evaluate this integral for the following two regions:

REGION I: $x^2 + y^2 \gg w_0^2$. In this region, the intensity of the laser beam is considered to be zero. The effect of laser on the photochemical reaction is dominant in the region bounded by $x'^2 + y'^2 \leq w_0^2$. The integrand of Eq. (5.75f) is almost zero because

$$\exp\left[-\frac{2(x'^2 + y'^2)}{w_0^2}\right] \approx 0 \quad \text{when} \quad x'^2 + y'^2 > w_0^2$$

For nonzero and significant values of the integrand, we need to have $x'^2 + y'^2 \leq w_0^2$. Under this condition, the denominator of the integrand can be approximated as

$$\sqrt{(x' - x)^2 + (y' - y)^2} = \sqrt{x^2 + y^2 + (x'^2 + y'^2 - 2xx' - 2yy')}$$
$$\approx \sqrt{x^2 + y^2}$$

With this approximation, Eq. (5.75f) can be written as

$$j_{f_{\text{ball}}} = \frac{\gamma n \sigma P_t}{2\pi^2 w_0^2 E \sqrt{x^2 + y^2}} \int_{-\infty}^{\infty} \exp\left(-2\frac{x'^2}{w_0^2}\right) dx' \int_{-\infty}^{\infty} \exp\left(-2\frac{y'^2}{w_0^2}\right) dy'$$

$$= \frac{\gamma n \sigma P_t}{E} \frac{1}{4\pi \sqrt{x^2 + y^2}} \quad (5.75g)$$

REGION II: $x^2 + y^2 < w_0^2/2$. To evaluate the integral of Eq. (5.75f) in this region, we will shift the origin to the point of interest, $P$ in Fig. 5.4, by using the following transformations:

$$x' - x = \rho' \cos \phi$$

and

$$y' - y = \rho' \sin \phi$$

where $0 \leq \rho' < \infty$, and $0 \leq \phi \leq 2\pi$. Equation (5.75f) can now be written as

$$j_{f_{\text{ball}}} = \frac{\gamma n \sigma P_t}{2\pi^2 w_0^2 E} \int_0^{2\pi} d\phi \int_0^{\infty}$$
$$\times \frac{\exp\{-(2/w_0^2)[\rho'^2 + 2(x\cos\phi + y\sin\phi)\rho' + x^2 + y^2]\}}{\rho'} \rho' \, d\rho'$$

$$= \frac{\gamma n \sigma P_t}{4\pi\sqrt{2\pi} w_0 E} \exp\left[-\frac{2(x^2+y^2)}{w_0^2}\right] \int_0^{2\pi} \exp\left(\frac{2}{w_0^2}\mu^2\right) \text{erfc}\left(\frac{\sqrt{2}}{w_0}\mu\right) d\phi$$
(5.75h)

where $\mu = x\cos\phi + y\sin\phi$. The integral of Eq. (5.75h) can be evaluated by taking the series expansions of the exponential and complementary error functions, that is,

$$\exp(\alpha) = 1 + \alpha + \frac{\alpha^2}{2!} + \frac{\alpha^3}{3!} + \cdots$$

and

$$\text{erfc}(\alpha) = 1 - \frac{2}{\sqrt{\pi}}\left(\alpha - \frac{\alpha^3}{3 \cdot 1!} + \frac{\alpha^5}{5 \cdot 2!} - \frac{\alpha^7}{7 \cdot 3!} + \cdots\right)$$

When $x^2 + y^2 < w_0^2/2$, we have $\mu < w_0/\sqrt{2}$, and under this condition we retain only the first two terms in the above series expansions to obtain

$$\exp\left(\frac{2}{w_0^2}\mu^2\right) \text{erfc}\left(\frac{\sqrt{2}}{w_0}\mu\right) \approx \left(1 + \frac{2}{w_0^2}\mu^2\right)\left(1 - \frac{2}{\sqrt{\pi}}\frac{\sqrt{2}}{w_0}\mu\right)$$

$$\approx 1 - \frac{2\sqrt{2}}{\sqrt{\pi}w_0}\mu + \frac{2}{w_0^2}\mu^2 \quad (5.75\text{i})$$

where the term containing $\mu^3$ has been neglected in order to be consistent with the approximation of $\text{erfc}(\sqrt{2}\mu/w_0)$ that has been used above without considering the term containing $\mu^3$. Substituting Eq. (5.75i) into Eq. (5.75h), we obtain

$$\begin{aligned} j_{f_{\text{ball}}} &\approx \frac{\gamma n\sigma P_t}{4\pi\sqrt{2\pi w_0 E}} \exp\left[-\frac{2(x^2+y^2)}{w_0^2}\right] \int_0^{2\pi} \rho_{xy}\, d\phi \\ &\approx \frac{\gamma n\sigma P_t}{2\sqrt{2\pi w_0 E}} \exp\left[-\frac{2(x^2+y^2)}{w_0^2}\right]\left[1+\frac{x^2+y^2}{w_0^2}\right] \end{aligned} \qquad (5.75j)$$

where

$$\rho_{xy} = 1 - \frac{2\sqrt{2}}{\sqrt{\pi w_0^2}}(x\cos\phi + y\sin\phi) + \frac{2}{w_0^2}(x\cos\phi + y\sin\phi)^2$$

Chen (1987) has given the following approximate expression that is valid for all values of $x$ and $y$ in the ranges $-\infty < x < \infty$ and $-\infty < y < \infty$:

$$j_{f_{\text{ball}}} \approx \frac{\gamma n\sigma P_t}{2E\sqrt{2\pi[w_0^2 + 2(x^2+y^2)]}} \frac{2w_0^4 + (x^2+y^2)^2}{2w_0^4 + \sqrt{\pi}(x^2+y^2)^2} \qquad (5.75k)$$

### 5.7.2. Gas Phase Decomposition: Diffusive Deposition Model

When the total pressure inside the deposition chamber is high such that $\bar{\lambda} < 2w_0$, collision effects have to be considered in order to model the movement of various species. The diffusion equation can be used to analyze the redistribution of the product species during laser photochemical deposition. Under steady state conditions and in the absence of convection, the governing equation for the diffusion of the product species can be written from Eq. (5.1) as follows:

$$D\nabla^2 n_p(\mathbf{r}) = S(\mathbf{r}) \qquad (5.76a)$$

where $-\infty < x < \infty$, $-\infty < y < \infty$, and $0 \leqslant z < \infty$, and the formation of the product species per unit time per unit volume, $S(\mathbf{r})$, is given by Eq. (5.73). To solve this equation, Chen (1987) considered the following expression for the particle flux at the substrate surface, that is,

$$j_{f_{\text{diff}}} = D\frac{\partial n_p}{\partial z} = \tfrac{1}{4}\gamma n_p \bar{v} \quad \text{at} \quad z = 0 \qquad (5.76b)$$

PHOTOLYTIC LCVD MODELING

where $\bar{v}$ is the average velocity of the product species and the diffusion coefficient $D$ is given by [Mayer and Mayer (1940)]

$$D = \tfrac{1}{3}\bar{v}\bar{\lambda} \tag{5.76c}$$

At the center of the laser beam, he has assumed that

$$\frac{\partial n_p}{\partial z} \approx \frac{\bar{n}_p}{w_0} \tag{5.76d}$$

where $\bar{n}_p$ is the average number density of the product species in the beam. Combining Eqs. (5.76b, c, d), we obtain

$$n_p = \frac{4}{3}\frac{\bar{\lambda}}{\gamma w_0}\bar{n}_p \quad \text{at} \quad z = 0 \tag{5.76e}$$

which can be used as the boundary condition to solve Eq. (5.76a). However, we note that if $\gamma \approx 1$ and $\bar{\lambda}/w_0 \ll 1$, we have $n_p \ll \bar{n}_p$, and under these conditions we can then write

$$n_p = 0 \quad \text{at} \quad z = 0 \tag{5.76f}$$

The expression (5.76f) is used as the boundary condition to solve Eq. (5.76a). As in the previous section, we assume that the distribution of the precursor molecules, $n(\mathbf{r}')$, is constant, and under this assumption the solution of Eq. (5.76a) can be written as

$$n_p(\mathbf{r}) = \frac{n\sigma}{D}\int G(\mathbf{r}' - \mathbf{r})I(\mathbf{r}')dV' \tag{5.76g}$$

where the Green function $G(\mathbf{r}' - \mathbf{r})$ is given by

$$G(\mathbf{r}' - \mathbf{r}) = \frac{1}{4\pi}\left\{\frac{1}{[(x'-x)^2 + (y'-y)^2 + (z'-z)^2]^{1/2}} - \frac{1}{[(x'-x)^2 + (y'-y)^2 + (z'+z)^2]^{1/2}}\right\} \tag{5.76h}$$

which will be derived later in this section. We note that

$$\left.\frac{\partial G(\mathbf{r}' - \mathbf{r})}{\partial z}\right|_{z=0} = \frac{1}{4\pi}\left\{\frac{z' - z}{[(x' - x)^2 + (y' - y)^2 + (z' - z)^2]^{3/2}}\right.$$

$$\left. + \frac{z' + z}{[(x' - x)^2 + (y' - y)^2 + (z' + z)^2]^{3/2}}\right\}_{z=0}$$

$$= \frac{1}{2\pi}\frac{z'}{[(x' - x)^2 + (y' - y)^2 + z'^2]^{3/2}} \quad (5.76\text{i})$$

Now the diffusive deposition rate is given by

$$j_{f_{\text{diff}}} = D\left.\frac{\partial n_p(\mathbf{r})}{\partial z}\right|_{z=0} = n\sigma \int I(\mathbf{r}')\left[\frac{\partial G(\mathbf{r}' - \mathbf{r})}{\partial z}\right]_{z=0} dV'$$

$$= \frac{n\sigma}{2\pi}\int_{-\infty}^{\infty} dx' \int_{-\infty}^{\infty} dy' \int_{0}^{\infty} I(x', y', z') \frac{z'}{[(x' - x)^2 + (y' - y)^2 + z'^2]^{3/2}} dz$$

$$(5.76\text{j})$$

Comparing Eqs. (5.75b) and (5.76j), we obtain

$$j_{f_{\text{diff}}} = (2/\gamma)J_{f_{\text{ball}}} \quad (5.76\text{k})$$

which shows that the diffusive deposition rate is twice the ballistic deposition rate when the sticking coefficient $\gamma$ is unity.

### 5.7.3. Similarity between the Ballistic and Diffusive Deposition Models

Equation (5.76k) shows that the particle fluxes at the substrate surface, that is, the deposition rates obtained from the ballistic and diffusive deposition models are different from each other. However, the governing equations for these two models are similar to one another. To show this similarity, let us first consider Eq. (5.75d). Taking the gradient of Eq. (5.75d) and then the divergence of the resulting expression, we can write

$$\nabla^2 F(\mathbf{r}) = n\sigma \int_{-\infty}^{\infty} dx' \int_{-\infty}^{\infty} dy' \int_{0}^{\infty} I(x', y', z')\nabla^2\left(\frac{1}{4\pi R}\right) dz' \quad (5.77\text{a})$$

where $\nabla^2 = \partial/\partial x^2 + \partial/\partial y^2 + \partial/\partial z^2$ in Cartesian coordinates, $\mathbf{r} = \hat{i}x + \hat{j}y + \hat{k}z$, and $R = [(x' - x)^2 + (y' - y)^2 + (z' - z)^2]^{1/2}$.

# PHOTOLYTIC LCVD MODELING

To integrate Eq. (5.77a), we will show that $1/4\pi R$ is a solution of the following equation:

$$\nabla^2 G(\mathbf{r}' - \mathbf{r}) = \delta(\mathbf{r}' - \mathbf{r}) \tag{5.77b}$$

in a spherically symmetric region. Here $G(\mathbf{r}' - \mathbf{r})$ is the Green function, and $\delta(\mathbf{r}' - \mathbf{r})$ is the Dirac delta function. Let us use the change of variables,

$$\xi = x' - x$$
$$\eta = y' - y$$
$$\zeta = z' - z$$

to shift the point $(x, y, z)$ to the origin. Now Eq. (5.77b) can be written as

$$\nabla^2 G = \frac{\partial^2 G}{\partial \xi^2} + \frac{\partial^2 G}{\partial \eta^2} + \frac{\partial^2 G}{\partial \zeta^2} = \delta(\xi)\delta(\eta)\delta(\zeta) \tag{5.77c}$$

Owing to spherical symmetry, we can integrate Eq. (5.77c) by noting that

$$\xi = \rho \sin\theta \cos\phi, \qquad \eta = \rho \sin\theta \sin\phi, \qquad \zeta = \rho \cos\theta,$$
$$\rho = (\xi^2 + \eta^2 + \zeta^2)^{1/2}, \qquad G = G(\rho), \qquad \delta(\xi)\delta(\eta)\delta(\zeta) = \delta(\rho)/4\pi\rho^2$$

where $\theta$ and $\phi$ represent the polar and azimuthal angles, respectively. Equation (5.77c) can now be written as

$$\frac{1}{\rho^2} \frac{d}{d\rho}\left(\rho^2 \frac{dG}{d\rho}\right) = \frac{\delta(\rho)}{4\pi\rho^2} \tag{5.77d}$$

which has to be solved under the following conditions:

$$\lim_{\rho \to \infty} G(\rho) = 0 \tag{5.77e}$$

$$\lim_{\rho \to \infty} \frac{dG}{d\rho} = 0 \tag{5.77f}$$

to determine $G$. To obtain the other boundary condition at $\rho = 0$, we integrate Eq. (5.77d) from 0 to $\infty$, and use the boundary condition (5.77f), which yields

$$\lim_{\rho \to 0} \left( \rho^2 \frac{dG}{d\rho} \right) = -\frac{1}{4\pi} \tag{5.77g}$$

From Eq. (5.77d), we can write

$$\frac{1}{\rho^2} \frac{d}{d\rho} \left( \rho^2 \frac{dG}{d\rho} \right) = 0 \quad \text{for} \quad \rho > 0 \tag{5.77h}$$

Solving Eq. (5.77h) and satisfying the boundary conditions (5.77e and g), we obtain

$$G(\rho) = \frac{1}{4\pi\rho} \tag{5.77i}$$

or

$$G(\mathbf{r'} - \mathbf{r}) = \frac{1}{4\pi(\xi^2 + \eta^2 + \zeta^2)^{1/2}}$$

$$= \frac{1}{4\pi[(x' - x)^2 + (y' - y)^2 + (z' - z)^2]^{1/2}}$$

$$= \frac{1}{4\pi R} \tag{5.77j}$$

Combining Eqs. (5.77b) and (5.77j), we can write

$$\nabla^2 \left( \frac{1}{4\pi R} \right) = \delta(\mathbf{r'} - \mathbf{r}) = \delta(x' - x)\delta(y' - y)\delta(z' - z) \tag{5.77k}$$

Substituting Eq. (5.77k) into Eq. (5.77a) and carrying out the integration, we obtain

$$\nabla^2 F(\mathbf{r}) = n\sigma I(\mathbf{r}) \tag{5.77l}$$

which is similar to Eq. (5.76a). Table 5.1 illustrates the similarities among the various expressions used in the ballistic and diffusive deposition models.

Table 5.1. Similarities between the Ballistic and Diffusive Deposition Models

| Expression | Ballistic model | Diffusion model |
|---|---|---|
| Governing equations | $\nabla^2 F(\mathbf{r}) = n\sigma I(\mathbf{r})$ [From Eq. (5.77l)] | $D\nabla^2 n_p(\mathbf{r}) = S(\mathbf{r}) = n\sigma I(\mathbf{r})$ [From Eq. (5.76a)] |
| Fluxes | $j_{f\text{ball}} = \gamma[\partial F(\mathbf{r})/\partial z]_{z=0}$ [From Eq. (5.75c)] | $j_{f\text{diff}} = D[\partial n_p(\mathbf{r})/\partial z]_{z=0}$ [From Eq. (5.76j)] |

### 5.7.4. Determination of the Time-Independent Green Function

In Section 5.7.2, we solved Eq. (5.76a) by using a Green function given by Eq. (5.76h). We will now derive an expression for the Green function $G(\mathbf{r} - \mathbf{r}')$, which is a solution of the following partial differential equation:

$$\nabla^2 G(\mathbf{r} - \mathbf{r}') = \frac{\partial^2 G}{\partial x^2} + \frac{\partial^2 G}{\partial y^2} + \frac{\partial^2 G}{\partial z^2} = \delta(x - x')\delta(y - y')\delta(z - z') \quad (5.78a)$$

where $-\infty < x < \infty$, $-\infty < y < \infty$, and $0 \leq z < \infty$. Note that the Green function $G(\mathbf{r}' - \mathbf{r})$ which appears in Eq. (5.76g) is determined by using the reciprocity theorem which states that $G(\mathbf{r}' - \mathbf{r}) = G(\mathbf{r} - \mathbf{r}')$. Since $G(\mathbf{r} - \mathbf{r}')$ has been used to solve Eq. (5.76a), it must satisfy the homogeneous counterpart of the boundary conditions that have been applied to solve for $n_p(\mathbf{r})$ from Eq. (5.76a). Therefore, the boundary conditions for $G(\mathbf{r} - \mathbf{r}')$ can be written as

$$G = \begin{cases} 0 & \text{as } x \to \pm\infty & (5.78b) \\ 0 & \text{as } y \to \pm\infty & (5.78c) \\ 0 & \text{as } z \to \infty & (5.78d) \\ 0 & \text{at } z = 0 & (5.78e) \end{cases}$$

We will use the Fourier transform for an infinite medium in the $x$ and $y$ directions to solve Eq. (5.78a). In the $z$ direction, we will use Fourier sine transform for a semi-infinite medium because $\sin z = 0$ at $z = 0$, which complies with the boundary condition (5.78e). If the Fourier transform exists for a given function $f(x)$, we can write the following formulas:

FOURIER TRANSFORM FOR AN INFINITE MEDIUM:

$$\bar{f}(k_x) = \int_{-\infty}^{\infty} f(x) \exp(ik_x x) dx \quad (5.78f)$$

INVERSE TRANSFORM FOR EQ. (5.78f):

$$f(x) = \frac{1}{2\pi} \int_{-\infty}^{\infty} \bar{f}(k_x) \exp(-ik_x x) dk_x \qquad (5.78g)$$

FOURIER SINE TRANSFORM FOR A SEMI-INFINITE MEDIUM:

$$\bar{f}(k_x) = \int_0^{\infty} f(x) \sin(k_x x) dx \qquad (5.78h)$$

INVERSE TRANSFORM FOR EQ. (5.78h):

$$f(x) = \frac{2}{\pi} \int_0^{\infty} \bar{f}(k_x) \sin(k_x x) dk_x \qquad (5.78i)$$

With these equations, the Fourier transform that is used for the Green function $G(\mathbf{r} - \mathbf{r}')$ can be written as follows:

FOURIER TRANSFORM:

$$\bar{G}(k_x, k_y, k_z) = \int_{-\infty}^{\infty} dx \int_{-\infty}^{\infty} dy \int_0^{\infty} G(\mathbf{r} - \mathbf{r}') \exp[i(k_x x + k_y y)] \sin(k_z z) dz \qquad (5.78j)$$

INVERSE TRANSFORM FOR EQ. (5.79j):

$$G(\mathbf{r} - \mathbf{r}') = \frac{1}{2\pi^3} \int_{-\infty}^{\infty} dk_x \int_{-\infty}^{\infty} dk_y \int_0^{\infty} \bar{G}(k_x, k_y, k_z)$$
$$\times \exp[-i(k_x x + k_y y)] \sin(k_z z) dk_z \qquad (5.78k)$$

Applying the Fourier transform (5.78j) to Eq. (5.78a), and satisfying the boundary conditions, we obtain

$$\bar{G}(k_x, k_y, k_z) = \frac{1}{k_x^2 + k_y^2 + k_z^2} \exp[i(k_x x' + k_y y')] \sin(k_z z') \qquad (5.78l)$$

Substituting Eq. (5.78l) into Eq. (5.78k), and noting that

$$\sin(k_z z) \sin(k_z z') = \tfrac{1}{2}[\cos k_z(z - z') - \cos k_z(z + z')]$$

# PHOTOLYTIC LCVD MODELING

we can write

$$G(\mathbf{r} - \mathbf{r}') = \frac{1}{4\pi^3} \int_{-\infty}^{\infty} dk_x \int_{-\infty}^{\infty} dk_y \int_{0}^{\infty} \left\{ \frac{\exp\{-i[k_x(x - x') + k_y(y - y')]\}}{k_x^2 + k_y^2 + k_z^2} \right.$$

$$\left. \times [\cos k_z(z - z') - \cos k_z(z + z')] \right\} dk_z \quad (5.78m)$$

We note that

$$\cos k_z(z - z') = \text{Re}\{\exp[-ik_z(z - z')]\}$$
$$\cos k_z(z + z') = \text{Re}\{\exp[-ik_z(z + z')]\}$$
$$(\text{Vector}) \; \mathbf{k} = \hat{i}k_x + \hat{j}k_y + \hat{k}k_z \quad (\text{see Fig. (5.5a))},$$
$$(\text{Vector}) \; \mathbf{R}_1 = \hat{i}(x - x') + \hat{j}(y - y') + \hat{k}(z - z') \quad (\text{see Fig. 5.5a})$$
$$(\text{Vector}) \; \mathbf{R}_2 = \hat{i}(x - x') + \hat{j}(y - y') + \hat{k}(z + z') \quad (\text{see Fig. 5.5a})$$
$$k = |\mathbf{k}| = (k_x^2 + k_y^2 + k_z^2)^{1/2}$$
$$R_1 = [(x - x')^2 + (y - y')^2 + (z - z')^2]^{1/2}$$

and

$$R_2 = [(x - x')^2 + (y - y')^2 + (z + z')^2]^{1/2}$$

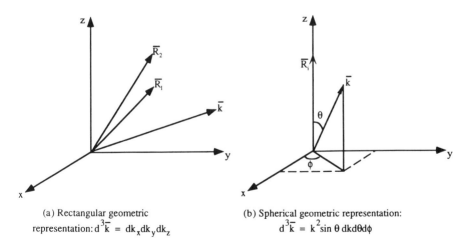

(a) Rectangular geometric representation: $d^3\overline{k} = dk_x dk_y dk_z$

(b) Spherical geometric representation: $d^3\overline{k} = k^2 \sin\theta \, dk d\theta d\phi$

Figure 5.5. Coordinate transformation to evaluate the integrals in Eq. (5.78n).

which can be used to rewrite Eq. (5.78m) as follows:

$$G(\mathbf{r} - \mathbf{r}') = \frac{1}{4\pi^3} \text{Re} \int_{-\infty}^{\infty} \int_{-\infty}^{\infty} \int_{0}^{\infty} \frac{\exp(-i\mathbf{k}\cdot\mathbf{R}_1)}{k^2} dk_x dk_y dk_z$$

$$- \frac{1}{4\pi^3} \text{Re} \int_{-\infty}^{\infty} \int_{-\infty}^{\infty} \int_{0}^{\infty} \frac{\exp(-i\mathbf{k}\cdot\mathbf{R}_2)}{k^2} dk_x dk_y dk_z \quad (5.78n)$$

To evaluate these two integrals, we will express $dk_x dk_y dk_z$ in spherical coordinates as shown in Fig. 5.5b, where the vector $\mathbf{R}_i$ has been aligned with the $z$-axis for $i = 1$ and 2. Therefore, Eq. (5.78n) can be written as

$$G(\mathbf{r} - \mathbf{r}') = \frac{1}{4\pi^3} \text{Re} \int_{0}^{2\pi} d\phi \int_{0}^{\infty} dk \int_{0}^{\pi/2} \exp(-ikR_1 \cos\theta) \sin\theta d\theta$$

$$- \frac{1}{4\pi^3} \text{Re} \int_{0}^{2\pi} d\phi \int_{0}^{\infty} dk \int_{0}^{\pi/2} \exp(-ikR_2 \cos\theta) \sin\theta d\theta$$

$$= \frac{1}{2\pi^2} \text{Re} \int_{0}^{\infty} dk \int_{0}^{1} \exp(-ikR_1 \mu) d\mu$$

$$- \frac{1}{2\pi^2} \text{Re} \int_{0}^{\infty} dk \int_{0}^{1} \exp(-ikR_2 \mu) d\mu, \quad \text{where } \mu = \cos\theta$$

$$= \frac{1}{2\pi^2} \int_{0}^{\infty} \frac{\sin(kR_1)}{kR_1} dk - \frac{1}{2\pi^2} \int_{0}^{\infty} \frac{\sin(kR_2)}{kR_2} dk \quad (5.78o)$$

To evaluate the two integrals in Eq. (5.78o), let us consider the following integral:

$$I_R(a) = \int_{0}^{\infty} \frac{\sin kR}{k} \exp(-ak) dk, \quad \text{where } I_R(\infty) = 0$$

$$\frac{dI_R}{da} = -\int_{0}^{\infty} \sin(kR) \exp(-ak) dk = \text{Im} \int_{0}^{\infty} \exp[-(a+iR)k] dk = -\frac{R}{a^2 + R^2}$$

Thus, $I_R(a) = \pi/2 - \tan^{-1}(a/R)$, which is obtained by solving the above differential equation and applying the condition $I_R(\infty) = 0$.

Now we can write

$$\int_{0}^{\infty} \frac{\sin kR}{k} dk = I_R(0) = \frac{\pi}{2} \quad (5.78p)$$

Combining Eqs. (5.78o and p), we obtain

$$G(\mathbf{r} - \mathbf{r}') = \frac{1}{4\pi}\left(\frac{1}{R_1} - \frac{1}{R_2}\right)$$

where the variables $\mathbf{r}$ and $\mathbf{r}'$ can be interchanged to write

$$G(\mathbf{r}' - \mathbf{r}) = \frac{1}{4\pi}\left\{\frac{1}{[(x'-x)^2 + (y'-y)^2 + (z'-z)^2]^{1/2}} - \frac{1}{[(x'-x)^2 + (y'-y)^2 + (z'+z)^2]^{1/2}}\right\} \quad (5.78q)$$

which is identical to Eq. (5.76h).

### 5.7.5. Adsorbed Phase Decomposition: Effect of Chemisorbed Phase

As discussed in Chapter 3, adsorption plays an important role in laser photochemical deposition. In particular, chemisorption causes a strong adherence of the film to the substrate due to the formation of chemical bonds between the atoms of the film and the substrate. However, the maximum thickness of the chemisorbed film is one monolayer [Chen (1987)], and the sticking coefficient for chemisorption is usually much smaller than unity, which results in a slow chemisorbed film deposition rate. Chen (1987) has modeled the deposition due to chemisorbed phase decomposition, which will be discussed below. For this purpose, let us represent the chemisorption process by the following expression:

$$A + {\overset{|}{\underset{(I)}{-S-}}} \rightarrow {\overset{A}{\underset{(II)}{\overset{|}{-S-}}}} \quad (5.79a)$$

where $A$ is the atom or molecule that is being adsorbed at the surface site (I) to form the chemisorbed product (II). The rate of formation of the chemisorbed species is given by

$$\frac{dN_c}{dt} = k_c J^* N_s^* \quad (5.79b)$$

where $k_c$ is the chemisorbed rate constant. Taking $k_c = \gamma_c/N_s$, we can rewrite

[Chen (1987)] Eq. (5.79b) as

$$\frac{dN_c}{dt} = \gamma_c J^* \frac{N_s - N_c}{N_s}$$

or

$$\frac{d\theta_c}{dt} = \frac{\gamma_c J^*}{N_s}(1 - \theta_c) \tag{5.79c}$$

To solve Eq. (5.79c), we note that $\theta_c = 0$ at $t = 0$ for a clean surface. Under this initial condition, the solution of Eq. (5.79c) is given by

$$\theta_c = 1 - \exp\left(-\frac{\gamma_c J^*}{N_s} t\right) \tag{5.79d}$$

where the time constant $\tau_c$ to produce a chemisorbed film is given by

$$\tau_c = \frac{N_s}{\gamma_c J^*} \tag{5.79e}$$

To evaluate $\theta_c$ from Eq. (5.79d), we need the values of $\gamma_c$, $J^*$, and $N_s$: The value of $N_s$ can be estimated by assuming that the density of the adsorbed film is equal to the density of the adsorbed species in its liquid state. Thus, the total number of molecules per unit volume is given by $\rho_1 A_v / M_p$. Assuming that the molecules are arranged in a cube with $x$ molecules along each edge, we find that $x = (\rho_1 A_v / M_p)^{1/3}$. So, the areal density is given by

$$N_s = x^2 = \left(\frac{\rho_1 A_v}{M_p}\right)^{2/3} \tag{5.79f}$$

The number of precursor molecules impinging on the surface per unit time per unit surface area is given by [Mayer and Mayer (1940)]

$$J^* = \frac{p_r}{\sqrt{2\pi m k_B T}} = 3.537 \times 10^{22} \times \frac{p_r}{\sqrt{M_p T}} \, \text{cm}^{-2}\,\text{s}^{-1} \tag{5.79g}$$

where $p_r$, $M_p$, $T$ are in Torrs, g/mol, and K, respectively; $\gamma_c$ can be calculated from Eq. (5.79e) if $\tau_c$ is known. Usually, $\tau_c$ is about 0.1 s or longer.

To determine the deposition rate due to chemisorbed phase decomposition, we have to consider the photochemical reaction in the chemisorbed phase, which leads to the following kinetic equation for chemisorption:

$$\frac{dN_c}{dt} = \frac{\gamma_c}{N_s} J^*(N_s - N_c) - N_c \sigma_c(\lambda) I^* \tag{5.79h}$$

## 5.7.6. Adsorbed Phase Decomposition: Effect of Physisorbed Phase

Similar to chemisorption, physisorption is also important in laser photochemical deposition, although the physisorbed film only adheres weakly to the substrate. The binding energies of the physisorbed and chemisorbed species are about $k_B T$ and much larger than $k_B T$, respectively. In Chapter 3, we discussed the differences between physisorption and chemisorption, and Chen (1987) has presented a model for deposition due to physisorbed phase decomposition. The mechanism of physisorption is different from that of chemisorption. Only one layer, that is, a monolayer, is formed from chemisorption, whereas, multiple layers are formed as a result of physisorption. Therefore, physisorption involves the bonding of a molecule with the substrate surface as well as with another molecule of the same kind that is already adsorbed on the surface. Based on this consideration, Brunauer, Emmett, and Teller derived the following expression for surface coverage under the equilibrium condition,

$$\theta_p^* = \frac{C}{(p_0/p_r - 1)[1 - (C - 1)p_r/p_0]} \tag{5.80a}$$

which is known as the BET isotherm [Brunauer et al. (1938), Young and Crowell (1962)]. Here $C$ is a constant which is given by

$$C = \exp\left(\frac{E_0 - E_1}{k_B T}\right) \tag{5.80b}$$

where $E_0$ is the binding energy of a molecule which is in the first layer and attached to the solid surface, and $E_1$ is the binding energy of a molecule which is in the upper layer and attached to a molecule below it. The optical absorption in the physisorbed film, which satisfies the BET-type adsorption isotherm [Eq. (5.80a)], is given by [Chen and Osgood (1983)]

$$A(p_r, \lambda) = A_0(\lambda) \frac{C}{(p_0/p_r - 1)[1 - (C - 1)p_r/p_0]} = A_0(\lambda)\theta_p^* \tag{5.80c}$$

The kinetic equation for physisorption can be written as

$$\frac{dN_p}{dt} = J_p^* - N_p \sigma_p(\lambda) I^* - N_p P_e \tag{5.80d}$$

where we have assumed that each photon absorption results in a photochemical decomposition. On the right-hand side of Eq. (5.80d), the first term represents the number of molecules physisorbed per unit area per unit time due to the impingement of the gas molecules on the surface, and the second and third terms represent the losses of the physisorbed molecules due to the photochemical decomposition and evaporation, respectively. It should be noted that the absorption cross section of the physisorbed species can be different from that of the gas phase molecules because the chemical bonds of the molecules are different in these two phases. However, the absorption of a single monolayer of the physisorbed film, $A_0(\lambda)$, is usually measured [Chen and Osgood (1983)]. Noting that $A_0(\lambda) = N_s \sigma_p(\lambda)$, Eq. (5.80d) can be written as

$$\frac{dN_p}{dt} = J_p^* - \theta_p A_0(\lambda) I^* - \theta_p N_s P_e$$

$$= J_p^* - \theta_p \frac{A(p_r, \lambda)}{\theta_p^*} I^* - \theta_p N_s P_e \qquad (5.80e)$$

Also, we note that the physisorbed film is in dynamic equilibrium with the surrounding gas, and in the absence of any chemical reaction the number of molecules physisorbed must be equal to the number of molecules evaporated from the surface, that is,

$$j_p^* = \theta_p^* N_s P_e \quad \text{at equilibrium} \qquad (5.80f)$$

At the steady state, $dN_p/dt = 0$, and Eq. (5.80e) becomes

$$\theta_p = \frac{\theta_p^*}{1 + R_p^*} \qquad (5.80g)$$

where $R_p^* = A(p_r, \lambda) I^* / J_p^*$.

The physisorbed phase decomposition rate is then given by

$$\frac{dn_{pp}}{dt} = \frac{A(p_r, \lambda) I^*}{1 + R_p^*} \qquad (5.80h)$$

with $R_p^*$ the ratio of the reaction rate in the physisorbed phase at equilibrium coverage to the physisorption rate. Based on the values of $R_p^*$, the deposition process can be divided into the following two regimes:

REACTION-CONTROLLED REGIME: This regime is attained when $R_p^* \ll 1$, which simplifies Eq. (5.80h) to the following form:

$$\frac{dn_{p_p}}{dt} = A(p_r, \lambda)I^* \tag{5.80i}$$

which shows that the deposition rate is linearly proportional to the laser intensity, and that there will be no saturation of the deposition rate.

CHEMISORPTION-CONTROLLED REGIME: This regime is attained when $R_p^* \gg 1$, which simplifies Eq. (5.80h) to the following form:

$$\frac{dn_{p_p}}{dt} = J_p^* \tag{5.80j}$$

which shows that the deposition rate does not depend on the laser intensity, and that it is linearly proportional to the physisorption rate.

## 5.8. THE GREEN FUNCTION APPROACH

Green's function approach provides an efficient way of solving partial differential equations. However, to obtain the Green function for a given problem, we have to solve a partial differential equation (see Section 5.7.4 or 5.8.2) which is similar to the original problem, but easier to solve.

The Green function is usually denoted by $G(\mathbf{r}, t|\mathbf{r}', t')$ or $G(\mathbf{r}, t; \mathbf{r}', t')$ or $G(\mathbf{r} - \mathbf{r}', t - t')$ to represent the effect of an impulsive point source of unit strength, which is located at the point $(\mathbf{r}', t')$, at the observation point $(\mathbf{r}, t)$. An important property of the Green function is that it obeys the reciprocity theorem, that is, $G(\mathbf{r}, t; \mathbf{r}', t') = G(\mathbf{r}', -t'; \mathbf{r}, -t)$. The theory of Green's function can be found in Roach (1982). Green's function has been used to model laser-induced surface reactions as well as for measuring the surface diffusivity [Zeiger and Ehrlich (1989), Zeiger et al. (1985), Tsao et al. (1985)].

### 5.8.1. Effects of Surface Diffusion and Adsorption on Photolytic Surface Reactions

In Section 5.7, we discussed the gas phase diffusion and photochemical reaction under steady state conditions, and also analyzed photochemical decomposition in the chemisorbed and physisorbed phases in the absence of surface diffusion. Tsao et al. (1985) have presented a model for laser-induced surface photolysis by considering the change of surface concentration with time and surface diffusion. The governing equation for this problem, which is similar to the problem studied by Freeman and Doll (1983), is given by Tsao et al. (1985):

$$\frac{\partial n_a(\mathbf{r}, t)}{\partial t} = D_s \nabla^2 n_a(\mathbf{r}, t) - k_d n_a(\mathbf{r}, t) + k_a P_0 [N_s - n_a(\mathbf{r}, t)] - I(\mathbf{r}, t) n_a(\mathbf{r}, t) \sigma_a$$

(5.81a)

On the right-hand side of Eq. (5.81a), the first, second, third, and fourth terms represent the surface diffusion, desorption, Langmuir-type adsorption, and photochemical reaction rates per unit surface area, respectively. For Eq. (5.81a), the boundary conditions are

$$n_a(x, y, z, t) = 0 \quad \text{as } x \to \pm\infty, \quad y \to \pm\infty, \quad \text{or} \quad z \to \pm\infty$$

(5.81b)

in Cartesian coordinates, and the initial condition is

$$n_a(x, y, z, 0) = n_0$$

(5.81c)

For simplicity, we rewrite Eq. (5.81a) in the following form:

$$\frac{\partial n_a}{\partial t} = D_s \nabla^2 n_a - S_1 n_a - S_2(\mathbf{r}, t)$$

(5.81d)

where $S_1 = k_d + k_a P_0$, and $S_2(\mathbf{r}, t) = I(\mathbf{r}, t) n_a(\mathbf{r}, t) \sigma_a - k_a P_0 N_s$.

The two-dimensional Green function for this equation is given by

$$G(\mathbf{r}, t; \mathbf{r}', t') = \frac{1}{4\pi D_s(t - t')} \exp\left[-\frac{(\mathbf{r} - \mathbf{r}')^2}{4D_s(t - t')}\right] \exp[-S_1(t - t')] \quad \text{for} \quad t > t'$$

(5.81e)

which is derived in the next section.

With this Green's function, the two-dimensional solution of Eq. (5.81d) can be written as

$$n_a(\mathbf{r}, t) = \int_{-\infty}^{\infty}\int_{-\infty}^{\infty} [n_a(\mathbf{r}', t')G(\mathbf{r}, t; \mathbf{r}', t')]_{t'=0} dx' dy'$$

$$- \int_{-\infty}^{\infty}\int_{-\infty}^{\infty}\int_{0}^{t} S_2(\mathbf{r}', t')G(\mathbf{r}, t; \mathbf{r}', t')dx' dy' dt'$$

$$= n_0 \exp(-S_1 t) + \frac{k_a P_0 N_s}{S_1}[1 - \exp(-S_1 t)]$$

$$- \int_{-\infty}^{\infty}\int_{-\infty}^{\infty}\int_{0}^{t} I(\mathbf{r}', t')n_a(\mathbf{r}', t')\sigma_a G(\mathbf{r}, t; \mathbf{r}', t')dx' dy' dt'$$

(5.81f)

Equation (5.81f) shows that Green's function can be used to convert a differential equation, such as Eq. (5.81a), into an integral equation. Although both Eqs. (5.81a) and (5.81f) require a numerical technique to solve for $n_a(\mathbf{r}, t)$, the integral equations are preferable for numerical computation because the numerical integration and differentiation involve smoothing and chopping discretization errors, respectively [Carnahan et al. (1969)]. In any case, after $n_a(\mathbf{r}, t)$ is determined the deposition rate can be obtained from the following expression:

$$\frac{dn_{pa}}{dt} = \gamma I(\mathbf{r}, t) n_a(\mathbf{r}, t) \sigma_a \qquad (5.81g)$$

### 5.8.2. Determination of the Time-Dependent Green Function

The governing equation to determine the Green function for Eq. (5.81d) is

$$\frac{\partial G(\mathbf{r}', -t'; \mathbf{r}, -t)}{\partial t} - D_s \nabla^2 G(\mathbf{r}', -t'; \mathbf{r}, -t) + S_1 G(\mathbf{r}', -t'; \mathbf{r}, -t) = \delta(\mathbf{r} - \mathbf{r}')\delta(t - t')$$

(5.82a)

where

$$\nabla^2 \equiv \frac{\partial}{\partial x^2} + \frac{\partial}{\partial y^2} + \frac{\partial}{\partial z^2}$$

$$\delta(\mathbf{r} - \mathbf{r}') = \delta(x - x')\delta(y - y')\delta(z - z')$$

with $-\infty < x < \infty$, $-\infty < y < \infty$, and $-\infty < z < \infty$ in Cartesian coordinates, and $0 \leqslant t < \infty$.

The Green function $G(\mathbf{r}, t; \mathbf{r}', t')$ which appears in Eq. (5.81f) is determined from $G(\mathbf{r}', -t'; \mathbf{r}, -t)$ by using the reciprocity theorem, which states that $G(\mathbf{r}, t; \mathbf{r}', t') = G(\mathbf{r}', -t'; \mathbf{r}, -t)$. $G(\mathbf{r}', -t'; \mathbf{r}, -t)$ satisfies the homogeneous counterpart of the boundary and initial conditions that have been applied to solve for $n_a(\mathbf{r}, t)$ from Eq. (5.81a), that is,

$$G(x', y', z', -t'; x, y, z, -t) = 0, \quad \text{as } x \to \pm\infty, \quad y \to \pm\infty, \quad \text{or} \quad z \to \pm\infty$$
(5.82b)

in Cartesian coordinates, and

$$G(x', y', z', -t^-; x, y, z, -t) = 0, \quad \text{for} \quad -t' < -t \quad \text{or} \quad t < t' \quad (5.82c)$$

The initial condition (5.82c) is based on the consideration that the effect of a source is zero at $(x', y', z', -t')$ before it is turned on at $(x, y, z, -t)$. We will use a Laplace transform [Eqs. (5.44a, b)] and the infinite-medium Fourier transform [Eqs. (5.78f, g)] to solve Eq. (5.82a) by defining the following variable:

LAPLACE AND FOURIER TRANSFORMS:

$$\bar{G} = \int_0^\infty dt \exp(-st) \int_{-\infty}^\infty \int_{-\infty}^\infty \int_{-\infty}^\infty G \exp[i(k_x x + k_y y + k_z z)] \, dx \, dy \, dz$$
(5.82d)

where $\bar{G} \equiv \bar{G}(x', y', z', -t'; k_x, k_y, k_z, -s)$, and $G \equiv G(x', y', z', -t'; x, y, z, -t)$.

INVERSE TRANSFORM FOR EQ. (5.82d):

$$G = \frac{1}{2\pi i}\int_{\beta-i\infty}^{\beta+i\infty} \exp(st) \, dx \, \frac{1}{(2\pi)^3} \int_{-\infty}^\infty \int_{-\infty}^\infty \int_{-\infty}^\infty \bar{G} \exp[-i(k_x x + k_y y + k_z z)] \, dk_x \, dk_y \, dk_z$$
(5.82e)

Applying the Laplace and Fourier transforms, and the boundary and initial conditions to Eq. (5.82a), we obtain

$$\bar{G} = \frac{\exp(-st')}{s + D_s(k_x^2 + k_y^2 + k_z^2) + S_1} \exp[i(k_x x' + k_y y' + k_z z')] \quad (5.82f)$$

The inverse transform of Eq. (5.82f) yields

$$G = \frac{1}{(2\pi)^3} \int_{-\infty}^{\infty} \int_{-\infty}^{\infty} \int_{-\infty}^{\infty} \exp[-s_1(t-t')] E_1 E_2 \, dk_x \, dk_y \, dk_z \quad \text{for } t > t' \quad (5.82\text{g})$$

$$= 0 \quad \text{for } t < t' \quad (5.82\text{h})$$

where

$$E_1 = \exp[-D_s(k_x^2 + k_y^2 + k_z^2)(t - t')]$$

and

$$E_2 = \exp\{-i[k_x(x - x') + k_y(y - y') + k_z(z - z')]\}$$

As in Section 5.7.4, we define the vectors $\mathbf{k} = \hat{i}k_x + \hat{j}k_y + \hat{k}k_z$, and $\mathbf{R} = \hat{i}(x - x') + \hat{j}(y - y') + \hat{k}(z - z')$, and carry out the integration in Eq. (5.82g) in spherical coordinates, that is,

$$G = \frac{1}{(2\pi)^3} \int_0^{2\pi} d\phi \int_0^{\infty} k^2 \, dk \int_0^{\pi} \exp[-S_1(t - t')]$$
$$\times \exp[-D_s k^2 (t - t')] \exp(-i\mathbf{k} \cdot \mathbf{R}) \sin\theta \, d\theta$$

$$= \frac{\exp[-S_1(t-t')]}{(2\pi)^2} \int_0^{\infty} k^2 \, dk \int_{-1}^{1} \exp[-D_s k^2 (t-t')] \exp(-ikR\mu) \, d\mu$$

where $\mathbf{k} \cdot \mathbf{R} = kR\cos\theta$ and $\mu = \cos\theta$

$$= \frac{\exp[-S_1(t-t')]}{2\pi^2} \int_0^{\infty} \exp[-D_s k^2 (t-t')] \frac{\sin kR}{R} k \, dk$$

$$= -\frac{\exp[-S_1(t-t')]}{4\pi^2 R} \text{Re} \frac{\partial}{\partial R} \int_{-\infty}^{\infty} \exp[-D_s k^2 (t-t') + ikR] \, dk$$

$$= -\frac{\exp[-S_1(t-t')]}{4\pi^2 R} \text{Re} \frac{\partial}{\partial R} \left\{ \exp\left[-\frac{R^2}{4D_s(t-t')}\right] \right.$$
$$\left. \times \int_{-\infty}^{\infty} \exp\left[-D_s(t-t')\left(k + \frac{iR}{2D_s(t-t')}\right)^2\right] dk \right\}$$

$$= \frac{\exp[-S_1(t-t')]}{[4\pi D_s(t-t')]^{3/2}} \exp\left[-\frac{(x-x')^2 + (y-y')^2 + (z-z')^2}{4D_s(t-t')}\right] \quad (5.82\text{i})$$

The exponent of $4\pi D_s(t - t')$ in Eq. (5.82i) depends on the dimensionality of the problem. It is 3/2, 1, and 1/2 for the three-, two-, and one-dimensional problems, respectively. For the one-dimensional problem, Eq. (5.82i) becomes

$$G = \frac{1}{[4\pi D_s(t - t')]^{1/2}} \exp\left[-\frac{(x - x')^2}{4D_s(t - t')}\right] \exp[-S_1(t - t')]$$

and for the two-dimensional problem, the Green function is given by

$$G = \frac{1}{4\pi D_s(t - t')} \exp\left[-\frac{(x - x')^2 + (y - y')^2}{4D_s(t - t')}\right] \exp[-S_1(t - t')]$$

$$= \frac{1}{4\pi D_s(t - t')} \exp\left[-\frac{(\mathbf{r} - \mathbf{r}')^2}{4D_s(t - t')}\right] \exp[-S_1(t - t')] \quad (5.82j)$$

which is identical to Eq. (5.81e).

## 5.9. MONTE CARLO METHOD

So far, we have discussed several macroscopic models where the gas is treated as a continuum, and the continuum equations for the conservations of mass, momentum, and energy are considered to hold good. Using such continuum models, the macroscopic velocity, pressure, temperature, and density can be determined at any time and point in the gas phase. However, at low pressures, the gas phase attains a particulate structure in which the particles are individual molecules or clusters of several molecules, and does not behave like a continuum medium. In such situations, the microscopic or molecular models are useful, where molecular dynamics or Monte Carlo simulation techniques are usually used to determine the position and velocity of each particle at all times. It should be noted that the basic conservation equations are applicable to both the macroscopic and microscopic models, and that they require expressions for the fluxes of mass, momentum, and energy in terms of other macroscopic quantities, which are known as the constitutive relations, to form a determinate set of equations. For example, the fluxes of mass, momentum (shear stress), and energy are related to the gradients of concentration, velocity, and temperature by Fick's diffusion, Newton's viscosity, and Fourier's heat condition laws, respectively. When these macroscopic quantities vary very rapidly over a very short distance, that is, when their gradients are so large that their scale length and the mean free path are almost equal, the continuum models fail.

Bird (1976) has used the Knudsen number Kn, which is a ratio of the mean free path $\bar{\lambda}$ to the characteristic dimension $L$, that is, $Kn = \bar{\lambda}/L$, to determine the applicability of the macroscopic (continuum) and microscopic (molecular) models. If $Kn \ll 1$, the continuum models are considered to be valid. On the other hand, the molecular approach is applicable when $Kn \geqslant 1$, that is, when the mean free path is very large or the characteristic dimension is very small. Since the mean free path is inversely proportional to the density or pressure, the condition $Kn \geqslant 1$ arises at low pressures or densities when microscopic modeling is required. Also, in many processes, especially during LCVD, the characteristic dimension is very small because the diameter of the laser beam is very small, which can lead to the condition $Kn \geqslant 1$, when the molecular approach is required to model and understand the process mechanisms.

As noted earlier, the molecular dynamics and Monte Carlo simulation techniques are usually used for microscopic modeling. Yardley *et al.* (1981) have used the molecular dynamics technique to model the laser photodissociation of iron pentacarbonyl. A discussion on the direct simulation Monte Carlo (DSMC) method and its application to molecular gas dynamics can be found in Bird (1976). Coronell and Jensen (1992) have used the DSMC technique to model the transition regime flows in low-pressure conventional CVD reactors. Zeiri *et al.* (1991) have developed a model, which is discussed in this section, by applying the Monte Carlo method to simulate the deposition rate during photolytic LCVD under the following assumptions:

1. The laser intensity is very low so that the rise in temperature of the substrate at the irradiated spot is very small. This implies that the product is formed due to gas phase decomposition of the precursor molecules instead of the pyrolytic process.
2. The diffusions of both the precursor and product species are important. The carrier (buffer) gas is always uniformly distributed inside the deposition chamber so that its diffusion need not be considered.
3. Due to the cylindrical symmetry of the Gaussian laser beam, a cylindrical volume would be a natural choice for the computational domain to carry out Monte Carlo simulation. However, numerical calculation in such a three-dimensional domain would require extensive computational resources. For this reason, a rectangular slice of this volume is considered to model the photolytic LCVD process by considering two-dimensional meshes (see Fig. 5.6).
4. The ideal gas law and the kinetic theory of gases are applicable to the precursor and product species during the deposition process.

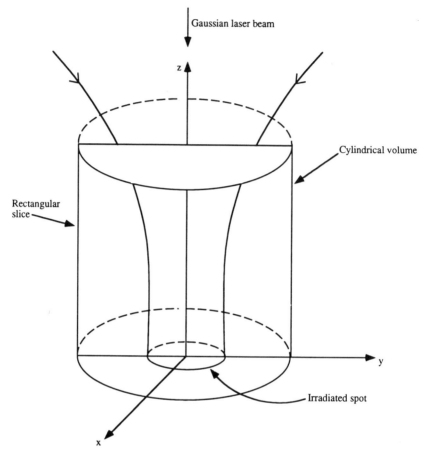

Figure 5.6. Geometry for the Monte Carlo simulation of photolytic laser chemical vapor deposition.

5. A hard sphere approximation is adequate to model the collisions between various particles.
6. Usually, the motion of a large number of particles ($\sim 10^4$ to $10^5$) needs to be followed for successful simulation of a typical LCVD process, which is difficult to accomplish in practice. For this reason, the precursor and product species are considered to obey Brownian statistics [Chandrasekhar (1943)], and the buffer (carrier) gas remains uniformly distributed during the simulation.

Figure 5.6 shows the geometric configuration used for simulating the deposition process. The $xy$-plane lies on the substrate where the deposition takes place. The Gaussian laser beam propagates in the negative direction of the $z$-axis. One of the sides of the rectangular slice considered in this model coincides with the $y$-axis and another side coincides with the $z$-axis as shown in Fig. 5.6. Since the two-dimensional slice represents the computational region according to assumption (3), we have to determine the total number of molecules per unit area of this slice. For this purpose, we use assumption (4) to obtain the total number of particles (precursor and buffer gas) per unit volume of the cylinder before the photochemical reaction begins, $n = P_0/(k_B T)$.

Assuming that the molecules are arranged in a cube with $x$ molecules along each edge, we find that $x = n^{1/3}$, and so the total number of particles (precursor and buffer gas) in each mesh of the rectangular slice is given by

$$n^* = S_m x^2 = S_m (P_0/k_B T)^{2/3} \tag{5.83a}$$

where $S_m$ is the area of each mesh. If $\Delta y$ and $\Delta z$ are the mesh sizes in the $y$ and $z$ directions, respectively, $S_m = \Delta y \Delta z$. The time step $\Delta t$ for the simulation is obtained from the following expression:

$$\Delta t = (\Delta l)^2 / 4 \bar{D} \tag{5.83b}$$

where $\Delta l$ is the average of the cell sizes in the $y$ and $z$ direction, and $\bar{D}$ is the weighted average diffusion coefficient in the gas phase inside the deposition chamber, that is,

$$\bar{D} = \frac{1}{\rho} \sum_{i=1}^{S} n_i m_i D_i \tag{5.83c}$$

Here $\rho$ is the density of the gas phase, $S$ is the total number of species present in the deposition chamber, and $n_i$, $m_i$, $D_i$ are the number density, mass, and diffusion coefficient of the $i$th species, respectively. If the deposition chamber contains only one type each of the precursor, product, and buffer gas molecules, $S = 3$. Then $\rho$ and $D_i$ are given by [Mayer and Mayer (1940)]

$$\rho = \sum_{i=1}^{S} n_i m_i \tag{5.83d}$$

and

$$D_i = \tfrac{1}{3} \bar{v}_i \lambda_i \tag{5.83e}$$

where $\bar{v}_i$ and $\lambda_i$, the average thermal speed and the mean free path in the mixture of the precursor, product, and buffer gas molecules for the $i$th species, respectively, are given by [Bird (1976), pp. 59, 70]:

$$\bar{v}_i = (8k_BT/\pi m_i)^{1/2} \tag{5.83f}$$

and

$$\lambda_i = \frac{\bar{v}_i}{\Sigma_{j=1}^S \sqrt{\pi/2}(d_i + d_j)^2 n_j[2k_BT(m_i + m_j)/m_im_j]^{1/2}} \tag{5.83g}$$

where $d_i$ is the effective diameter of the $i$th species [Bird (1976), p. 14].

After determining $\Delta t$, Zeiri et al. (1991) carried out the Monte Carlo simulation through the following sequence of steps for each $\Delta t$:

STEP 1: Calculate the total number of product molecules produced due to the photochemical reaction in each of two-dimensional cells at the beginning of each time step by using the following expression:

$$n_p = I_c n_r \sigma \tag{5.83h}$$

where $n_p$ and $n_r$ are the total number of the product and reactant molecules, respectively, in each cell, and $I_c$ is the laser intensity at the center of the cell. Here $n_r$ represents the total number of precursor molecules in a cell before the photochemical reaction. It should be noted that the attenuation of the laser beam as it propagates toward the substrate can be neglected if the partial pressure of the laser-absorbing gas is very low.

STEP 2: Move the precursor and product particles to new positions by using the following Brownian particle distribution function [Chandrasekhar (1943; see assumption (6)].

$$W(\mathbf{r}, t; \mathbf{r}_0, \mathbf{u}_0)$$
$$= \left\{\frac{m\beta^2}{2\pi k_BT[2\beta t - 3 + 4\exp(-\beta t) - \exp(-2\beta t)]}\right\}^{3/2}$$
$$\times \exp\left\{-\frac{m\beta^2|\mathbf{r} - \mathbf{r}_0 - \mathbf{u}_0[1 - \exp(-\beta_i t)]/\beta|^2}{2k_BT[2\beta t - 3 + 4\exp[-\beta t] - \exp(-2\beta t)]}\right\}$$

(5.83i)

where $\beta = k_BT/mD$, and for the $i$th species, $m = m_i$ and $d = d_i$; that is, $m$ and $d$ are the mass and diffusion coefficient of a particle, respectively; $\mathbf{r}_0$ and $\mathbf{u}_0$ are the position and velocity vectors of the particle, respectively, at time $t = 0$, and **r** is its position vector at any time $t$. To use Eq. (5.83i), we have to know $\mathbf{r}_0$ and $\mathbf{u}_0$; $\mathbf{r}_0$ is randomly chosen by assuming that the initial distribution of the particles in uniform, and $\mathbf{u}_0$ is obtained by randomly sampling the following canonical distribution function:

$$f(u) = \left(\frac{m}{2\pi k_B T}\right)^{1/2} \exp\left(-\frac{mu^2}{2k_B T}\right) \qquad (5.83j)$$

During the Monte Carlo simulation, a lot of time is consumed in generating the representative values of various variables from a prescribed distribution function by using random numbers. For this reason, efficient sampling techniques are required to reduce the computational time. The method of sampling in a given distribution function is discussed in the next section.

STEP 3: For the product species, if $z < 0$, that is, if the final location of the species is below the substrate surface, use linear interpolation between this final and the previous positions to obtain a new location for the species on the substrate surface ($z = 0$) where the species is assumed to stick to the surface. For the precursor species, if $z < 0$, that is, if the final location of the species is below the substrate surface, the species is assumed to be reflected from the surface. It should be noted that the sticking coefficients of the precursor and product species have been essentially assumed to be zero and unity, respectively, which holds good for many cases. For example, it has been shown [Foord and Jackman (1986), Ho (1988)] that various metal carbonyls are not adsorbed on Si surfaces at room temperature, and that the metal products stick to the Si surface with nearly unit sticking probability.

STEP 4: Apply the reflective boundary condition along the $z$-axis at $x = 0$ due to the axial symmetry of the cylindrical volume (see Fig. 5.6).

STEP 5: We have taken care of the boundary conditions at the boundaries $z = 0$, and $x = 0$ in Steps 3 and 4, respectively. At the other two sides, that is, at the top and left of the rectangular slice (see Fig. 5.6), the concentrations are assumed to be the same as the concentrations beyond the boundaries. Due to this boundary condition, when a particle leaves a cell through any point in these two boundaries, a new particle is placed at the same boundary point.

STEP 6: At the end of the time step $\Delta t$, the number of the deposited particles is counted, and the numbers of the precursor and product particles are updated in each cell. Here, the collisions among the precursor and product particles, which can lead to the loss of particles from the cell, are neglected, which is a reasonable approximation at low partial pressures of these species. Only the buffer gas is assumed to collide with these low partial pressure species.

After Step 6, the simulation is repeated starting from Step 1 for the next time step $\Delta t$.

STATISTICAL SAMPLING OF DISTRIBUTION FUNCTIONS

Most of the computational time in Monte Carlo simulations is spent in sampling prescribed distribution functions to obtain reasonable values for various variables in order to represent the physical process that is under study in a realistic way. For this reason, an efficient technique to sample from a given distribution function is required.

Let $f(x)$ be a normalized distribution, such as the normalized velocity, or concentration, or temperature distribution, function of the variate $x$; $f(x)$ is known as the probability density function (pdf), and $f(x)dx$ represents the probability that a value of $x$ lies between $x$ and $x + dx$, which is denoted by $P(x \leqslant X \leqslant x + dx)$. Therefore, we can write,

$$P(x \leqslant X \leqslant x + dx) = f(x)dx \tag{5.84a}$$

If the value of $x$ ranges from $a$ to $b$, that is, $a \leqslant x \leqslant b$, the pdf, $f(x)$ must satisfy the following condition:

$$\int_a^b f(x)dx = 1 \tag{5.84b}$$

and the cumulative distribution function $F(x)$, which represents the probability that a random selection of $x$, say $X$, lies between $a$ and $x$, is defined by

$$F(x) = P(a \leqslant X \leqslant x) = \int_a^x f(x')dx' \tag{5.84c}$$

It should be noted that

$$f(x) = dF(x)/dx \geqslant 0 \tag{5.84d}$$

To sample from a given distribution function, we equate $F(x)$ to the random number $R_n$, where $0 \leqslant R_n \leqslant 1$, that is,

$$F(x) = R_n = \int_a^x f(x')dx' \tag{5.84e}$$

SAMPLING FROM A UNIFORM DISTRIBUTION: For a uniform distribution, the probability density function is

$$f(x) = k \tag{5.84f}$$

where $a \leq x \leq b$, and $k$ is a constant. Since $f(x)$ must satisfy the condition (5.84b), we have

$$k = \frac{1}{b-a} \tag{5.84g}$$

From Eq. (5.84e), we have

$$R_n = \int_a^x \frac{1}{b-a} dx' = \frac{x-a}{b-a}$$

or

$$x = a + R_n(b-a) \tag{5.84h}$$

where the variate $x$ is expressed in terms of the random number $R_n$. However, in some cases, $f(x)$ can be so complicated that the integral in Eq. (5.84e) cannot be evaluated analytically or $x$ cannot be expressed in terms of $R_n$. In such cases, special techniques have to be used as discussed below.

SAMPLING FROM A MAXWELLIAN VELOCITY DISTRIBUTION: In Eq. (5.83j), we have used the following Maxwell velocity distribution:

$$f(u) = (\delta/\sqrt{\pi})\exp(-\delta^2 u^2) \tag{5.84i}$$

to carry out the Monte Carlo simulation, where $\delta = (m/2k_B T)^{1/2}$ and $-\infty < u < \infty$. Equation (5.84i) satisfies the condition (5.84b). From Eq. (5.84e), we obtain

$$R_n = \int_{-\infty}^{\infty} \frac{\delta}{\sqrt{\pi}} \exp(-\delta^2 u'^2) du' = \frac{1}{2}[1 + \text{erf}(\delta u)] \tag{5.84j}$$

which shows that $u$ cannot be expressed in terms of $R_n$ and, therefore, the sampling for $u$ will require numerical techniques to solve the transcendental Eq. (5.84j). However, this difficulty can be overcome by noting that we need two velocities, $v_y$ and $v_z$ in the $y$ and $z$ directions, respectively, to determine

$\mathbf{u}_0$ for the Monte Carlo simulation in the previous section, and that the Maxwell velocity distribution for two thermal velocity components in an equilibrium gas is given by

$$f(v_y, v_z) = \frac{\delta^2}{\pi} \exp[-\delta^2(v_y^2 + v_z^2)] \tag{5.84k}$$

where $-\infty < v_y < \infty$ and $-\infty < v_z < \infty$. Equation (5.84k) satisfies the condition (5.84b), and Eq. (5.84e) yields

$$R_n = \int_{-\infty}^{v_y} \int_{-\infty}^{v_z} \frac{\delta^2}{\pi} \exp[-\delta^2(v_y'^2 + v_z'^2)] dv_y' dv_z' \tag{5.84l}$$

We will integrate Eq. (5.84l) in polar coordinates by setting $v_y' = u' \cos\theta$ and $v_z' = u' \sin\theta$, which transform Eq. (5.84l) into the following form:

$$R_n = \frac{\delta^2}{\pi} \int_0^{2\pi} d\theta \int_0^u \exp(-\delta^2 u'^2) u' du' = 1 - \exp(-\delta^2 u^2) \tag{5.84m}$$

or

$$u = \frac{[-\ln(1 - R_n)]^{1/2}}{\delta} \tag{5.84n}$$

Also, we note from Eq. (5.84m) that $\theta$ is uniformly distributed between 0 and $2\pi$, and therefore $\theta = 2\pi R_n'$ according to Eq. (5.84h), where $R_n'$ is a random number such that $0 \leq R_n' \leq 1$. By selecting the random numbers $R_n$ and $R_n'$, the values of $u$ and $\theta$ can now be calculated, and then $v_y$ and $v_z$ can be obtained from the expressions, $v_y = u \cos\theta$ and $v_z = u \sin\theta$. An efficient technique to sample complicated distribution functions can be found in Metropolis et al. (1953).

ACCEPTANCE–REJECTION METHOD: In Eq. (5.84j), we have shown that $u$ cannot be expressed in terms of $R_n$. In such cases, or when the integral in Eq. (5.84e) cannot be evaluated analytically, the acceptance–rejection method provides a convenient way of sampling from such complicated functions. This method involves the following steps:

STEP 1: Let $f(x)$ be a normalized distribution function of the variate $x$, where the value of $x$ ranges from $a$ to $b$, that is, $a \leq x \leq b$. Also, let $f_{max}$ be the maximum value of $f(x)$. Define the ratio

$$f^*(x) = f(x)/f_{max} \tag{5.84o}$$

STEP 2: Assume that $x$ is uniformly distributed between $a$ and $b$, and use Eq. (5.84h) to determine $x$. Let this value of $x$ be $x'$.

STEP 3: Determine $f^*(x')$, and select a random number $R'_n$, where $0 \leq R'_n \leq 1$.

STEP 4: Accept $x'$ as a representative value of $x$ if $f^*(x') > R'_n$, and reject $x'$ if $f^*(x') < R'_n$.

EXAMPLE

Consider the distribution function given by Eq. (5.84i), where $f_{max} = \delta/\sqrt{\pi}$, and $f^*(u) = \exp(-\delta^2 u^2)$. Although, $-\infty < u < \infty$, we will assume that $-3/\delta \leq u \leq 3/\delta$, because the fraction of values lying outside this range is $1 - \text{erf}(3) = 0.000022$, which is very small. Therefore, we have $a = -3/\delta$ and $b = 3/\delta$, and according to Steps 2 and 3,

$$u' = -(3 + 6R_n)/\delta$$
$$f^*(u') = \exp[-(-3 + 6R_n)^2]$$

If $R'_n$ represents another random number, $u'$ is accepted if $f^*(u') > R'_n$, and $u'$ is rejected if $f^*(u') < R'_n$.

## 5.10. MULTIPHOTON PROCESSES

Two important features of the models that have been discussed so far are that the models require the value of the dissociation cross section, and that the photochemical reactions have been assumed to occur due to single-photon absorption phenomena. However, in many cases, photochemical reactions occur due to the absorption of two or more photons [Karny et al. (1978), Duncan et al. (1979), Gerrity et al. (1980), Fisanick et al. (1981), Fuchs et al. (1988), Kato and Takeuchi (1992)], and in such cases, the reaction rate is not linearly proportional to the laser intensity. In multiphoton dissociation, the dissociation probability $\sigma(r,z)$ depends on the laser fluence $\phi(r,z)$, and the spatial variation of the fluence plays an important role in modeling the overall reaction yields, or in determining the dissociation probability from experimental data through deconvolution. Kato and Takeuchi (1992) have used the following analytical method to model the reaction yields due to the multiphoton dissociation.

DISSOCIATION PROBABILITY

In multiphoton dissociation, $\sigma(r,z)$ is assumed to depend on $\phi(r,z)$ according to the following power-law model:

$$\sigma(r,z) = (\phi(r,z)/\phi_c)^a \quad \text{for} \quad \phi(r,z) < \phi_c \quad (5.85a)$$
$$= 1 \quad \text{for} \quad \phi(r,z) \geq \phi_c \quad (5.85b)$$

The reaction volume, that is, the volume of the gas where photochemical reaction occurs, is given by

$$V_r = \int_{V_0} \sigma(r, z) dV \qquad (5.85c)$$

These expressions are used below to determine the reaction yields.

LINEAR (SINGLE-PHOTON) ABSORPTION

We assume that the intensity of the laser beam is radially uniform so that we can write $\phi(z) = \phi(r, z)$, and that the Beer–Lambert law is applicable in the linear absorption regime, that is,

$$-d\phi/dz = n\sigma_A \phi$$

or

$$\phi(z) = \phi(0) \exp(-n\sigma_A z) \qquad (5.85d)$$

Let us define the dimensionless fluence $\phi_1$ and reaction volume $V_1$ by the following expressions:

$$\phi_1 = \phi(0)/\phi_c \qquad (5.85e)$$

$$V_1 = V_r/V_f \qquad (5.85f)$$

where the laser beam focal volume $V_f$ is defined as

$$V_f = \pi r_0^2 / n\sigma_A \qquad (5.85g)$$

Since the transmittance $T$ can be measured experimentally, we will express $n\sigma_A$ in terms of $T$ by noting that

$$T = \frac{\phi(0) \exp(-n\sigma_A L)}{\phi(0)}$$

that is

$$n\sigma_A = \frac{\ln(1/T)}{L} \qquad (5.85h)$$

and therefore, $V_f$ is given by

$$V_f = \frac{\pi r_0^2 L}{\ln(1/T)} \qquad (5.85i)$$

Noting that $\phi(r, z) = \phi(z)$, we will encounter three situations depending on the values of $\phi(0)$, $\phi(L)$, and $\phi_c$:

PHOTOLYTIC LCVD MODELING    373

SITUATION I: $\phi(0) < \phi_c$, that is, $\phi_1 < 1$. In this situation, Eq. (5.85a) holds good over the entire length of the laser beam $L$, that is, in the range $0 \leq z \leq L$. Thus, $V_1$ can be written as

$$V_1 = \frac{1}{V_f} \int_0^L (\phi(z)/\phi_c)^a \pi r_0^2 dz = \frac{\ln(1/T)}{L} \phi_1^a \int_0^L \exp\left[-\frac{\ln(1/T)}{L} za\right] dz$$

$$= \frac{\phi_1^a}{a}(1 - T^a) \qquad (5.85j)$$

SITUATION II: $\phi(0) \geq \phi_c > \phi(L)$, that is, $\phi(0) \geq \phi_c > \phi(0)\exp(-n\sigma_A L)$, or $1 \leq \phi_1 < 1/T$. Let us assume that $\phi(z) \geq \phi_c$ in the region $0 \leq z \leq z_1$, and $\phi(z) < \phi_c$ in the region $z_1 \leq z \leq L$, such that $\phi(z) = \phi_c$ at $z = z_1$, where $z_1$ is a point in the region $0 \leq z_1 \leq L$. Therefore, Eq. (5.85c) can be written as

$$V_r = \pi r_0^2 \left[\int_0^{z_1} \sigma(r,z)dz + \int_{z_1}^L \sigma(r,z)dz\right]$$

$$= \pi r_0^2 \left[\int_0^{z_1} dz + \int_{z_1}^L \phi_1^a \exp(-n\sigma_A za)dz\right]$$

$$= \pi r_0^2 \left[z_1 + \phi_1^a \frac{\exp(-n\sigma_A z_1 a) - \exp(-n\sigma_A La)}{n\sigma_A a}\right] \qquad (5.85k)$$

However, but we note that $\phi(z_1) = \phi_c = \phi(0)\exp(-n\sigma_A z_1)$, which implies that $\exp(-n\sigma_A z_1 a) = (1/\phi_1)^a$, and $z_1 = (\ln \phi_1)/(n\sigma_A)$. Also, we have $T = \exp[-n\sigma_A L]$, and $n\sigma_A = \ln(1/T)/L$. Combining these relationships with Eqs. (5.85f and k), we obtain

$$V_1 = \ln \phi_1 + \frac{1}{a}(1 - T^a \phi^a) \qquad (5.85l)$$

SITUATION III: $\phi(0) > \phi(L) \geq \phi_c$, that is, $1/T \leq \phi_1$. In this situation, Eq. (5.85b) holds good over the entire length of the laser beam $L$. Therefore, $V_1$ can be expressed as

$$V_1 = \frac{\ln(1/T)}{\pi r_0^2 L} \int_0^L \pi r_0^2 dz = \ln(1/T) \qquad (5.85m)$$

Eq. (5.85j, l, or m) can be used to determine the reaction yield.

### Nonlinear (Multiphoton) Absorption

As in the case of single-photon absorption, we assume that the intensity of the laser beam is radially uniform so that we can write $\phi(r, z) = \phi(z)$. When multiphoton absorption occurs, the Beer–Lambert law does not hold good, and the fluence is assumed to vary with $z$ according to the following expression:

$$-\frac{d\phi}{dz} = \Lambda \phi^m \tag{5.86a}$$

where $m < 1$, and $\Lambda$ is the absorption coefficient. It should be noted that $\Lambda = n\sigma_A(n)$, where the absorption cross section $\sigma_A(n)$ depends on the number density of the reactant molecules $n$, in the case of multiphoton absorption processes. Lyman et al. (1986) have shown that $m = \frac{2}{3}$ for many molecules under several experimental conditions. The solution of Eq. (5.86a) yields

$$\phi^{1-m}(z) = \phi^{1-m}(0) - (1 - m)\Lambda z \tag{5.86b}$$

The transmittance $T$ is given by

$$T = \frac{\phi(L)}{\phi(0)} = [1 - (1 - m)\Lambda L \phi^{m-1}(0)]^{1/(1-m)} \tag{5.86c}$$

As before, we define the dimensionless fluence $\phi_2$ and reaction volume $V_2$ by the following expressions:

$$\phi_2 = \phi(0)/\phi_c \tag{5.86d}$$

$$V_2 = V_r/V_f \tag{5.86e}$$

where the laser beam focal volume is given by

$$V_f = \frac{\pi r_0^2}{\Lambda \phi^{m-1}(0)} = \frac{\pi r_0^2 L(1 - m)}{1 - T^{1-m}}. \tag{5.86f}$$

Now $V_2$ can be determined for the following three situations:

SITUATION I: $\phi(0) < \phi_c$, that is, $\phi_2 < 1$. In this situation, we can use Eqs. (5.85a, c, 5.86b, d, e, and f) to obtain

$$V_2 = \frac{\Lambda \phi^{m-1}(0)}{\pi r_0^2} \int_0^L \frac{1}{\phi_c^a} [\phi^{1-m}(0) - (1-m)\Lambda z]^{a'(1-m)} \pi e_0^2 dz$$

$$= \Lambda \phi^{m-1}(0) \phi_2^a \int_0^L \left[1 - \frac{1-m}{\phi^{1-m}(0)} \Lambda z\right]^{a'(1-m)} dz$$

$$= \frac{\phi_2^a}{1-m+a} (1 - T^{1-m+a}) \qquad (5.86g)$$

SITUATION II: $\phi(0) \geq \phi_c > \phi(L)$, that is, $1 \leq \phi_2 < 1/T$. As in the case of single-photon absorption, we can write

$$V_r = \int_{V_0} \sigma(r,z) dV = \int_0^{z_1} \sigma \pi r_0^2 dz + \int_{z_1}^L \sigma \pi r_0^2 dz$$

$$= \pi r_0^2 \left[z_1 + \int_{z_1}^L \phi_2^a \left(1 - \frac{1-m}{\phi^{1-m}(0)} \Lambda z\right)^{a/(1-m)} dz\right] \qquad (5.86h)$$

Noting that

$$\phi^{1-m}(z_1) = \phi_c^{1-m} = \phi^{1-m}(0) - (1-m)\Lambda z_1$$

that is,

$$\phi_2^{m-1} = \frac{1 - (1-m)\Lambda z_1}{\phi^{1-m}(0)},$$

or

$$z_1 = \frac{(1 - \phi_2^{m-1})\phi^{1-m}(0)}{\Lambda(1-m)},$$

we obtain

$$V_2 = \frac{V_r}{V_f} = \frac{\Lambda \phi^{m-1}(0)}{\pi r_0^2} V_r$$

$$= \frac{1}{1-m+a} (\phi_2^{m-1} - \phi_2^a T^{1-m+a}) + \frac{1}{1-m} (1 - \phi_2^{m-1}) \qquad (5.86i)$$

SITUATION III: $\phi(0) > \phi(L) \geqslant \phi_c$, that is, $1/T \leqslant \phi_2$. In this situation, Eq. (5.85b) holds good over the entire length of the laser beam $L$. Equations (5.85b, c, 5.86e and f) yield

$$V_2 = \frac{V_r}{V_f} = \frac{1 - T^{1-m}}{\pi r_0^2 L(1-m)} \int_0^L \pi r_0^2 \, dz$$

$$= \frac{1 - T^{1-m}}{1 - m} \tag{5.86j}$$

Equations (5.86g, i, or j) can be used to determine the reaction yield. For a Gaussian laser beam, where the intensity of the beam is not radially uniform, expressions for the reaction volume can be found in Kato and Takeuchi (1992).

## NOMENCLATURE

| | |
|---|---|
| $A(p_r, \lambda)$ | Absorption of laser by the physisorbed film as a function of $p_r$ and $\lambda$ |
| $A_0(\lambda)$ | Absorption of laser by a single monolayer of physisorbed species |
| $A_c(\lambda)$ | Laser absorption factor for chemisorbed phase decomposition $\{I(r,z) = I(r,0) \exp[-A_c(\lambda)()]\}$ |
| $A_v$ | Avogadro's number |
| $b$ | Distance of the point of interest from the origin along $x$-axis on the substrate surface |
| $D$ | Diffusion coefficient of the atoms of interest (in Section 5.3) |
| $D$ | Diffusion coefficient of the reactant molecules (in Section 5.5) |
| $D_s$ | Surface diffusivity of the precursor molecules |
| $D_{i1s}$ | Diffusion coefficient of the $i$th species in the solid phase in region 1 |
| $D_{ir}$ | Diffusion coefficient in the gas phase for the $i$th species in the $r$th region; $D_{ir}$ is a function of the concentration $N_{1r}, N_{2r}, N_{3r}, N_{4r}, \ldots, N_{i*r}$, and temperature $T$, that is, $D_{ir} \equiv D_{ir}(N_{1r}, N_{2r}, \ldots, N_{i*r}, T)$, where $i*$ is the total number of species inside the deposition chamber |
| Dm | Damkohler number, the ratio of the reaction speed to the diffusive speed $(\sigma I^* w_0 / D / w_0)$. |
| $E$ | Energy of a photon of the laser beam |
| $H$ | Height of region 1 or 2 in Fig. 4.1a |
| $H_1(t)$ | Location of the film–gas interface near $z = H$ in region 1 |
| $H_2(t)$ | Location of the film–gas interface near $z = H$ in region 2 |
| $i$ | Imaginary number |
| $\hat{i}$ | Unit vector along the $x$-axis |

|  |  |
|---|---|
| $I(\mathbf{r})$ | Laser intensity at any position **r** (number of photons per unit area per unit time) |
| $I(\mathbf{r}, t)$ | Time-dependent $I(\mathbf{r})$ |
| $I^*$ | Uniform laser intensity (number of photons per unit area per unit time) |
| $I_0$ | Laser intensity at $r = 0$ (number of photons per unit area per unit time) |
| $\bar{I}$ | Uniform intensity within the diameter $\sqrt{2}w_0$ (number of photons per unit area per unit time) |
| $\hat{j}$ | Unit vector along the $y$-axis |
| $j_d$ | Diffusive flux of the atoms of interest at the surface of the reaction zone (surface of the cylinder in Fig. 5.1) |
| $j_f$ | Particle flux (normal component of the number of the atoms of interest impinging per unit time per unit area of the substrate surface) due to the gas-phase dissociation |
| $j_{f_{\text{ball}}}$ | Particle flux $j_f$ due to ballistic motion |
| $j_{f_{\text{diff}}}$ | Particle flux $j_f$ due to diffusive motion |
| $j_{f_a}$ | Film growth rate (number of atoms deposited per unit area per unit time) due to the decomposition of adsorbed molecules |
| $j_p$ | Volumetric production rate (number of atoms of interest per unit volume per unit time) |
| $j_s$ | Surface source term (number of the atoms of interest crossing a unit area per unit time in all possible directions at the surface of the reaction zone) |
| $J^*$ | Surface collision rate (number of molecules impinging on the surface per unit time per unit surface area) |
| $J_p^*$ | Number of molecules physisorbed per unit time per unit surface area |
| $\hat{k}$ | Unit vector along the $z$-axis |
| $k_a$ | Adsorption rate constant |
| $k_B$ | Boltzmann constant |
| $k_d$ | Desorption rate constant |
| $l$ | Characteristic distance traveled by a metal atom in the gas phase before it is removed by a gas-phase reaction |
| $L$ | Length of the laser beam in the irradiated gas volume (in Section 5.10) |
| $m$ | Mass of one molecule of the precursor |
| $M$ | Molecular weight of the film material (desired product species) |
| $M_p$ | Molecular weight of the precursor |
| $n$ | Number density of the atoms of interest (number of the atoms of interest per unit volume) (in Section 5.3) |
| $n$ | Number density of the reactant molecules at any time $t$ (in Sections 5.5, 5.7, and 5.10) |

| | |
|---|---|
| $\hat{n}$ | Unit vector normal to the film-gas interface |
| $n_0$ | Initial density of the adsorbed precursor molecules before the laser was turned on |
| $n_0$ | Number density of the reactant molecules (number of reactant molecules per unit volume) (in Sections 5.3 and 5.4) |
| $n_0$ | Number density of the reactant molecules at time $t = 0$ (in Section 5.5) |
| $n_1$ | Number density of the reactant molecules in region I, $0 \leqslant w_0$ (Fig. 5.3) |
| $n_2$ | Number density of the reactant molecules in region II, $w_0 \leqslant r < \infty$ or $w_0 \leqslant r \leqslant R_i$ (Fig. 5.3) |
| $n_a$ | Density of adsorbed precursor molecules (number of adsorbed parent molecules per unit surface area) |
| $n_p$ | Number density of the product species due to the gas-phase decomposition |
| $n_{p_a}$ | Density of the product species due to the adsorbed phase decomposition (number of product species generated by the adsorbed phase reaction per unit surface area) |
| $n_{p_c}$ | Density of the product species due to chemisorbed phase decomposition (number of product species generated by the chemisorbed phase reaction per unit surface area) |
| $n_{p_p}$ | Density of the product species due to physisorbed phase decomposition (number of product species generated by the physisorbed phase reaction per unit surface area) |
| $N_{Ar}$ | Number density of the reactant species, $A$ in the $r$th region [see Eq. (5.4)] |
| $N_c$ | Density of chemisorbed species (number of chemisorbed species per unit surface area) |
| $N_{i1f}$ | Number density of the $i$th species in the film in region 1 |
| $N_{i1s}$ | Number density of the $i$th species in the solid phase in region 1 |
| $N_{ir}$ | Number density (number of the $i$th species per unit volume) of the $i$th species in the $r$th region (see Fig. 4.1a) |
| $N_{irj}$ | Number density of the $i$th species at the $j$th location in the $r$th region; $r = 1$ and $j = H$ refer to the location $z = H$; $j = 0, j = H$, and $j = r$ refer to the locations $z = 0$, $z = H$, and $r = r_2$, respectively, when $r = 2$ |
| $N_p$ | Density of physisorbed species (number of physisorbed species per unit surface area) |
| $N_s$ | Density of total (occupied and unoccupied) adsorption sites at the surface or density of one complete monolayer (number of adsorption sites per unit surface area before the occurrence of any adsorption) |

| | |
|---|---|
| $N_{ir}^*$ | Distribution of the number density of the $i$th species in the $r$th region at time $t = 0$. |
| $N_s^*$ | Density of unoccupied adsorption sites at any time $t$ |
| $p_0$ | Vapor pressure of the precursor molecules at the deposition temperature |
| $p_e$ | Probability that a physisorbed molecule will evaporate per unit time |
| $p_r$ | Partial pressure of the reactant (precursor) molecules inside the deposition chamber |
| $P_t$ | Total laser power |
| $P_0$ | Ambient pressure inside the deposition chamber |
| $r$ | Radical variable in cylindrical coordinates |
| $r_0$ | A length scale in the hemispheric model ($r_0 = 2D/\eta u$) |
| $r_0$ | Radius of the laser beam (in Section 5.10) |
| $r_1$ | Radius of region 1 as well as the radius of the interface of regions 1 and 2 |
| $r_2$ | Radius of the outer boundary of region 2; region 2 is considered to be cylindrical |
| $\dot{r}_{irs}$ | Number of the $i$th species produced per unit area per unit time at the solid interface in the $r$th region |
| $R_1$ | Distance between the point of interest on the substrate surface and a differential element |
| $\hat{R}_1$ | Unit vector along the distance $R_1$, pointing from the point of interest on the substrate surface to a differential element |
| $R_2(t)$ | Location of the film–gas interface near $r = r_2$ in region 2 (in Section 5.2) |
| $R_i$ | Radius of the inner wall of the deposition chamber |
| $s$ | Laplace transform variable [see Eqs. (5.44a, b)] |
| $S_{pir}$ | Number of the $i$th species produced per unit volume per unit time in the $r$th region |
| $t$ | Time variable |
| $T$ | Laser beam transmittance (in Section 5.10) |
| $T$ | Temperature of the gas phase inside the deposition chamber |
| $u$ | Root mean square velocity of the reactant molecules (in Section 5.5) |
| $\bar{v}$ | Average velocity |
| $V_0$ | Volume of the irradiated gas (in Section 5.10) |
| $V_f$ | Laser beam focal volume (volume of the laser beam over the length that decreases the fluence to $1/e$) |
| $V_r$ | Reaction volume (volume of the gas where photochemical reaction occurs) |
| $w_0$ | Radius of the laser beam at its waist |

$z$            Axial variable in cylindrical coordinates
$Z(t)$       Location of the film–gas interface near the $z = 0$ plane in region 1.
$Z_0(t)$      Location of the film–gas interface near $z = 0$ in region 2.

GREEK SYMBOLS

$\gamma$            Sticking coefficient
$\gamma_c$           Chemisorption sticking coefficient
$\gamma_{irj}$        Sticking coefficient of the $i$th species at the $j$th film–gas interface in the $r$th region. $j = 0$ and $j = H$ refer to $z = Z(t)$ and $z = H_1(t)$ locations, respectively, when $r = 1$. $j = 0, j = H$, and $j = r$ refer to $z = Z_0(t)$, $z = H_2(t)$, and $r = R_2(t)$ locations, respectively, when $r = 2$
$\delta(\alpha)$        Dirac delta function ($\alpha = \xi, \eta, \zeta, (x - x'), (y - y'), (z - z')$, etc.)
$\eta$            Collisional reaction efficiency, that is, the fraction of surface collisions that induce the chemical reaction
$\theta$            Angle between the vector that connects the point of interest to the differential element, and the normal to the substrate surface at the point of interest (throughout chapter except Section 5.2)
$\theta$            Angular variable in cylindrical coordinates (in Section 5.2)
$\theta_p^*$          Surface coverage due to physisorption under equilibrium condition
$\theta_c$           Surface coverage due to chemisorption ($\theta_c = N_c/N_s$)
$\theta_p$           Surface coverage due to chemisorption ($\theta_p = N_p/N_s$)
$\lambda$            Laser wavelength
$\bar{\lambda}$            Mean free path of the product species
$\rho$            Density of the film (throughout chapter except Section 5.7.3)
$\rho_l$           Density of the precursor in its liquid state
$\sigma$            Gas-phase dissociation cross section in units of area per unit molecule
$\sigma(r, z)$    Dissociation probability (in Section 5.10)
$\sigma_A$           Photon absorption cross section of species $A$ in units of area per unit of $A$
$\sigma_a$           Adsorbed phase dissociation cross section in units of area per unit molecule
$\sigma_c(\lambda)$      Absorption cross section of the precursor in the chemisorbed phase
$\sigma_p(\lambda)$      Absorption cross section of the precursor in the physisorbed phase
$\tau_c$           Time constant to produce a chemisorbed film
$\phi$            Azimuthal angle (angular variable in the plane of the substrate surface)
$\phi(0)$       Fluence at $z = 0$, $\phi(0) = \phi(r, 0)$

$\phi(r, z)$     Laser fluence, $\phi = I(r, t)t$, (in Section 5.10)
$\phi_0(r, t)$     Photon flux (number of photons passing per unit area per unit time) at the bottom surface of the window in Fig. 4.1a, at any radial position $r$ and time $t$
$\phi_c$     Critical fluence for complete dissociation
$\phi_{max}$     Maximum azimuthal view angle (see Fig. 5.1) at the point of interest on the substrate surface
$\Phi_r$     Quantum efficiency or quantum yield
$\psi$     Photon flux vector (number of photons passing per unit area per unit time in a particular direction)

## REFERENCES

Bellman, R., Kalaba, R. E., and Lockett, J. A. (1966), *Numerical Inversion of the Laplace Transform: Applications to Biology, Economics, Engineering, and Physics,* American Elsevier, New York.

Bilenchi, R., Gianinoni, I., Musci, M. (1982), *J. Appl. Phys.* **53**, 6479.

Bird, G. A. (1976), *Molecular Gas Dynamics,* Clarendon, Oxford.

Bird, R. B., Stewart, W. E., and Lightfoot, E. N. (1960), *Transport Phenomena,* Wiley, New York.

Baunauer, S., Emmett, P. H., and Teller, E. (1938), *J. Am. Chem. Soc.* **60**, 309.

Byrd, P. F., and Friedman, M. D. (1954), *Handbook of Elliptic Integrals for Engineers and Physicists,* Springer-Verlag, Berlin, pp. 9, 176, 297.

Carnahan, B., Luther, H. A., and Wilkes, J. O. (1969), *Applied Numerical Methods,* Wiley, New York, pp. 69, 342, 433.

Carslaw, H. S., and Jaeger, J. C. (1959), *Conduction of Heat in Solids,* 2nd Ed., Clarendon, Oxford, p. 494.

Chandrasekhar, S. (1943), Stochastic Problems in Physics and Astronomy, *Rev. Mod. Phys.* **15**, 1–89.

Chen, C. J. (1987), Kinetic Theory of Laser Photochemical Deposition, *J. Vac. Sci. Technol.* **A5**, 3386.

Chen, C. J., and Osgood Jr., R. M. (1983), *Chem. Phys. Lett.* **98**, 363.

Coronell, D. G., and Jensen, K. F. (1992), Analysis of Transition Regime Flows in Low Pressure Chemical Vapor Deposition Reactors using the Direct Simulation Monte Carlo Method, *J. Electrochem. Soc.* **139**, 2264.

Duncan, M. A., Dietz, T. G., and Smalley, R. E. (1979), Efficient Multiphoton Ionization of Metal Carbonyls Cooled in a Pulsed Supersonic Beam, *Chem Phys.* **44**, 415.

Ehrlich, D. J., Osgood, R. M., and Deutsch, T. F. (1980), *IEEE J. Quant. Electron.* **QE-16**, 1233.

Ehrlich, D. J., and Osgood, R. M. (1981), *Chem. Phys. Lett.* **79**, 381.

Ehrlich, D. J., and Tsao, J. Y. (1983), A Review of Laser Microchemical Processing, *J. Vac. Sci. Technol.* **B1**, 969.

Fisanick, G. J., Gedanken, A., Eichelberger, IV, T. S., Kuebler, N. A., and Robin, M. B. (1981), Multiphoton Ionization Spectroscopy of Organometallics: The $Cr(CO)_6$, $Cr(CO)_3C_6H_6$, $Cr(C_6H_6)_2$ Series, *J. Chem. Phys.* **75**, 5215.

Flint, J. H., Meunier, M., Adler, D., and Haggerty, J. S. (1984), a-Si:H Films Produced from Laser-Heated Gases: Process Characteristics and Film Properties, in: *Laser-Assisted Deposition, Etching, and Doping*, SPIE Proc. Vol. 459, S. D. Allen, ed., SPIE—The International Society for Optical Engineering, Washington, pp. 66–70.

Foord, J. S., and Jackson, J. B. (1986), *Surf. Sci.* **171**, 197.

Freeman, D. L., and Doll, J. D. (1983), *J. Chem. Phys.* **78**, 6002.

Fuchs, C., Boch, E., Fogarassy, E., Aka, B., and Siffert, P. (1988), Two-Photon Absorption Cross-Section for Silane under Pulsed ArF (193 nm) Excimer Laser Irradiation, in: *Laser and Particle Beam Chemical Processing for Microelectronics, Materials Research Soc. Symp. Proc.*, Vol. 101, Ehrlich, D. J., Higashi, G. S., and Oprysko, M. M., eds., Materials Research Society, Pittsburgh, pp. 361–365.

Gattuso, T. R., Meunier, M., Adler, D., and Haggerty, J. S. (1983), IR Laser-Induced Deposition of Silicon Thin Films, in: *Laser Diagnostics and Photochemical Processing for Semiconductor Devices*, Materials Research Soc. Symp. Proc., Vol. 17, Osgood, R. M., Brueck, S. R. J., and Schlossberg, H. R., eds., North-Holland, Amsterdam, pp. 215–222.

Gerrity, D. P., Rothberg, L. J., and Vaida, V. (1980), Multiphoton Ionization of Metal Atoms Produced in the Photodissociation of Group VI Hexacarbonyls, *Chem. Phys. Lett.* **74**, 1.

Gradshteyn, I. S., and Ryzhik, I. M. (1980), *Table of Integrals, Series, and Products*, Academic, New York, 66.

Guest, P. G. (1961), The Solid Angle Subtended by a Cylinder, *Rev. Sci. Instr.* **32**, 164.

Ho, W. (1988), Comments, *Cond. Mat. Phys.* **13**, 293.

Kar, A., and Mazumder, J. (1988), One-Dimensional Finite-Medium Diffusion Model for Extended Solid Solution in Laser Cladding of Hf on Nickel, *Acta. Metal.* **36**, 701.

Karny, Z., Naaman, R., and Zare, R. N. (1978), Production of Excited Metal Atoms by UV Multiphoton Dissociation of Metal Alkyl and Metal Carbonyl Compounds, *Chem. Phys. Lett.* **59**, 33.

Kato, S., and Takeuchi, K. (1992), Infrared Multiphoton Dissociation by an Unfocussed Beam in an Optically Thick Medium: An Analytical Method for Reaction Yields, *Appl. Opt.* **31**, 2825.

Krchnavek, R. K., Gilgen, H. H., and Chen, J. C., Shaw, P. S., Lieata, T. J., and Osgood, Jr., R. M. (1987), *J. Vac. Sci. Technol.* **B5**, 20.

Lyman, J. L., Quigley, G. P., and Judd, O. P. (1986), Single-Infrared-Frequency Studies of Multiple-Photon Excitation and Dissociation of Polyatomic Molecules, in: *Multiple-Photon Excitation and Dissociation of Polyatomic Molecules*, Cantrell, C. D., ed., Vol. 35 of Topics in Current Physics, Springer-Verlag, Berlin, pp. 9–94.

Masket, A. V. (1957), Solid Angle Contour Integrals, Series, and Tables, *Rev. Sci. Instr.* **28**, 191.

Mayer, J. E., and Mayer, M. G. (1940), *Statistical Mechanics*, Wiley, New York.

Metropolis, N., Rosenbluth, A. W., Rosenbluth, M. N., Teller, A. H., and Teller, E. (1953), Equation of State Calculations by Fast Computing Machines, *J. Chem. Phys.* **21**, 1087.

Meunier, M., Gattuso, T. R., Adler, D., Haggerty, J. S. (1983), *Appl. Phys. Lett.* **43**, 273.

Morgan, K. Z., and Emerson, L. C. (1967), Dose from Extended Sources of Radiation, in: *Principles of Radiation Protection: A Textbook of Health Physics*, Morgan, K. G., and Turner, J. E., eds., Wiley, New York, pp., 268–300.

Özisik, M. N., and Murray, R. L. (1974), On the Solution of Linear Diffusion Problems with Variable Boundary Parameters, *ASME J. Heat Transfer*, **96C**, 48.

Roach, G. F. (1982), *Green's Functions*, 2nd Ed., Cambridge University Press, London.

Rockwell, III, T., ed. (1956), *Reactor Shielding Design Manual*, Van Nostrand, New York, pp. 400–404.

Siegel, R., and Howell, J. R. (1981), *Thermal Radiation Heat Transfer*, McGraw-Hill, New York.

Sneddon, I. N. (1972), *The Use of Integral Transforms*, McGraw-Hill, New York.

Tsao, J. Y., and Ehrlich, J. (1984a), Recent Advances in UV Laser Photodeposition, in: *Laser-Controlled Chemical Processing of Surfaces*, Materials Research Soc. Symp. Proc., Vol. 29, Johnson, A. W., Ehrlich, D. J., and Schlossberg, H. R., eds., North-Holland, Amsterdam, pp. 115–126.

Tsao, J. Y., and Ehrlich, D. J. (1984b), Surface and Gas Processes in Photodeposition in Small Zones, in: *Laser-Assisted Deposition, Etching, and Doping*, SPIE Proc. Vol. 459, S. D. Allen, ed., SPIE—The International Society for Optical Engineering, Washington, pp. 2–8.

Tsao, J. Y., Zeiger, H. J., and Ehrlich, D. J. (1985), Measurement of Surface Diffusion by Laser-Beam-Localized Surface Photochemistry, *Surf. Sci.* **160**, 419.

West, G. A., and Gupta, A. (1984), Laser-Induced Chemical Vapor Deposition of Silicon Nitride Films, in: *Laser-Controlled Chemical Processing of Surfaces*, Materials Research Soc. Symp. Proc., Vol. 29, Johnson, A. W., Ehrlich, D. J., and Schlossberg, H. R., eds., North-Holland, Amsterdam, pp. 61–66.

Wood, T. H., White, J. C., and Thacker, B. A. (1983a), Ultraviolet Photodecomposition for Metal Deposition: Gas Versus Surface Phase Processes, *Appl. Phys. Lett.* **42**, 408.

Wood, T. H., White, J. C., and Thacker, B. A. (1983b), UV Photodecomposition for Metal Deposition: Gas vs. Surface Phase Processes, in: *Laser Diagnostics and Photochemical Processing for Semiconductor Devices*, Materials Research Soc. Symp. Proc., Vol. 17, Osgood, R. M., Brueck, S. R. J., and Schlossberg, H. R., eds., North-Holland, Amsterdam, pp. 35–41.

Yardley, J. T., Gitlin, B., Nathanson, G., and Rosan, A. M. (1981), Fragmentation and Molecular Dynamics in the Laser Photodissociation of Iron Pentacarbonyl, *J. Chem. Phys.* **74**, 370.

Yener, Y., and Özisik, M. N. (1974), On the Solution of Unsteady Conduction in Multiregion Finite Media with Time Dependent Heat Transfer Coefficient, *Proc. 5th International Heat Transfer Conference,* Vol. I, American Institute Chemical Engineers, New York, pp. 188–192.

Young, D. M., and Crowell, A. D. (1962), *Physical Adsorption of Gases*, Butterworth, London.

Zeiger, H. J., Tsao, J. Y., and Ehrlich, D. J. (1985), Technique for Measuring Surface Diffusion by Laser-Beam-Localized Surface Photochemistry, *J. Vac. Sci. Technol.* **B3**, 1436.

Zeiger, H. J., and Ehrlich, D. J. (1989), Lateral Confinement of Microchemical Surface Reactions: Effects on Mass Diffusion and Kinetics, *J. Vac. Sci. Technol.* **B7**, 466.

Zeiger, H. J., Ehrlich, D. J., and Tsao, J. Y. (1989), Transport and Kinetics, in: *Laser Microfabrication: Thin Film Processes and Lithography*, Ehrlich, D. J., and Tsao, J. Y., eds., Academic, New York, pp. 285–330.

Zeiri, Y., Atzmony, U., and Bloch, J. (1991), Monte Carlo Simulation of Laser Induced Chemical Vapor Deposition, *J. Appl. Phys.* **69**, 4110.

# APPENDIX

## A.1. DEFINITIONS OF ENERGY DENSITY, IRRADIANCE, AND INTENSITY

ENERGY DENSITY $W$ refers to the amount of energy per unit volume, that is, $J\,m^{-3}$.

IRRADIANCE $\bar{I}$ refers to the amount of energy flowing per unit area per unit time, that is, $J\,m^{-2}\,s^{-1}$ [Allmen (1987), Eckbreth (1988)].

INTENSITY $\bar{I}^*$ refers to the amount of energy flowing per unit area per unit time per unit solid angle, that is, $J\,m^{-2}\,s^{-1}\,sr^{-1}$ [Allmen (1987)].

THE SYMBOLS $W_v$, $\bar{I}_v$, and $I_v^*$ refer, respectively, to $W$, $\bar{I}$, and $I^*$ per unit oscillation frequency interval. The units of $W_v$, $\bar{I}_v$, and $I_v^*$ are $J\,m^{-3}\,Hz^{-1}$, $J\,m^{-2}\,s^{-1}\,Hz^{-1}$, and $J\,m^{-2}\,s^{-1}\,sr^{-1}\,Hz^{-1}$, respectively.

THE SYMBOLS $W_\omega$, $\bar{I}_\omega$, and $I_\omega^*$ refer, respectively, to $W$, $\bar{I}$, and $I^*$ per unit interval of the cylic or angular frequency of oscillation. Here $\omega = 2\pi v$, and the units of $W_\omega$, $\bar{I}_\omega$, and $I_\omega^*$ are $J\,m^{-3}\,s$, $J\,m^{-2}\,s^{-1}\,s$, and $J\,m^{-2}\,s^{-1}\,sr^{-1}\,s$, respectively.

It should be noted that the oscillation frequency $v$ and the cyclic or angular frequency $\omega = 2\pi v$ of oscillations are expressed [Matveev (1986)] in units of Hz and $s^{-1}$, respectively. Actually, $\omega$ should be expressed in units of $rad\,s^{-1}$. The expressions for the radiation absorption and emission rates (see Table 3.9) may be written by using any one of the above nine variables, and the units and magnitudes of the rate constants will vary depending on which variable is used in the rate expressions.

Since Einstein (1916, 1967, 1972) derived the rate expressions in terms of radiation density of frequency $v$, we have denoted the Einstein $B$ coefficients, such as $B_{lu}$ and $B_{ul}$, as those rate constants that appear in the rate expressions written in terms of $W_v$ as shown in Table 3.9. A similar convention for the Einstein coefficients has been used by Eastham (1989). However, Shimoda (1983) has written the rate expressions in terms of $W_\omega$ and defined the associated rate constants as the Einstein coefficients. Eckbreth (1988) has used the following absorption rate expression:

$$\text{Absorption rate} = \frac{B_{lu}\bar{I}_v}{c} N_1$$

$$= \frac{B_{lu}W_v c}{c} N_1, \quad \text{since } \bar{I}_v = W_v c$$

$$= B_{lu}W_v N_1$$

which is same as the expression given in Table 3.9. Calvert and Pitts (1966) have expressed the absorption rate in terms of intensity, and pointed out that the rate constant that is associated with their absorption rate expression is not the original Einstein coefficient.

As the terms irradiance and intensity are used interchangeably in the literature of laser-aided materials processing, the term intensity is used in place of irradiance in many places in this book.

## A.2. THERMAL STRESS ANALYSIS

In pyrolytic LCVD, the temperature is very high at the center and very low at the edge of the film, especially for a Gaussian laser beam. Due to this large temperture variation, the film will experience thermal stresses that can crack it. When there is relative motion between the laser beam and substrate, the film is usually deposited in the form of a stripe while for a stationary laser beam and substrate, the film is usually deposited in the form of a dot. To illustrate the thermal stress analysis technique, we will consider a thin circular film deposited on a substrate of a different material. Since the film is thin, we assume that the variation of temperature along the thickness of the film is negligible.

The strain in the film is due to the (a) thermal stresses in the film, (b) thermal expansion of the film, and (c) thermal expansion of the substrate.

# APPENDIX

Since the thermal expansion of the film and substrate are different, the substrate will induce a stress at the lower surface of the film, that is, at the interface of the film and substrate. Due to this coupling between the film and substrate, the stress equations have to be solved in the film and substrate regions. However, to simplify the analysis, we assume that as the film is thin the stresses and strains do not vary along its thickness. The strain in the film is $\alpha_f T$ due to the thermal expansion of the film, and

$$(\alpha_s T - \alpha_f T) \frac{E_s}{E_f}$$

due to the thermal expansion of the substrate. Therefore, we can write the following expressions for strain:

$$\begin{Bmatrix} \text{Radial strain in the film} \\ \text{due to thermal stresses} \end{Bmatrix} = \varepsilon_r - \alpha_f T - (\alpha_s T - \alpha_f T) \frac{E_s}{E_f}$$

$$= \varepsilon_r - \alpha T = \frac{1}{E_f}(\sigma_r - \nu_f \sigma_\theta) \quad (A.1)$$

where $\alpha = \alpha_f + (\alpha_s - \alpha_f) E_s / E_f$.

$$\begin{Bmatrix} \text{Tangential strain in the film} \\ \text{due to thermal stresses} \end{Bmatrix} = \varepsilon_\theta - \alpha_f T - (\alpha_s T - \alpha_f T) \frac{E_s}{E_f}$$

$$= \varepsilon_\theta - \alpha T = \frac{1}{E_f}(\sigma_\theta - \nu_f \sigma_r) \quad (A.2)$$

The governing equations are the balance of forces in the radial and tangential directions, which can be writtenh as follows:

FORCE BALANCE IN THE $r$-DIRECTION:

$$\partial \sigma_r \partial r + \frac{1}{r} \frac{\partial \tau_{\theta r}}{\partial \theta} + \frac{\sigma_r - \sigma_\theta}{r} + F_r = 0 \quad (A.3)$$

FORCE BALANCE IN THE $\theta$-DIRECTION:

$$\frac{1}{r} \frac{\partial \sigma_\theta}{\partial \theta} + \frac{\partial \tau_{r\theta}}{\partial r} + \frac{2\tau_{r\theta}}{r} F_\theta = 0 \quad (A.4)$$

We assume that there are no body forces, that is, $F_r = F_\theta = 0$. Since $\tau_{\theta r} = \tau_{r\theta}$, and $\tau_{r\theta} = 0$ due to the symmetry of the deformation [Timoshenko and Goodier (1982)], Eq. (B.4) implies that $\sigma_\theta$ is a function of $r$ only. Thus, Eq.

(B.3) can be written as

$$\frac{\partial \sigma_r}{\partial r} + \frac{\sigma_r - \sigma_\theta}{r} = 0 \tag{A.5}$$

Also, we note that

$$\varepsilon_r = du/dr \tag{A.6}$$

and

$$\varepsilon_\theta = u/r \tag{A.7}$$

Equations (A.1) through (A.7) can be combined [Timoshenko and Goodier (1982)] to write

$$\frac{d}{dr}\left[\frac{1}{r}\frac{d(ru)}{dr}\right] = (1 + v_f)\alpha \frac{dT}{dr} \tag{A.8}$$

which has been solved in Timoshenko and Goodier (1982) and Burgreen (1971) under the boundary conditions that $u = 0$ at $r = 0$, and $\sigma_r = 0$ at $r = r_0$ to obtain

$$\sigma_r = \alpha E_f \left(\frac{1}{r_0^2}\int_0^{r_0} Tr\, dr - \frac{1}{r^2}\int_0^r Tr'\, dr'\right) \tag{A.9}$$

$$\sigma_\theta = \alpha E_f \left(-T + \frac{1}{r_0^2}\int_0^{r_0} Tr\, dr - \frac{1}{r^2}\int_0^r Tr'\, dr'\right) \tag{A.10}$$

EXPRESSIONS FOR $\sigma_r$ AND $\sigma_\theta$ IN TERMS OF AVERAGE TEMPERATURE: Let us define the average temperatures $\bar{T}$ and $\bar{T}_r$ by the following expressions:

$$\bar{T} = \frac{\int_0^{r_0} Tr\, dr}{\int_0^{r_0} r\, dr} = \frac{2}{r_0^2}\int_0^{r_0} Tr\, dr \tag{A.11}$$

$$\bar{T}_r = \frac{\int_0^r Tr'\, dr'}{\int_0^r r'\, dr'} = \frac{2}{r^2}\int_0^r Tr'\, dr' \tag{A.12}$$

From Eqs. (A.9) through (A.12), we obtain

$$\sigma_r = \tfrac{1}{2}\alpha E_f(\bar{T} - \bar{T}_r) \tag{A.13}$$

$$\sigma_\theta = \tfrac{1}{2}\alpha E_f(-2T + \bar{T} - \bar{T}) \tag{A.14}$$

EXPRESSIONS FOR $\sigma_r$ AND $\sigma_\theta$ AT THE CENTER OF THE FILM: It should be noted [Timoshenko and Goodier (1982)] that

$$\lim_{r \to 0} \frac{1}{r^2} \int_0^r Tr' \, dr' = \tfrac{1}{2} T_c, \tag{A.15}$$

which is identical to the expression that can be obtained from Eq. (7.12) by taking the limit as $r \to 0$. From Eqs. (A.9) through (A.11), and Eq. (A.15), we obtain

$$\sigma_r(0) = \tfrac{1}{2}\alpha E_f(\bar{T} - T_c) \tag{A.16}$$

$$\sigma_\theta(0) = \tfrac{1}{2}\alpha E_f(-2T_c + \bar{T} + T_c)$$

$$= \tfrac{1}{2}\alpha E_f(\bar{T} - T_c) \tag{A.17}$$

EXPRESSIONS FOR THE MAXIMUM VALUES OF $\sigma_r$ AND $\sigma_\theta$: The maximum value of $\sigma_r$ is obtained when $d\sigma_r/dr = 0$. By applying this condition to Eq. (A.5), we obtain $\sigma_r = \sigma_\theta$ as a condition for $\sigma_r$ to be maximum. From Eqs. (A.16) and (A.17), we find that $\sigma_r = \sigma_\theta$ at the center of the film. Therefore, $\sigma_{r_{\max}}$ and $\sigma_{\theta_{\max}}$, which are the maximum values of $\sigma_r$ and $\sigma_\theta$, respectively, are given by

$$\sigma_{r_{\max}} = \sigma_r(0) = \tfrac{1}{2}\alpha E_f(\bar{T} - T_c) \tag{A.18}$$

$$\sigma_{\theta_{\max}} = \sigma_\theta(0) = \tfrac{1}{2}\alpha E_f(\bar{T} - T_c) \tag{A.19}$$

It should be noted that the expression (A.19) is obtained by applying the condition $d\sigma_\theta/dr = 0$ to Eq. (4.10) to establish that $\sigma_\theta(r)$ becomes maximum at $r = 0$.

## A.3. VOLUMETRIC ABSORPTION RATE

We want to derive an expression for the loss or absorption of photons per unit volume per unit time. For this purpose, we consider a control volume of sides $\Delta x$, $\Delta y$, $\Delta z$ as shown in Fig. A.1. Let $N$ be the number of photons absorbed per unit volume per unit time in this control volume, and

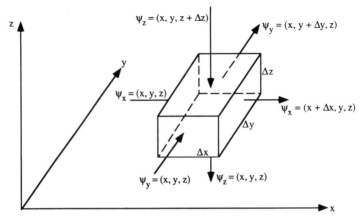

Figure A.1. Control volume $\Delta x \Delta y \Delta z$ fixed in space through which photons are flowing.

$\psi$ be the photon flux vector that is given by

$$\psi = \hat{i}\psi_x + \hat{j}\psi_y - \hat{k}\psi_z$$

where $\hat{i}, \hat{j}, \hat{k}$ are unit vectors, and $\psi_x$, $\psi_y$, and $\psi_z$ are the components of $\psi$ in the $x$, $y$, $z$ directions, respectively.

Taking a balance of the number of photons in this control volume, we can write

$$N\Delta x \Delta y \Delta z = [\psi_x(x, y, z) - \psi_x(x + \Delta x, y, z)]\Delta y \Delta z$$
$$+ [\psi_y(x, y, z) - \psi_y(x, y + \Delta y, z)]\Delta x \Delta z$$
$$+ [\psi_z(x, y, z + \Delta z) - \psi_z(x, y, z)]\Delta x \Delta y$$

Dividing both sides by $\Delta x \Delta y \Delta z$, and taking the limit as $\Delta x \to 0$, $\Delta y \to 0$, and $\Delta z \to 0$, we obtain

$$N = -\left(\frac{\partial \psi_x}{\partial x} + \frac{\partial \psi_y}{\partial y} - \frac{\partial \psi_z}{\partial z}\right)$$
$$= -\left(\hat{i}\frac{\partial}{\partial x} + \hat{j}\frac{\partial}{\partial y} + \hat{k}\frac{\partial}{\partial z}\right) \cdot (\hat{i}\psi_x + \hat{j}\psi_y - \hat{k}\psi_z)$$
$$= -\nabla \cdot \psi$$

The above expression shows that the volumetric rate of photon absorption is given by the negative value of the divergence of the photon flux vector.

APPENDIX 391

Table A.1. Physical Constants and Conversion Factors

| Physical constants | mks | cgs |
|---|---|---|
| Speed of light in vacuum $c$ | $2.9979 \times 10^8$ m/s | $2.9979 \times 10^{10}$ cm/s |
| Planck's constant $h$ | $6.6256 \times 10^{-34}$ J s/quantum | $6.6256 \times 10^{-27}$ erg s/quantum |
| Avogadro's number $Av$ | $6.023 \times 10^{23}$ mol$^{-1}$ | $6.023 \times 10^{23}$ mol$^{-1}$ |
| Boltzmann's constant $k_B = R/Av$ | $1.3804 \times 10^{-23}$ J/K | $1.3804 \times 10^{-16}$ erg/K |
| Universal gas constant $R$ | 8.314 J/mol K | $8.314 \times 10^7$ ergs/mol K |
| Magnitude of electron charge $|e|$ | $1.602 \times 10^{-19}$ C | $4.806 \times 10^{-10}$ esu |
| 1 eV of temperature $k_B T$ | 11600 K | 11600 K |
| Energy of a photon of 1 μm wavelength $E = hc/\lambda$ ($\lambda = 1\,\mu$m) | $1.9863 \times 10^{-19}$ J = 1.2399 eV | $1.9863 \times 10^{-12}$ erg |

Conversion Factors:
1 erg = $10^{-7}$ J
1 cal = 4.1868 J
1 eV = $1.602 \times 10^{-19}$ J
    = $1.602 \times 10^{-12}$ erg
    = 23060 cal/mol
    = $1.6021 \times 10^{-12}$ erg/molecule
    = 8065.7 cm$^{-1}$

## NOMENCLATURE

$E_f$    Young's modulus of the film
$E_s$    Young's modulus of the substrate
$F_r$    Radial body force (force per unit volume in the $r$-direction)
$F_\theta$    Tangential body force (force per unit volume in the tangential or circumferential or $\theta$-direction in polar coordinates)
$r$    Radial variable
$r_0$    Radius of the film
$T(r)$    Temperature distribution in the film
$T_c$    Temperature at the center of the film
$u$    Displacement

**GREEK SYMBOLS**

$\alpha_f$    Linear thermal expansion coefficient of the film
$\alpha_s$    Linear thermal expansion coefficient of the substrate
$\varepsilon_r$    Radial strain (strain in the $r$-direction)
$\varepsilon_\theta$    Tangential strain (strain in the $\theta$-direction)
$\theta$    Angular variable in polar coordinates

$v_f$    Poisson's ratio for the film
$\sigma_r$    Normal stress component in the radial direction
$\sigma_\theta$    Normal stress component in the circumferential direction
$\tau_{r\theta}$    Shear stress in the $\theta$-direction on a plane perpendicular to the $r$-direction
$\tau_{\theta r}$    Shear stress in the $r$-direction on a plane perpendicular to the $\theta$ direction

## REFERENCES

Allmen, M. V. (1987) *Laser-Beam Interactions with Materials: Physical Principles and Application*, Springer-Verlag, New York, p. 8.

Burgreen, D. (1971), *Elements of Thermal Stress Analysis*, C. P. Press, New York, pp. 216–218.

Calvert, J. G., and Pitts, J. N., Jr. (1966), *Photochemistry*, Wiley, New York, p. 54.

Eastham, D. (1989), *Atomic Physics of Lasers*, Tayer and Francis, Philadelphia, pp. 34, 35.

Eckbreth, A. C. (1988), *Laser Diagnostics for Combustion Temperature and Species*, Abacus, Massachusetts, pp. 36, 306.

Einstein, A. (1916), Strahlungs-emission und-absorption nach der quantentheorie, *Verth. Dtsch. Phys. Ges.* **18**, 318.

Einstein, A (1967) *On the Quantum Theory of Radiation*, A reprint in *The Old Quantum Theory*, Haar, D. T., Pergamon, New York, pp. 167–183; or in: *Laser Theory*, Barnes, F. S., ed. (1972), IEEE Press, New York, pp. 5–21.

Matveev, A. N. (1986), *Electricity and Magnetism*, Translated from the Russian by Wadhwa, R. and Deineko, N., Mir, Moscow, pp. 399, 441.

Shimoda, K. (1983), *Introduction to Laser Physics*, 2nd Ed., Springer-Verlag, New York, pp. 78, 79.

Timoshenko, S. P., and Goodier, J. N. (1982), *Theory of Elasticity*, 3rd Ed., McGraw-Hill, New Delhi, pp. 441–443.

# INDEX

Absorption
    cross-section, 145, 146
    coefficient, 65, 70
    rate, 25
Activation energy, 112, 113
Adsorption isotherm, 184
Ampére's law, 61
Anharmonic, 28
Anomalous
    dispersion, 76
    skin effect, 70, 84
Arrhenius
    constant, 112
    law, 109
Aspect ratio, 38
Attenuation, 66
    coefficient, 147

Barn, 24
Beam divergence, 8
Beer's law, 85
Beer–Lambert law, 145, 147, 297
Boltzmann plot, 203
Brownian statistics, 364

Chemical thermometers, 30
Chemiluminescence, 131
Chemisorbed phase, 353
Chemisorption, 181, 183
Collision rate with a surface, 111
Collision theory, 110
Collisional quenching, 127
Complex
    dielectric constant, 64
    refractive index, 64
Compton effect, 124, 125

Conduction band, 133
Confocal parameter, 6, 10
Corpuscular, 123
Coulomb's law, 61
Cumulative distribution function, 368
Cutting, 3
CVD, 14, 15, 16

Depth of focus, 8
Dielectric medium, 62
Diffusion coefficient, 117
Diffusive flux, 303
Dipole moment, 60, 72
Dissociation
    cross section, 150
    probability, 371
Drilling, 3
Dufour effect, 221

Einstein, 124
    coefficients, 192
Electronic spectra, 60
Enthalpy, 105
Equilibrium constant, 105
Exciton band, 133

F-number, 8, 24
Faraday's law, 61
Fick's law, 219, 297
Fluence, 25, 150
Fluorescence, 127
Fluorescent efficiency, 201
Focal plane, 9
Forced diffusion, 219
Fourier transform, 229, 349
Fourier's law, 221

Franck–Condon diagrams, 132, 135
Free energy, 105
Fresnel
    reflection coefficients, 91
    relation, 69, 98
    transmission coefficients, 91
FWHH, 194
FWHM, 194

Gibb's function, 105
Green function, 345

Hagen–Rubens relations, 70, 71, 98
Hardening, 1, 2
HAZ, 2
Heisenberg's uncertainty principle, 195
Hole burning, 30
Huyghen's principle, 123

Integral transform, 229, 333
Internal conversion (IC), 130
Intersystem crossing (ISC), 130
Isotope shift, 23, 24

Jablonsky energy level diagrams, 132

Laplace transform, 313, 324
Large molecules, 30
Laser-induced fluorescence, 188
LCVD, 1
Le Chatelier's principle, 107
LIF, 188, 199
Linear absorption, 371
Local thermodynamic equilibrium, 218
Luminescence, 127

Maxwell's equations, 60, 71
MBE, 15
Mean free time, 78
MOCVD, 14, 15
Molal enthalpy, partial, 221
Molar enthalpy, partial, 221
Molecularity, 103
Monte Carlo method, 362
Multiphoton absorption, 85, 373

Natural frequency, 71
Nonequilibrium process, 29

Nonlinear
    absorption, 373
    susceptibility, 86
Nonthermal process, 24, 28
Normal
    dispersion, 76
    incidence, 7

Oblique incidence, 8
Optical
    electrons, 71
    properties, 60
Order of a reaction, 103, 113

p-Polarization, 89
Particle
    flux, 303–304
    theory, 125
PECVD, 16
Phosphorescence, 127
Photochemical reactions, laws of, 142
Photoelectric effect, 124, 125
Photolysis, 128
Photosensitization, 30
Physisorbed phase, 355
Physisorption, 181, 183
Planck, 124
    quantum theory, 124
    radiation law, 194
Plasma frequency, 76
Polarizability, 72
Polarization, 72
Potential surface, 109
Predissociation, 150, 175, 178
Pre-exponential factor, 113, 115
Pressure diffusion, 219
Probability density function (pdf), 367

Quantum yield, 131

Rayleigh range, 6, 10, 302
Reaction
    quotient, 107
    rate constant, 103
Reciprocity theorem, 357
Red-shift, 25, 32
Refractive index, 63
Relaxation time, 62, 78
Resonance frequency, 73, 76

# INDEX

Rotational transitions, 60
RRKM, 26

$s$-Polarization, 89
Selection rules, 138
Self-focusing, 85
Sensitizer, 30
Single-photon absorption, 371
Skin depth, 65
Skin effect, 65, 70, 84
Soret effect, 219
Spin rule, 178
Spot size, 5
Stern–Vollmer
  factor, 201
  relation, 174
Sticking
  coefficient, 299, 301, 302, 337, 346, 353, 367
  probability, 304, 367
Stoichiometric coefficient, 103
Susceptibility, 86

TE, 89
TEA, 31

TEM, 5
Thermal
  conductivity, 117
  diffusion, 219
  monitor, 30
  process, 28
  runaway, 81, 85
  self-focusing, 85
TM, 89
Transverse
  electric, 89
  magnetic, 89
Two-photon absorption, 86
  absorption constant, 149

Vibrational spectra, 60
Viscosity, 116
VLSI, 21

Wave
  theory, 125
  fronts, 66
Welding, 1, 3